Lecture Notes in Computer Science 8959

Commenced Publication in 1973
Founding and Former Series Editors:
Gerhard Goos, Juris Hartmanis, and Jan van Leeuwen

T0213877

Sumit Ganguly   Ramesh Krishnamurti (Eds.)

Algorithms and Discrete Applied Mathematics

First International Conference, CALDAM 2015
Kanpur, India, February 8-10, 2015
Proceedings

 Springer

Volume Editors

Sumit Ganguly
Indian Institute of Technology
Department of Computer Science and Engineering
Kanpur 208016, India
E-mail: sganguly@cse.iitk.ac.in

Ramesh Krishnamurti
Simon Fraser University
School of Computing Science
Burnaby, BC V5A 1S6, Canada
E-mail: ramesh@sfu.ca

ISSN 0302-9743   e-ISSN 1611-3349
ISBN 978-3-319-14973-8   e-ISBN 978-3-319-14974-5
DOI 10.1007/978-3-319-14974-5
Springer Cham Heidelberg New York Dordrecht London

Library of Congress Control Number: 2014959099

LNCS Sublibrary: SL 1 – Theoretical Computer Science and General Issues

Typesetting: Camera-ready by author, data conversion by Scientific Publishing Services, Chennai, India

Printed on acid-free paper

Springer is part of Springer Science+Business Media (www.springer.com)

Preface

The First International Conference on Algorithms and Discrete Applied Mathematics was held during February 8–10, 2015, at the Indian Institute of Technology, Kanpur, India. This event was organized by the Department of Computer Science and Engineering, Indian Institute of Technology Kanpur. The workshop covered a diverse range of topics on algorithms and discrete mathematics, comprising computational geometry, algorithms including approximation algorithms, graph theory and computational complexity. This volume contains 26 contributed papers presented during CALDAM 2015. There were 58 submissions from 10 countries. These submissions were carefully reviewed by the Program Committee members with the help of external reviewers. Pavol Hell and C.R. Subramanian delivered excellent invited talks whose abstracts are included in this volume.

We would like to thank the authors for contributing high-quality research papers to the workshop. We express our thanks to the Program Committee members and the external reviewers for their active participation in reviewing the papers. We thank Springer for publishing the proceedings in the reputed *Lecture Notes in Computer Science* series. We thank our invited speakers Pavol Hell and C.R. Subramanian. We thank the Organizing Committee chaired by Surender Baswana and the CSE IIT Kanpur technical team B.M. Shukla, Meeta Bagga, Nagendra Yadav and Adarsh Jagannatha from CSE IIT Kanpur, for the smooth functioning of the workshop. We thank the chair of the Steering Committee, Subir Ghosh, for his active help, support and guidance throughout. We thank our sponsor Google Inc. for their financial support. Finally, we thank the EasyChair conference management system which was very effective in handling the entire reviewing process.

November 2014

Sumit Ganguly
Ramesh Krishnamurti

Organization

Program Committee

Amitabha Bagchi	Indian Institute of Technology, Delhi, India
Amitava Bhattacharya	Tata Institute of Fundamental Research, Bombay, India
Bostjan Bresar	University of Maribor, Slovenia
Sunil Chandran	Indian Institute of Science, Bangaluru, India
Manoj Changat	University of Kerala, India
Sandip Das	Indian Statistical Institute, Kolkata, India
Giuseppe Di Battista	Universita' Roma Tre, Rome, Italy
Sumit Ganguly (Co-Chair)	Indian Institute of Technology, Kanpur, India
Daya Gaur	University of Lethbridge, Canada
Partha P. Goswami	University of Calcutta, India
Sathish Govindarajan	Indian Institute of Science, Bengaluru, India
Prahladh Harsha	Tata Institute of Fundamental Research, Bombay, India
R. Inkulu	Indian Institute of Technology, Guwahati, India
Matthew Katz	Ben-Gurion University, Israel
Ken-Ichi Kawarabayashi	National Institute of Informatics, Japan
Shuji Kijima	Kyushu University, Japan
Sandi Klavzar	University of Ljubljana, Slovenia
Ramesh Krishnamurti (Co-Chair)	Simon Fraser University, Canada
Andrzej Lingas	Lund University, Sweden
Anil Maheshwari	Carleton University, Canada
Bojan Mohar	Simon Fraser University, Canada
Ian Munro	University of Waterloo, Canada
N.S. Narayanaswamy	Indian Institute of Technology, Madras, Chennai, India
Sudebkumar P. Pal	Indian Institute of Technology, Kharagpur, India
Abhiram Ranade	Indian Institute of Technology, Bombay, India
Bhawani S. Panda	Indian Institute of Technology, Delhi, India
Kunihiko Sadakane	National Institute of Informatics, Japan
Shakhar Smorodinsky	Ben-Gurion University, Israel
Angelika Steger	ETH Zurich, Switzerland
C.R. Subramanian	Institute of Mathematical Sciences, Chennai, India
Pavel Valtr	Charles University Prague, Czech Republic
David Wood	Monash University, Australia

Additional Reviewers

Abu-Affash, A. Karim
Angelini, Patrizio
Arun-Kumar, S.
Babu, Jasine
Basavaraju, Manu
Benkoczi, Robert
Da Lozzo, Giordano
Di Donato, Valentino
Floderus, Peter
Ganian, Robert
Gonzalez Yero, Ismael
Hell, Pavol
Iranmanesh, Ehsan
Jakovac, Marko
Kabanets, Valentine
Khodamoradi, Kamyar
Krohn, Erik
Levcopoulos, Christos
Manokaran, Rajsekar
Meierling, Dirk
Mishra, Tapas Kumar

Morgenstern, Gila
Mukherjee, Joydeep
Muthu, Rahul
Nilsson, Bengt J.
Otachi, Yota
Pal, Sudebkumar
Pradhan, Dinabandhu
Rafiey, Arash
Ramakrishna, Gadhamsetty
Ranu, Sayan
Rizzi, Romeo
Roy, Bodhayan
Saumell, Maria
Singh, Nitin
Tewari, Raghunath
Thiagarajan, Karthick
Vadivel Murugan, Sunitha
Varma, Girish
Varma, Nithin
Venkateswaran, Jayendran
Żyliński, Paweł

Invited Talk (Abstracts)

Obstruction Characterizations
in Graphs and Digraphs

Pavol Hell

School of Computing Science, Simon Fraser University, Canada

Abstract. Some of the nicest characterizations of graph families are stated in terms of obstructions – forbidden induced subgraphs or other substructures. A typical example characterizes interval graphs by the absence of asteroidal triples and induced cycles of length greater than three. I will discuss similar obstruction characterizations for classes of digraphs. The obstructions are novel, but similar in spirit to asteroidal triples. Surprisingly, these obstructions permit new characterizations even for undirected graphs. In particular, I will describe the first obstruction characterization of circular arc graphs, and a corresponding certifying polynomial time recognition algorithm for this graph class. The digraph results are joint with Arash Rafiey, Jing Huang, and Tomás Feder, and the circular arc graph characterization and algorithm results are joint with Juraj Stacho and Mathew Francis.

Probabilistic Arguments in Graph Coloring

C.R. Subramanian

The Institute of Mathematical Sciences,
Taramani, Chennai - 600 113, India
crs@imsc.res.in

Abstract. Probabilistic arguments has come to be a powerful way to obtain bounds on various chromatic numbers. It plays an important role not only in obtaining upper bounds but also in establishing the tightness of these upper bounds. It often calls for the application of various (often simple) ideas, tools and techniques (from probability theory) like moments, concentration inequalities, known estimates on tail probabilities and various other probability estimates. A number of examples illustrate how this approach can be a very useful tool in obtaining chromatic bounds. For many of these bounds, no other approach for obtaining them is known so far. In this talk, we illustrate this approach with some specific applications to graph coloring. On the other hand, several specific applications of this approach have also motivated and led to the development of powerful tools for handling discrete probability spaces. An important tool (for establishing the tightness results) is the notion of random graphs. We do not necessarily present the best results (obtained using this approach) since the main purpose is to provide an introduction to the power and simplicity of the approach. Many of the examples and the results are already known and published in the literature and have also been improved further.

Table of Contents

Computational Complexity

Algorithms

Probabilistic Arguments in Graph Coloring (Invited Talk)

C.R. Subramanian

The Institute of Mathematical Sciences,
Taramani, Chennai - 600 113, India
`crs@imsc.res.in`

Abstract. Probabilistic arguments has come to be a powerful way to obtain bounds on various chromatic numbers. It plays an important role not only in obtaining upper bounds but also in establishing the tightness of these upper bounds. It often calls for the application of various (often simple) ideas, tools and techniques (from probability theory) like moments, concentration inequalities, known estimates on tail probabilities and various other probability estimates. A number of examples illustrate how this approach can be a very useful tool in obtaining chromatic bounds. For many of these bounds, no other approach for obtaining them is known so far. In this talk, we illustrate this approach with some specific applications to graph coloring. On the other hand, several specific applications of this approach have also motivated and led to the development of powerful tools for handling discrete probability spaces. An important tool (for establishing the tightness results) is the notion of random graphs. We do not necessarily present the best results (obtained using this approach) since the main purpose is to provide an introduction to the power and simplicity of the approach. Many of the examples and the results are already known and published in the literature and have also been improved further.

1 Introduction

We mostly focus on simple undirected graphs $G = (V, E)$. A proper k-coloring of G is any map $f : V \to [k]$ which satisfies $f(u) \neq f(v)$ for every $uv \in E$. Here, $[k]$ denotes $\{1, \ldots, k\}$. The least value of k for which G admits a proper k-coloring is known as the chromatic number of G and is denoted by $\chi(G)$. The chromatic index of G (denoted by $\chi'(G)$) is the least value of k such that there is a map $f : E \to [k]$ satisfying $f(uv) \neq f(uw)$ for every $uv, uw \in E$ with $v \neq w$. It is well-known (from the theorems of Brooks and Vizing) that $\chi(G) \leq d + 1$ and $d \leq \chi'(G) \leq d + 1$ for any graph G with maximum degree d.

Several variants of vertex/edge colorings have also been introduced and studied as these notions model several situations arising in practice. Some of these variants impose further restrictions on the standard notion of colorings mentioned before, while others generalize this standard notion. For example, an acyclic vertex k-coloring is a proper k-coloring in which every cycle of G is colored with three or more colors. Eqivalently, the union of any two color classes

S. Ganguly and R. Krishnamurti (Eds.): CALDAM 2015, LNCS 8959, pp. 1–8, 2015.

induces a forest. The acyclic chromatic number $a(G)$ is the least k for which G admits such a coloring. When the union is further restricted to be a star forest, it is called a star coloring with star chromatic number being the corresponding invariant. Acyclic and star colorings were introduced by Grünbaum [2]. Acyclic and star colorings model certain partition problems arising in sparse matrix computation [26]. Further examples of such restrictions could be proper colorings in which the union of any two color classes induces (i) a partial 2-tree or (ii) a planar graph, etc.

An example of a generalization is a list coloring f in which each $u \in V$ is provided with a list L_u of colors and f satisfies $f(u) \in L_u$ for each u and $f(u) \neq f(v)$ for every $uv \in E$. Such a coloring is referred to as a proper $\mathcal{L}$-coloring where $\mathcal{L} = \{L_u : u \in V\}$. The least value of k such that G admits a $\mathcal{L}$-coloring for every $\mathcal{L}$ satisfying $|L_u| \geq k$ (for every $u \in V$) is known as the choice number or list chromatic number of G and is denoted by $ch(G)$. Clearly, $\chi(G) \leq ch(G)$ for any G.

List colorings naturally model scheduling problems where one needs to find a conflict-free schedule of resources to users where a resource can be assigned only to a user who is willing to accept that resource. List colorings also naturally arise in constructive proofs of chromatic bounds. To prove the existence of a k-coloring, one starts with an initial partial proper coloring which colors a subset of vertices and extends this partial coloring in an iterative fashion to a full coloring of V. Any such extension is essentially about proving the existence of a list coloring where any uncolored vertex is forbidden to use colors used on its colored neighbors.

Total coloring is another generalization. A total k-coloring is a k-coloring f of $V \cup E$ with colors from $[k]$ such that (i) f restricted to V (or E) is a proper vertex (or edge) coloring and (ii) $f(u) \neq f(uv)$ for every $uv \in E$. The least k for which G admits a total k-coloring is known as the total chromatic number of G and is denoted by $\chi_T(G)$. A trivial bound is $\chi_T(G) \leq \chi(G) + \chi'(G)$.

Given a coloring notion, one is interested in knowing whether the associated chromatic number can be bounded by a function g of one or more of other graph invariants like maximum degree $\Delta(G)$, maximum clique size $\omega(G)$ or the standard chromatic number $\chi(G)$? In addition, one is interested in actually obtaining such a function g and also in determining how tight the estimate provided by g is ? To establish tightness, one needs to obtain implicit or explicit examples of graphs where any such coloring will require a number of colors which is closer to the estimate provided by g. Probabilistic arguments is a very useful way to obtain bounds and also to establish their tightness.

1.1 Constrained Colorings

For an illustration, consider the example of b-frugal colorings. For a given $b \geq 1$, a proper k-coloring f of G is a b-frugal coloring if for every $u \in V$ and $j \in [k]$, at most b neighbors of u are colored j. The least k for which such a coloring exists, is known as the b-frugal chromatic number of G and is denoted by $\chi_b^{frug}(G)$. By introducing randomness, one can show that for every $b \geq 1$, we have

$$\chi_b^{frug}(G) \leq 16d^{b+1/b} \quad \ldots\ldots \quad (A)$$

for every G of maximum degree d. Hind, et.al. [8] obtained this bound with e^3 in place of 16. The idea is as follows : Define $k = 16d^{b+1/b}$. Choose uniformly at random a k-coloring $f : V \to [k]$. Define a collection of bad events as follows :

(i) For every $uv \in E$, $\mathcal{E}_{uv}$ happens if $f(u) = f(v)$.
(ii) For every $u, S \subseteq N(u)$ with $|S| = b + 1$, $\mathcal{E}_{u,S}$ happens if $f(v) = f(w)$ for every $v, w \in S$.

The random choice f is proper and b-frugal if none of these events (referred to as bad events) happens. Note that the claimed bound (A) is true if it can be shown that, **(C1)** : with positive probability f satisfies none of the bad events. **(C1)** can be established by applying a powerful probability tool known as Local Lemma (discovered by Erdös and Lovasz [6]) (see also [17]). This tool is very useful in situations where we have a collection of bad events each of which is independent of all but at most a "small" number of other events. The precise meaning of "small" will vary with events and are implicitly captured by a set of inequalities (involving probabilities of bad events and the number of influential events) which need to be satisfied.

The choice and definition of bad events play an important role in simplifying and shortening the proof arguments. Another important factor is in obtaining tight bounds on the number of events influencing a given bad event. Good estimates of probabilities also play an important role.

Using probabilitic arguments (particularly, Local Lemma), good upper bounds (many of them are also tight) have been obtained on various constrained chromatic numbers. For a sample, we suggest the reader to [7] (acyclic chromatic number), [14] (star chromatic numbers), [25, 10] (acyclic chromatic index), [13] (generalized acyclic chromatic number), [11] (k-intersection chromatic index), [5] (chromatic number bounds for graphs with sparse neighborhoods).

Recently, Aravind and Subramanian [22] have generalized the notions of acyclic chromatic number, frugal chromatic number, etc. to a generic notion of $(j, \mathcal{F})$-colorings and and obtained upper bounds (in terms of Δ) on the associated chromatic numbers. Here, $j \geq 2$ and $\mathcal{F}$ is a family of connected j-colorable graphs with each member being a graph on more than j vertices. A $(j, \mathcal{F})$-coloring of G is a proper vertex coloring such that the union of any j color classes induces a subgraph which is free of any copy of any member of $\mathcal{F}$. The $(j, \mathcal{F})$-chromatic number of G (denoted by $\chi_{j,\mathcal{F}}(G)$) is the least k used on any such coloring. This notion specializes to star colorings if we set $j = 2$ and $\mathcal{F} = \{P_4\}$. It specializes to b-frugal colorings if we set $j = 2$ and $\mathcal{F} = \{K_{1,b+1}\}$.

In a related work [21], they also considered the edge analogues of $(j, \mathcal{F})$-colorings and obtained nearly tight (within a constant multiplicative factor) upper bounds on the associated chromatic indices. For both vertex and edge colorings, it was also shown that several such restrictions (with each specified by a pair $(j_i, \mathcal{F}_i)$) can be simultaneously enforced on the colorings. As a sample consequence, it follows that graphs of maximum degree d can be properly edge colored with $O(d)$ colors so that all of the following restrictions can be simultaneously enforced : (i) union of any four matchings forms a partial 2-tree, (ii) union of any 16 matchings forms a 5-degenerate graph, (iii) union of any $\frac{k^2-k+2}{2}$ (for fixed k) matchings forms a k-colorable graph, etc. The upper bounds of [22, 21] were based on probabilistic arguments employing Local Lemma.

In a subsequent work [23], the authors improved the bound of [22] (for the special case of $j = 2$) to $O\left(d^{\frac{m}{m-1}}\right)$. Here, m denotes the minimum number of edges in any member of $\mathcal{F}$. This bound is also nearly tight in view of the lower bounds in derived in [22] for the case $j = 2$. In this work, a connection between $(j, \mathcal{F})$-chromatic numbers and oriented chromatic numbers was also established. It also presented an improvement of the previously best bound of $2d^2 2^d$ (due to Kostochka, Sopena and Zhu [16]) to $16kd2^d$ on the oriented chromatic number. Here, d refers to the maximum degree and k denotes the degeneracy.

The oriented chromatic number $\chi_o(G)$ of G is the least K such that for every orientation $\mathbf{G}$ of $E(G)$, there is an oriented graph (without self-loops) $\mathbf{H}$ on K vertices admitting a homomorphism $f : \mathbf{G} \to \mathbf{H}$. The improvement is basically due to a more careful analysis of the proof of the bound obtained in [16]. It works by proving the existence of a tournament T on K vertices in which for every $I \subseteq V(T), |I| = i \le d$ and for every $a \in \{IN, OUT\}^i$, there are at least $kd + 1$ vertices in $V(T) \setminus I$ all having a as the set of orientations into I. The existence was proved by choosing a random tournament on K vertices and showing that it satisfies the required conditions with a positive probability.

1.2   Randomized Coloring Procedure

Sometimes, to obtain strong and tight chromatic bounds, proving the existence of a coloring does not reduce to the probabilistic analysis of one random experiment. It may call for repeated random choices and analyses to arrive at the conclusion. In other words, it reduces to building a desired coloring in an incremental fashion by starting with an initial partial coloring and extending it to a full coloring. Each such extension will involve random choices and their analyses.

One paradigm that has been successfully employed in deriving several chromatic bounds is the following procedure (referred to as *randomized coloring procedure* in [12]) :

- Each uncolored vertex u picks a color randomly from the list of colors available to it. This coloring may not be proper. In that case, uncolor any recently colored vertex whose choice confcts with the random choice or pre-assigned color of any neighbor.

An application of this procedure initially helps us in obtaining a proper partial coloring that takes care of at least the colored vertices. This forces us to prune the list of colors available to uncolored vertices. However, one needs to ensure that each such list is of "sufficient" size to ensure that one can proceed further and obtain a full coloring of desired type. This calls for analyzing the randomized coloring procedure using some of the various concentration inequalities (like Azuma's martingale inequality, Talagrand's inequality) and tail bounds. Some striking applications of this procedure (and its variants) are the list chromatic index bounds (obtained by Kahn [3]), list chromatic bounds for triangle-free graphs (by Kim [4]), list coloring constants (by Reed and Sudakov [24]), total chromatic number bounds (by Hind, Molloy and Reed [9]), frugal coloring and weighted equitable coloring bounds (by Srinivasan and Pemmaraju [12]).

1.3   List Coloring

As mentioned before, we have $\chi(G) \leq ch(G)$ for any G. How large can $ch(G)$ be compared to $\chi(G)$. It can be shown that $ch(K_{n,n}) = \Theta(\ln n)$ and hence one cannot hope to bound $ch(G)$ by only a function of $\chi(G)$. One can easily see that $ch(G) \leq d(G) + 1 \leq \Delta(G) + 1$ for any G. Here, $d(G)$ denotes the degeneracy of G. As the example of $K_{n,n}$ shows, even these bounds can be quite far from $ch(G)$.

The following upper bound is well-known (as has been observed by several researchers using probabilistic arguments)

$$(B1 :)\ \ ch(G) \leq c\chi(G)(\ln n)$$

for any G, where $c > 0$ is a constant. An asymptotic improvement of this bound was obtained by Noga Alon in [1] for very special classes of graphs. It was shown that $ch(K_{r*m}) \leq cr(\ln m)$ for every $r, m \geq 2$. Here, K_{m*r} denotes a complete r-partite graph with each part being of size m. For those m satisfying $m = n^{o(1)}$ where $n = rm$, this leads to an asymptotic improvement. This bound was also established to be tight within a constant multiplicative factor for every $r, m \geq 2$. The proof is based on probabilistic arguments. Consider the unique optimal coloring $(V_1, \ldots, V_r)$ of K_{r*m}. If $r \leq \sqrt{n}$, then $m \geq r$ and hence $(B1)$ can be applied. When $r \geq m$, let $L = \cup_u L_u$ and consider the bipartition $L = L_1 \cup L_2$ formed by a uniformly random choice of $h : L \to \{1, 2\}$. Prune each L_u into $L_u \cap L_1$ or $L_u \cap L_2$ depending on whether $u \in \cup_{j \leq r/2} V_j$ or otherwise. It can be shown (using Chernoff bounds on tail probabilities) that this random pruning of lists does not reduce the list sizes by too much. Now it reduces to proving the bound for $K_{\frac{r}{2}*m}$. Continuing in this way, one reaches a stage where $(B1)$ can be applied. The assumption of equal part sizes played a very important role in this proof.

Later, Subramanian [18] extended the proof arguments of [1] to work for any G. Precisely, it was shown that

$$(B2 :)\ \ \ ch(G) \leq c\chi(G)\left(\ln \frac{n}{\chi} + 1\right)$$

for any G. The main difficulty in extending the arguments was the potentially highly non-uniform part sizes in any optimal coloring of G under consideration. Hence, to obtain a proof based on inductive reduction on the value of χ, a simple uniform bipartition of L will not work and one needs to choose a random bipartition which takes into account the sizes of the parts and which also ensures that the pruned lists are of required length (with positive probability) as determined by the upper bound one wants to establish. [18] shows how such a random bipartition can be achieved which finally establishes the bound $(B2)$. It was also shown how the proof arguments lead to an efficient algorithm producing a $\mathcal{L}$-coloring provided the lists are of sufficient length.

In a subsequent work [19], the author extended the result of [18] to obtain an analogue of $(B2)$ for the list version of hereditary colorings. For a given hereditary property $\mathcal{P}$, a $\mathcal{P}$-coloring is a partition of V into parts such that the subgraph induced by each part is a member of $\mathcal{P}$. The minimum k such that there exists $V = V_1 \cup \ldots \cup V_k$ where each $G[V_i] \in \mathcal{P}$ is known as the $\mathcal{P}$-chromatic number $\chi_\mathcal{P}(G)$. One can define list version of $\mathcal{P}$-colorings and [19] establishes that

$$(B3:) \quad ch_\mathcal{P}(G) \leq c\chi_\mathcal{P}(G)\left(\ln\frac{n}{\chi_\mathcal{P}} + 1\right)$$

for any G.

In a further subsequent work [20], this result is being further extended to list version of various constrained colorings of graphs and hypergraphs.

1.4   Establishing Tightness

We illustrate the application of probabilistic arguments for establishing the tightness of chromatic bounds with a simple example. Recall the bound $\chi_2^{frug}(G) \leq 16d^{3/2}$ for any G.

Consider the random graph $G \in \mathcal{G}(n,p)$ with $p = 6\left(\frac{\ln n}{n}\right)^{1/3}$. Define $\mu = (n-1)p \approx 6\left(n^{2/3}(\ln n)^{1/3}\right)$. It is well known (from an application of Chernoff bounds) that

$$(A): \quad \mathbf{Pr}\left(\frac{\mu}{2} \leq d \leq 2\mu\right) \to 1 \quad \text{as } n \to \infty$$

where d denotes the maximum degree of G.

Consider any fixed partition $(V_1, \ldots, V_s)$ of V into $s \leq n/3$ parts. Remove at most 2 least numbered vertices from each part so as to get a new collection $(U_1', \ldots, U_t')$ (where $t \leq s$) of non-empty parts with each of size divisible by 3 and each U_j being a subset of some V_i. The number of vertices removed is at most $2n/3$ and hence the new collection has at least $n/3$ vertices. Now, further dividing each part into sets of size 3 each, we get a collection $(W_1, \ldots, W_r)$ (for some $r \geq n/9$) of smaller parts totally which are such that union $W_i \cup W_j$ of any two smaller parts induces a subgraph with maximum degree less than 3. For any fixed pair this happens with probability at most $1 - p^3$. Since these events are independent for different pairs (involving disjoint edges), we deduce that

$$\mathbf{Pr}(\chi_2^{frug}(G) \leq n/3) \leq n^n \cdot (1-p^3)^{\binom{n/9}{2}} \leq e^{n(\ln n) - \frac{n^2}{162}p^3}$$
$$\leq e^{n(\ln n) - \frac{216n(\ln n)}{162}} = o(1)$$

Thus, $(B):$ $\quad \chi_2^{frug}(G) > n/3$ with a probability approaching 1 as $n \to 1$. Comparing (A) and (B), we deduce that for some postive constant C, there are infinitely many values of d for which there is a graph G having maximum degree d and $\chi_2^{frug}(G) \geq C \frac{d^{3/2}}{(\ln d)^{1/2}}$. This establishes the tightness of the upper bound within a multiplicative factor $(\ln d)^{1/2}$. This proof is a simplification of the proof of a more general theorem which can be found in [22], the latter proof using arguments similar to those employed in [7].

2   Conclusions

For a good introduction to the applications of probabilistic arguments in graph coloring and in combinatorics generally, the interested reader is referred to the book by Molloy and Reed [15] and also the one by Alon and Spencer [17], apart from the various references provided below.

References

[1] Alon, N.: Choice numbers of graphs: A probabilistic approach. Combinatorics, Probability and Computing 1(2), 107–114 (1992)

[2] Grünbaum, B.: Acyclic colorings of planar graphs. Israel Journal of Mathematics 14(3), 390–408 (1973)

[3] Kahn, J.: Asymptotically good list-colorings. Journal of Combinatorial Theory, Series A 73(1), 1–59 (1996)

[4] Kim, J.H.: On brooks' theorem for sparse graphs. Combinatorics, Probability and Computing 4(2), 97–132 (1995)

[5] Alon, N., Krivelevich, M., Sudakov, B.: Colouring graphs with sparse neighborhoods. Journal of Combinatorial Theory, Series B 77(1), 73–82 (1999)

[6] Erdös, P., Lovasz, L.: Problems and results on 3-chromatic hypergraphs and some related questions. In: Hajnal, A., Rado, R., Sos, V.T. (eds.) Infinite and Finite Series, pp. 609–628 (1975)

[7] Alon, N., McDiarmid, C., Reed, B.: Acyclic coloring of graphs. Random Structures and Algorithms 2(3), 277–288 (1991)

[8] Hind, H., Molloy, M., Reed, B.: Colouring a graph frugally. Combinatorica 17(4), 469–482 (1997)

[9] Hind, H., Molloy, M., Reed, B.: Total coloring with $\Delta + poly(\log \Delta)$ colors. SIAM Journal on Computing 28(3), 816–821 (1998)

[10] Muthu, R., Narayanan, N., Subramanian, C.R.: Improved bounds on acylic edge colouring. Discrete Mathematics 307(23), 3063–3069 (2007)

[11] Muthu, R., Narayanan, N., Subramanian, C.R.: On k-intersection edge colourings. Discussiones Mathematicae Graph Theory 29(2), 411–418 (2009)

[12] Pemmaraju, S., Srinivasan, A.: The randomized coloring procedure with symmetry-breaking. In: Aceto, L., Damgård, I., Goldberg, L.A., Halldórsson, M.M., Ingólfsdóttir, A., Walukiewicz, I. (eds.) ICALP 2008, Part I. LNCS, vol. 5125, pp. 306–319. Springer, Heidelberg (2008)

[13] Greenhill, C., Pikhurko, O.: Bounds on the generalized acyclic chromatic numbers of bounded degree graphs. Graphs and Combinatorics 21(4), 407–419 (2005)

[14] Fertin, G., Raspaud, A., Reed, B.: Star coloring of graphs. Journal of Graph Theory 47(3), 163–182 (2004)

[15] Molly, M., Reed, B.: Graph Colouring and the Probabilistic Method. Springer, Germany (2000)

[16] Kostochka, A.V., Sopena, E., Zhu, X.: Acyclic and oriented chromatic numbers of graphs. Journal of Graph Theory 24(4), 331–340 (1997)

[17] Alon, N., Spencer, J.H.: The Probabilistic Method, 3rd edn. John Wiley & Sons, Inc., New York (2008)

[18] Subramanian, C.R.: List set coloring: bounds and algorithms. Combinatorics, Probability and Computing 16(1), 145–158 (2007)

[19] Subramanian, C.R.: List hereditary colorings. In: Proceedings of the 2nd International Conference on Discrete Mathematics (ICDM), India, June 6-10. RMS Lecture Note Series, vol. 13, pp. 191–205. Ramanujan Mathematical Society (2010)

[20] Subramanian, C.R.: List hereditary colorings of graphs and hypergraphs (2014) (manuscript)

[21] Aravind, N.R., Subramanian, C.R.: Bounds on edge colorings with restrictions on the union of color classes. SIAM Journal of Discrete Mathematics 24(3), 841–852 (2010)

[22] Aravind, N.R., Subramanian, C.R.: Bounds on vertex colorings with restrictions on the union of color classes. Journal of Graph Theory 66(3), 213–234 (2011)

[23] Aravind, N.R., Subramanian, C.R.: Forbidden subgraph colorings and the oriented chromatic number. European Journal of Combinatorics 34, 620–631 (2013)

[24] Reed, B., Sudakov, B.: Asymptotically the list colouring constants are 1. Journal of Combinatorial Theory, Series B 86(1), 27–37 (2002)

[25] Alon, N., Sudakov, B., Zaks, A.: Acylic edge colorings of graphs. Journal of Graph Theory 37(3), 157–167 (2001)

[26] Gebremedhin, A., Tarafdar, A., Manne, F., Pothen, A.: New acyclic and star coloring algorithms with applications to Hessian computation. SIAM Journal on Scientific Computing 29, 1042–1072 (2007)

A PTAS for the Metric Case of the Minimum Sum-Requirement Communication Spanning Tree Problem

Santiago V. Ravelo and Carlos E. Ferreira

Instituto de Matemática e Estatística, Universidade de São Paulo, Brasil
{ravelo,cef}@ime.usp.br

Abstract. This work considers the metric case of the minimum sum-requirement communication spanning tree problem (SROCT), which is an NP-hard particular case of the minimum communication spanning tree problem (OCT). Given an undirected graph $G = (V, E)$ with non-negative lengths $\omega(e)$ associated to the edges satisfying the triangular inequality and non-negative routing weights $r(u)$ associated to nodes $u \in V$, the objective is to find a spanning tree T of G, that minimizes: $\frac{1}{2} \sum_{u \in V} \sum_{v \in V} (r(u) + r(v)) d(T, u, v)$, where $d(H, x, y)$ is the minimum distance between nodes x and y in a graph $H \subseteq G$. We present a polynomial approximation scheme for the metric case of the SROCT improving the until now best existing approximation algorithm for this problem.

1 Introduction

In this work we consider a particular case of the minimum communication spanning tree problem (OCT). The OCT was introduced by Hu in 1974. In the problem it is given an undirected graph $G = (V, E)$ with non-negative length $\omega(e)$ associated to each edge $e \in E$ and non-negative requirement $\psi(u, v)$ between each pair of nodes $u, v \in V$. The problem is to find a spanning tree T of G which minimizes the total communication cost: $C(T) = \sum_{u \in V} \sum_{v \in V} \psi(u, v) d(T, u, v)$, where $d(H, x, y)$ denotes the minimum distance between nodes x and y in the sub-graph H of G. ([1,2])

In [3] it was proved that the minimum routing cost spanning tree problem (MRCT) is NP-hard (by a reduction from the 3-exact cover problem (3-EC)). Observe that MRCT is a particular case of OCT where the requirement between all pair of nodes is equal to one ($\psi(u, v) = 1$ for all $u, v \in V$). In [4] a PTAS for the MRCT was given. The authors presented a reduction from the general to the metric case, which implies that MRCT with edge-lengths that satisfy the triangular inequality is also NP-hard. Also, in [4] an $O(\log^2(n))$-approximation was given for OCT applying a result from [5] which was later improved to a $O(\log(n))$-approximation by [6].

In [7], the minimum product-requirement communication spanning tree problem (PROCT) and the minimum sum-requirement communication spanning tree problem (SROCT) were introduced. In these problems each vertex $u \in V$

S. Ganguly and R. Krishnamurti (Eds.): CALDAM 2015, LNCS 8959, pp. 9–20, 2015.

has a non-negative routing weight $r(u)$. For PROCT the requirement is defined as $\psi(u,v) = \frac{1}{2}r(u)r(v)$, and for SROCT $\psi(u,v) = \frac{1}{2}(r(u) + r(v))$. Both problems are NP-hard. In [7] a 1.577-approximation algorithm for PROCT and a 2-approximation for SROCT are presented.

The approximation ratio for PROCT was improved in [8] where a PTAS was given. A particular case of SROCT is the weighted p-MRCT, were given an integer p, only p nodes of the graph will have a positive routing weight (i.e. the remaining nodes have zero weight). The particular case in which the p nodes have routing weight 1 is called p-MRCT. In [9] it was proved that 2-MRCT is NP-hard. It also was proved in [10] where PTASs for 2-MRCT and the metric case of weighted 2-MRCT were given.

To the best of our knowledge, there are no results improving the 2-approximation ratio for SROCT which is also the best known ratio for the metric case of SROCT (denoted by m-SROCT). Observe that this problem is also NP-hard, since MRCT is a particular case in which $r(u) = 1$ for all $u \in V$.

In this work we give a PTAS for m-SROCT improving the best previous known result for this problem. The idea of our algorithm was inspired in the previous PTASs for related problems such as MRCT and PROCT. This paper is organized as follows. In the next section we present some notation. In section 3 we show how to obtain an optimal k-star for SROCT in polynomial time for a fixed integer k. In section 4 we present a PTASfor the m-SROCT. Finally, in section 5 the conclusions and future work are given.

2   Definitions

Unless specified we consider all graphs as undirected graphs. Given a graph G we denote the set of its nodes by V_G and the set of its edges by E_G (when G is implicit by context we use V as V_G and E as E_G).

Definition 1. *Given a graph G with non-negative lengths associated to its edges, the* **length of a path** *in G is defined as the sum of the lengths of its edges (a path with no edges has length zero). The* **distance** *between node x and node y in H sub-graph of G is the length of a path with minimum length between x and y in H and is denoted by $d(H, x, y)$.*

Now we can define SROCT as:

Problem 1. SROCT - Sum-Requirement Communication Spanning Tree problem

Input: A graph G, a non-negative length function over the edges of G, $\omega : E \to \mathbb{Q}_+$ and a non-negative routing weight function over the nodes of G, $r : V \to \mathbb{Q}_+$.

Output: A spanning tree T of G which minimizes the total weighted routing cost:

$$C(T) = \sum_{u \in V} \sum_{v \in V} \tfrac{1}{2}(r(u) + r(v))d(T, u, v) = \sum_{u \in V} \sum_{v \in V} r(u)d(T, u, v).$$

Definition 2. *Given a graph G and a non-negative routing weight function over the nodes of G, $r : V \to \mathbb{Q}_+$, we denote $r(G) = \sum_{u \in V_G} r(u)$ and $n(G) = |V_G|$. When G is implicit by the context we use R to denote $r(G)$ and n to denote $n(G)$.*

This paper considers the m-SROCT, the metric case of SROCT, which is the particular case of SROCT where the graph G is complete and the length function over the edges satisfies the triangular inequality. In order to approximate an optimal solution of m-SROCT we introduce the concept of a k-star[1]:

Definition 3. *Given a graph G and a positive integer k, a k-**star** of G is a spanning tree of G with no more than k internal nodes (that is, at least $n - k$ leaves). A **core** of a k-star T of G is a tree resulting by eliminating $n - k$ leaves from T.*

Note that a k-star T can be represented by (τ, S), where τ is a core of T and $S = \{S_{u_1}, ..., S_{u_k}\}$ is a vector indexed by the nodes in τ where S_{u_i} is the set of leaves adjacent in T to $u_i \in V_\tau$ $(1 \leq i \leq k)$.

The problem of finding an optimal k-star for m-SROCT can be defined as:

Problem 2. Optimum k-star for m-SROCT

Input: A positive integer k and an instance of m-SROCT: a complete graph G, a non-negative length function over the edges of G which satisfies the triangular inequality, $\omega : E \to \mathbb{Q}_+$ and a non-negative routing weight function over the nodes of G, $r : V \to \mathbb{Q}_+$.

Output: A k-star T of G which minimizes the total weighted routing cost: $C(T) = \sum_{u \in V} \sum_{v \in V} r(u)d(T, u, v)$.

The next section shows an efficient algorithm to find an optimal k-star.

3   Optimal k-Star for m-SROCT

First we introduce the notion of configuration of a k-star:

Definition 4. *Given a k-star $T = (\tau, S)$ a **configuration** of T is (τ, L) where $L = \{l_{u_1}, ..., l_{u_k}\}$ is a vector of integers being $l_{u_i} = |S_{u_i}|$ $(1 \leq i \leq k)$. A configuration (τ, L) is **over** (k, G), where k is a positive integer and G is a graph, if τ is a tree of G with k nodes (that is, $\tau \subseteq G$ and $|V_\tau| = k$) and $\sum_{u \in V_\tau} l_u = n - k$.*

In [4] it was observed that given a complete graph G and a fixed positive integer k, the number of configurations over (k, G) is polynomial in n, resulting $O(k^k n^{2k-1})$. Then, given an instance $\langle G, \omega, r, k \rangle$ of the optimum k-star for m-SROCT, our proposal is to enumerate all possible configurations over (K, G),

---

[1] The definition of k-star used in this paper is the same used by [4,7,8], which is different from the usual definition of k-star in graph theory (a tree with k leaves linked to a single vertex of degree k).

finding an optimal k-star of each configuration, and finally select the best k-star among them.

We find an optimal k-star for an instance $\langle G, \omega, r, k \rangle$ of the optimum k-star for m-SROCT and a configuration (τ, L) over (k, G), reducing the problem to an uncapacitated minimum cost flow problem (UMCF).

Problem 3. UMCF - Uncapacitated Minimum Cost Flow problem

Input: A directed graph G, a cost function over the arcs $\omega : E \to \mathbb{Q}_+$ and a demand function over the nodes $r : V \to \mathbb{Z}$.

Output: An integer vector indexed by the arcs $X = (x_e)_{e \in E}$ which minimizes $C(X) = \sum_{e \in E} \omega(e) x_e$ and guaranties for each node $u \in V$:
$$\sum_{e \in \delta^+(u)} x_e - \sum_{e \in \delta^-(u)} x_e = r(u),$$
where $e \in \delta^+(w)$ and $e \in \delta^-(v)$ iff $e = \langle v, w \rangle$ ($\forall e \in E, v, w \in V$).

Proposition 1. *Given an instance $I = \langle G, \omega, r, k \rangle$ of the optimum k-star for m-SROCT and a configuration $c = (\tau, L)$ over (k, G), the problem of finding an optimal k-star with configuration c for I can be reduced in polynomial time to the UMCF with instance $I' = \langle G', \omega', r' \rangle$, where:*

- $V_{G'} = V_G$;
- $E_{G'} = \{(u, v) | u \in V_{G-\tau} \wedge v \in \tau\}$;
- $\omega'(u, v) = R\omega(u, v) + \sum_{w \in V_\tau} r(u) \left(d(\tau, v, w) + \omega(u, v) \right) (l_w + 1) - 2r(u)\omega(u, v)$;
- *if $u \in V_{G-\tau}$ then $r'(u) = -1$, otherwise $r'(u) = l_u$.*

The graph G' is a complete bipartite graph on the same node set V_G of G. The bi-partition is given by the nodes in τ and outside this set. The cost of arc $\langle u, v \rangle$ is equivalent to the value of assigning u as adjacent of v in a k-star with the given configuration. We have to consider the cost of sending the routing weight from u to all nodes of τ assuming that each node $w \in V_\tau$ receives $(l_w + 1)$ times the value $r(u)$ (considering the transmission to the node w and the leaves adjacent to it); also, we add the cost of sending the routing weight of the entire graph $(R - r(u))$ to node u, which must pass by node v. Finally the demands r' are set to ensure assignment between nodes out of τ and nodes in τ.

Proof. Since demands are integer we know that in any feasible solution the values x_e will be either zero or one. Moreover, exactly $n - k$ arcs of G' will have value 1. This guaranties that every feasible solution S' of the flow problem represents an assignment of leaves outside τ to be adjacent to nodes in τ for a k-star T of G with configuration (τ, L). Also, it is easy to see that any k-star T with configuration (τ, L) provides a feasible solution to the flow problem: connect node $u \in \tau$ to the l_u leaves adjacent to it in T.

Observe that[2]: $C(S') = C(T) - \sum_{u \in V_\tau} \sum_{v \in V_\tau} r(v)d(\tau, u, v)(l_u + 1)$, where $\sum_{u \in V_\tau} \sum_{v \in V_\tau} r(v)d(\tau, u, v)(l_u + 1)$ is the same for every solution with the same configuration. Then, an optimum of UMCF with instance I' is associated to an optimal k-star with configuration c of m-SROCT with instance I.

---

[2] A detailed proof of this fact can be found in the full version of the paper.

In order to obtain I' from I the cost of each arc in G' must be calculated. This can be done in $O((n-k)k^3)$. Defining the demands and the graph G' itself can be done in $O((n-k)k+n)$. Finally, obtaining the k-star T associated to a solution S' can be done in $O(n-k)$, while the complexity of calculating $C(T)$ would be $O(k^3)$. So, the reduction above can be done in $O(nk^3)$.

$\square$

It is well known that UMCF can be solved in $O(n\log(n)(nk+n\log(n))) = O(n^2\log^2(n))$ (e.g. [11]). Then, finding an optimal k-star for m-SROCT with fixed k can be done efficiently.

Lemma 1. *The optimum k-star for m-SROCT with fixed k can be solve in $O(n^{2k+1}\log^2(n))$.*

4   PTAS for m-SROCT

In this section we prove that for $0 < \delta \leq \frac{1}{2}$ there exists a k-star, with k depending on δ, which is a $\frac{1}{1-\delta}$-approximation of m-SROCT. For that, from now on, we will consider an instance I of m-SROCT. Remember that $n = n(G)$ and $R = r(G)$.

The idea of the proof is similar to those presented in [4], [7] and [8]. Given $0 < \delta \leq \frac{1}{2}$ and a spanning tree T of G, we show the existence of a set Y of internally disjoint paths whose union results in a sub-tree S of T, such that the communication cost of each component $B \in T - S$ is at most a small fraction of the communication cost of T, which implies that most of the communication cost of T passes by S. Also, we prove that the size of Y is limited by a function of δ and we show how to construct a k-star from Y, where the value of k depends on the size of Y. The communication cost of the k-star approximates the communication cost of T by a factor of $\frac{1}{1-\delta}$.

4.1   Notation

First, in order to present the results of this section, we need some notation, which generalizes the notation given in [4], [7] and [8]:

Definition 5. *Given a spanning tree T of G, a set of edges H of T and a node u of T, $VB(T, H, u)$ is the set of nodes in the component of $T - H$ containing the vertex u.*

Definition 6. *Given a spanning tree T of G, a path $P = u_1, ..., u_h$ of T, we denote by f_P (or f, when P is clear by the context) the first node of P and l_P (or l) the last node. We will use η to denote number of nodes and ρ to denote communication requirements. We define:*

For the number of nodes (figure 1 gives an example of these notation):

- *$\eta_P^f = |VB(T, E_P, f)|$ and $\eta_P^l = |VB(T, E_P, l)|$, are the number of nodes in the component of $T - E_P$ containing the first node of P and the component containing the last node of P,*

- $\eta_P^m = n - \eta_P^f - \eta_P^l$, is the number of nodes of all the component of $T - E_P$ containing an internal node of P,
- $N(P, v) = \sum_{i=2}^{h-1} |VB(T, E_P, u_i)| \, d(P, v, u_i)$, is the sum over each internal node u_i of P of the number of nodes in the component of $T - E_P$ containing u_i times the distance in P from u_i to node v,
- $N_f(P) = N(P, f)$, $N_l(P) = N(P, l)$, represent the sums over each internal node u_i of P of the number of nodes in the component of $T - E_P$ containing u_i times the distance in P from u_i to the first node of P and to the last node of P,
- $\eta_P^s = \max\{\eta_P^f, \eta_P^l\}$, $\eta_P^i = \min\{\eta_P^f, \eta_P^l\}$, represents the number of nodes in the greater component of $T - E_P$ containing an extremal node of P,
- if $\eta_P^f = \eta_P^s$ then $N(P) = N_f(P)$, else $N(P) = N_l(P)$.

Analogously, for the communication requirements:

- $\rho_P^f = r\,(VB(T, E_P, f))$, $\rho_P^l = r\,(VB(T, E_P, l))$, $\rho_P^m = R - \rho_P^f - \rho_P^l$,
- $R(P, v) = \sum_{i=2}^{h-1} r\,(VB(T, E_P, u_i))\, d(P, v, u_i)$, $R_f(P) = R(P, f)$, $R_l(P) = R(P, l)$,
- $\rho_P^s = \max\{\rho_P^f, \rho_P^l\}$, $\rho_P^i = \min\{\rho_P^f, \rho_P^l\}$,
- if $\rho_P^f = \rho_P^s$ then $R(P) = R_f(P)$, else $R(P) = R_l(P)$.

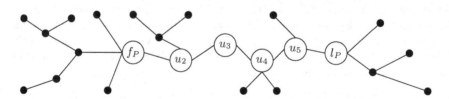

Fig. 1. Consider the above spanning tree T of a graph G, where all the edges have unitary weights and P is the path of T from node f_P to l_P. Observe that $VB(T, E_P, f_P)$ is the set of nodes to the left of f_P (including f_P), $VB(T, E_P, l_P)$ is the set of nodes to the right of l_P (including l_P), $VB(T, E_P, u_2)$ is the set of nodes containing u_2 and the three nodes above it, $VB(T, E_P, u_3) = \{u_3\}$, $VB(T, E_P, u_4)$ is the set of nodes containing u_4 and the two nodes below it, and $VB(T, E_P, u_5)$ is the set of nodes containing u_5 and the node above it. Then, $\eta_P^f = 9$, $\eta_P^l = 5$, $\eta_P^m = 10$, and thus $\eta_P^s = 9$ and $\eta_P^i = 5$. Also, $N_f(P) = |VB(T, E_P, u_2)| \times 1 + |VB(T, E_P, u_3)| \times 2 + |VB(T, E_P, u_4)| \times 3 + |VB(T, E_P, u_5)| \times 4 = 4 \times 1 + 1 \times 2 + 3 \times 3 + 2 \times 4 = 23$ and $N_l(P) = |VB(T, E_P, u_2)| \times 4 + |VB(T, E_P, u_3)| \times 3 + |VB(T, E_P, u_4)| \times 2 + |VB(T, E_P, u_5)| \times 1 = 4 \times 4 + 1 \times 3 + 3 \times 2 + 2 \times 1 = 27$, which yields $N(p) = 23$.

Now we introduce definitions for *separators*. A δ-separator is a sub-tree of a spanning tree T of G, whose deletion gives rise to components that are bounded (in the number of nodes, routing weight or both) by a factor δ of the total value (n or R). Formally:

Definition 7. *Given $0 < \delta \leq \frac{1}{2}$ and a spanning tree T of G, a sub-tree S of T is a δ-η-**separator** of T if every component B of $T - S$, satisfies $n(B) \leq \delta n$. If every component B of $T - S$, satisfies $r(B) \leq \delta R$, S is a δ-ρ-separator of T. If both conditions apply, S is a δ-$\eta\rho$-**separator** of T.*

Also, we define δ-$\eta\rho$-path and δ-$\eta\rho$-spine:

Definition 8. *Given $0 < \delta \leq \frac{1}{2}$ and a spanning tree T of G, a path P of T is a δ-$\eta\rho$-**path** of T if $\eta_P^m \leq \delta \frac{n}{6}$ and $\rho_P^m \leq \delta \frac{R}{6}$.*

Definition 9. *Given $0 < \delta \leq \frac{1}{2}$ and a spanning tree T of G, a set $Y = \{P_1, P_2, ..., P_l\}$ of δ-$\eta\rho$-paths internally-disjoint of T is a δ-$\eta\rho$-**spine**, if $S = \bigcup_{i=1}^{l} P_i$ is a minimal δ-$\eta\rho$-separator of T. $ext(Y)$ denotes the endpoints set of all paths in Y.*

4.2   Approximation Lemma

Using those definitions we prove that for any $0 < \delta \leq \frac{1}{2}$, any spanning tree T of G and any δ-$\eta\rho$-spine Y of T, there exists a $|ext(Y)|$-star with communication cost bounded by $\frac{1}{1-\delta} C(T)$. This lemma, together with lemma 4 are the basis of the main result of this work.

Lemma 2. *Given $0 < \delta \leq \frac{1}{2}$, a spanning tree T of G and a δ-$\eta\rho$-spine Y of T, there exists a $|ext(Y)|$-star X of G satisfying $C(X) \leq \frac{1}{1-\delta} C(T)$.*

In order to conclude that result, first we use the following three propositions[3].

Proposition 2. *Given $0 < \delta \leq \frac{1}{2}$ a δ-$\eta\rho$-path P of a δ-$\eta\rho$-spine of a spanning tree T of G, then:*

$$(R + n - \eta_P^m - \rho_P^m)\left(\eta_P^f \rho_P^l + \eta_P^l \rho_P^f\right) \omega(P)$$
$$+ \left(\eta_P^l + \rho_P^l\right)\left(\omega(P)\left(\eta_P^m \rho_P^f + \eta_P^f \rho_P^m\right) + RN_l(P) + nR_l(P)\right)$$
$$+ \left(\eta_P^f + \rho_P^f\right)\left(\omega(P)\left(\eta_P^m \rho_P^l + \eta_P^l \rho_P^m\right) + RN_f(P) + nR_f(P)\right)$$
$$\leq \frac{6 + 5\delta}{6}(R + n)\left(\eta_P^f \rho_P^l + \eta_P^l \rho_P^f + \eta_P^l \rho_P^m + \eta_P^m \rho_P^i\right)\omega(P)$$
$$+ (R + n)\left(\left(\eta_P^s - \eta_P^i\right)R(P) + \left(\rho_P^s - \rho_P^i\right)N(P)\right).$$

Proposition 3. *Given $0 < \delta \leq \frac{1}{2}$, a spanning tree T of G and a δ-$\eta\rho$-spine Y of T, there exists a $|ext(Y)|$-star X of G which satisfies:*

$$C(X) \leq \sum_{P \in Y} \left(\eta_P^f \rho_P^l + \rho_P^f \eta_P^l\right)\omega(P) + \min\{\Delta_{fl}(P), \Delta_{lf}(P)\}$$
$$+ R \sum_{u \in V_G} d(T, u, S) + (n - 2)\sum_{u \in V_G} r(u)d(T, u, S).$$

---

[3] The proofs of the propositions are intricate and do not bring any new insight into the problem, also they can be found in the full version of the paper.

Where $\Delta_{wz}(P) = \omega(P)\left(\eta_P^m \rho_P^w + \rho_P^m \eta_P^w\right) + RN_z(P) + (n-2)R_z(P)$, $w, z \in \{f, l\}$, and $S = \bigcup_{P \in Y} P$.

Observe that the result above shows the existence of a $|ext(Y)|$-star X of G (which comes from a δ-$\eta\rho$-spine Y of a spanning tree of G) and also gives us an upper bound for the associated communication cost of X. The next proposition gives us a lower bound for the communication cost of a spanning tree of G.

Proposition 4. *Given* $0 < \delta \le \frac{1}{2}$, *a spanning tree* T *of* G *and a* δ-$\eta\rho$-*spine* Y *of* T, *then:*

$$C(T) \ge \sum_{P \in Y} \left(\rho_P^l \eta_P^f + \rho_P^f \eta_P^l + \rho_P^i \eta_P^m + \eta_P^i \rho_P^m\right) \omega(P)$$

$$+ \sum_{P \in Y} \left(\eta_P^s - \eta_P^i\right) R(P) + \left(\rho_P^s - \rho_P^i\right) N(P)$$

$$+ (1 - \delta)\left(n \sum_{u \in V_G} r(u)d(T, u, S) + R \sum_{u \in V_G} d(T, u, S)\right).$$

Where $S = \bigcup_{P \in Y} P$.

Now we demonstrate the lemma 2, which states that if we are given a δ-$\eta\rho$-spine Y of a spanning tree T, then we can construct a star whose core lies on $ext(Y)$, such that its communication cost is bounded by $\frac{1}{1-\delta}C(T)$.

Proof. Let $S = \bigcup_{P \in Y} P$ be the minimal δ-$\eta\rho$-separator associated with Y and X the $|ext(Y)|$-star of G given by proposition 3, then:

$$C(X) \le \sum_{P \in Y} \left(\eta_P^f \rho_P^l + \rho_P^f \eta_P^l\right) \omega(P) + \min\{\Delta_{fl}(P), \Delta_{lf}(P)\}$$

$$+ R \sum_{u \in V_G} d(T, u, S) + (n-2) \sum_{u \in V_G} r(u)d(T, u, S)$$

$$\le \sum_{P \in Y} \frac{R + n - \rho_P^m - \eta_P^m}{R + n - \rho_P^m - \eta_P^m} \left(\eta_P^f \rho_P^l + \rho_P^f \eta_P^l\right) \omega(P) + \min\{\Delta_{fl}(P), \Delta_{lf}(P)\}$$

$$+ R \sum_{u \in V_G} d(T, u, S) + n \sum_{u \in V_G} r(u)d(T, u, S).$$

Since the minimum between two numbers is less than or equal to their weighted median, we have:

$$C(X) \le \sum_{P \in Y} \frac{R + n - \rho_P^m - \eta_P^m}{R + n - \rho_P^m - \eta_P^m} \left(\eta_P^f \rho_P^l + \rho_P^f \eta_P^l\right) \omega(P)$$

$$+ \sum_{P \in Y} \frac{\eta_P^l + \rho_P^l}{\eta_P^l + \rho_P^l + \eta_P^f + \rho_P^f} \Delta_{fl}(P) + \frac{\eta_P^f + \rho_P^f}{\eta_P^l + \rho_P^l + \eta_P^f + \rho_P^f} \Delta_{lf}(P)$$

$$+ R \sum_{u \in V_G} d(T, u, S) + n \sum_{u \in V_G} r(u)d(T, u, S).$$

Since every path P satisfies $R + n - \eta_P^m - \rho_P^m = \eta_P^f + \eta_P^l + \rho_P^f + \rho_P^l$, and every $P \in Y$ is a δ-$\eta\rho$-path, then by applying the result of Proposition 2 we conclude:

$$C(X) \le \sum_{P \in Y} \frac{\left(\frac{6+5\delta}{6}\right)(R+n)}{R+n-\rho_P^m-\eta_P^m} \left(\eta_P^f \rho_P^l + \rho_P^f \eta_P^l + \eta_P^i \rho_P^m + \rho_P^i \eta_P^m\right) \omega(P)$$

$$+ \sum_{P \in Y} \frac{n+R}{R+n-\rho_P^m-\eta_P^m} \left(\left(\eta_P^s - \eta_P^i\right) R(P) + \left(\rho_P^s - \rho_P^i\right) N(P)\right)$$

$$+ R \sum_{u \in V_G} d(T, u, S) + n \sum_{u \in V_G} r(u) d(T, u, S).$$

Notice that any δ-$\eta\rho$-path satisfies: $\rho_P^m + \eta_P^m \le \frac{\delta}{6} R + \frac{\delta}{6} n = \frac{\delta}{6}(n + R)$, then:

$$C(X) \le \sum_{P \in Y} \frac{\left(\frac{6+5\delta}{6}\right)(R+n)}{R+n-\frac{\delta}{6}(R+n)} \left(\eta_P^f \rho_P^l + \rho_P^f \eta_P^l + \eta_P^i \rho_P^m + \rho_P^i \eta_P^m\right) \omega(P)$$

$$+ \sum_{P \in Y} \frac{n+R}{R+n-\frac{\delta}{6}(R+n)} \left(\left(\eta_P^s - \eta_P^i\right) R(P) + \left(\rho_P^s - \rho_P^i\right) N(P)\right)$$

$$+ R \sum_{u \in V_G} d(T, u, S) + n \sum_{u \in V_G} r(u) d(T, u, S)$$

$$= \frac{6+5\delta}{6-\delta} \sum_{P \in Y} \left(\eta_P^f \rho_P^l + \rho_P^f \eta_P^l + \eta_P^i \rho_P^m + \rho_P^i \eta_P^m\right) \omega(P)$$

$$+ \frac{6}{6-\delta} \sum_{P \in Y} \left(\left(\eta_P^s - \eta_P^i\right) R(P) + \left(\rho_P^s - \rho_P^i\right) N(P)\right)$$

$$+ \frac{1}{1-\delta}(1-\delta)\left(R \sum_{u \in V_G} d(T, u, S) + n \sum_{u \in V_G} r(u) d(T, u, S)\right).$$

Since $\frac{6+5\delta}{6-\delta} = \frac{(6+5\delta)(1-\delta)}{(6-\delta)(1-\delta)} = \frac{6-\delta-5\delta^2}{(6-\delta)(1-\delta)} < \frac{6-\delta}{(6-\delta)(1-\delta)} = \frac{1}{1-\delta}$ and $\frac{6}{6-\delta} < \frac{6+5\delta}{6-\delta} < \frac{1}{1-\delta}$, by applying Proposition 4 we obtain: $C(X) \le \frac{1}{1-\delta} C(T)$. $\square$

4.3   Existence of Bounded δ-$\eta\rho$-Spine

In the next lemma we show that there exists a δ-$\eta\rho$-spine Y of T whose $|ext(Y)|$ is bounded by a function of δ.

Lemma 3. *Given $0 < \delta \le \frac{1}{2}$ and a spanning tree T of G, there exists a δ-$\eta\rho$-spine Y of T satisfying $|ext(Y)| \le 3\left(\left\lceil\frac{6}{\delta}\right\rceil^2 - 11\left\lceil\frac{6}{\delta}\right\rceil + 1\right)$.*

Proof. Consider a minimal δ-ρ-separator S_ρ of T and a minimal δ-η-separator S_η of T. If S_ρ and S_η have at least one node in common, then define $S' = S_\rho \cup S_\eta$ and obviously S' is a δ-$\eta\rho$-separator. If S_ρ and S_η have no nodes in common, then since both are trees, S_η must be included in a component of $T - S_\rho$. But, S_ρ

is a δ-ρ-separator of T, then every component of $T - S_\rho$ has weight bounded by δR, so the path P in T connecting S_ρ to S_η satisfies $\rho_P^m < \delta R$. Analogously, P also satisfies $\eta_P^m < \delta n$. Then, P can be divided into 6 paths each one with weight bounded by $\frac{\delta R}{6}$ and another 6 paths each one with number of nodes bounded by $\frac{\delta n}{6}$. Since each division uses 5 internal nodes, in the worst case, using 10 internal nodes we obtain a division of P in δ-$\eta\rho$-paths, and $S' = S_\rho \cup S_\eta \cup P$ is a δ-$\eta\rho$-separator.

By modifying[4] a proof of [4], [7] and [8], we prove that there exists an Y'_ρ and an Y'_η sets of internally-disjoint δ-$\eta\rho$-paths, satisfying $\cup_{P\in Y'_\rho} P = S_\rho$, $\cup_{P\in Y'_\eta} P = S_\eta$ and $\left|ext(Y'_\rho)\right|, \left|ext(Y'_\eta)\right| \leq \left\lceil \frac{6}{\delta} \right\rceil^2 - 11 \left\lceil \frac{6}{\delta} \right\rceil + 1$.

For each path $P \in Y'_\rho$ if P contains internal nodes in $ext(Y'_\eta)$, divide P on those nodes to create new internally-disjoint δ-$\eta\rho$-paths and put those new paths in Y_ρ, otherwise add P to Y_ρ. Analogously, define Y_η from Y'_η and $ext(Y'_\rho)$. Observe that no path of $Y_\eta \cup Y_\rho$ has an internal node in $ext(Y_\eta) \cup ext(Y_\rho)$, also Y_η and Y_ρ are sets of internally-disjoint δ-$\eta\rho$-paths such that $\cup_{P\in Y_\rho} P = S_\rho$, $\cup_{P\in Y_\eta} P = S_\eta$ and:

$$|ext(Y_\rho \cup Y_\eta)| \leq 2 \left(\left\lceil \frac{6}{\delta} \right\rceil^2 - 11 \left\lceil \frac{6}{\delta} \right\rceil + 1 \right).$$

Notice that, since $S_\eta \cup S_\rho$ is acyclic, each path of Y_ρ internally-intersects at most one path in Y_η and vice-verse.

If there are two paths $P_\eta \in Y_\eta$ and $P_\rho \in Y_\rho$ whose internal-intersection is not empty and their end-points do not belong to their intersection, then no other path of Y_η intersects any path of Y_ρ, and by removing from P_η the internal nodes of the intersection we add at most two new extremal points (the end-points of the intersection). Then $Y' = (Y_\eta - P_\eta) \cup Y_\rho \cup (P_\eta - (P_\eta \cap P_\rho))$ is a set of internally-disjoint δ-$\eta\rho$-paths which satisfies $\cup_{P\in Y'} P = S'$ and:

$$|ext(Y')| \leq 2 \left(\left\lceil \frac{6}{\delta} \right\rceil^2 - 11 \left\lceil \frac{6}{\delta} \right\rceil + 1 \right) + 2.$$

Otherwise, if no path of Y_η intersects any path of Y_ρ then, as seen before, there exists a path P connecting S_η to S_ρ that can be divided in at most 11 δ-$\eta\rho$-paths, and the union of those paths with Y_η and Y_ρ results in a set Y' of internally-disjoint δ-$\eta\rho$-paths such that $\cup_{P\in Y'} P = S'$ and:

$$|ext(Y')| \leq 2 \left(\left\lceil \frac{6}{\delta} \right\rceil^2 - 11 \left\lceil \frac{6}{\delta} \right\rceil + 1 \right) + 10.$$

The last possibility is that at least one path of Y_η internally-intersects a path of Y_ρ and each not-empty intersection between a path of Y_η and a path of Y_ρ contains at least one endpoint. Then, remove from each path in Y_η the internal nodes of the intersection with each path in Y_ρ (notice that a path of Y_η at most internally-intersects one path in Y_ρ). In this case the number of new extremal

---

[4] Such modification can be found in the full version of the paper.

points will be at most $|Y'_\eta|$ and the set Y' defined by the union of Y_ρ with the modified Y_η is a set of internally-disjoint δ-$\eta\rho$-paths that satisfies $\cup_{P\in Y'} P = S'$ and:

$$|ext(Y')| \leq 3\left(\left\lceil\frac{6}{\delta}\right\rceil^2 - 11\left\lceil\frac{6}{\delta}\right\rceil + 1\right).$$

Since, for $0 < \delta \leq \frac{1}{2}$: $\left(\left\lceil\frac{6}{\delta}\right\rceil^2 - 11\left\lceil\frac{6}{\delta}\right\rceil + 1\right) \geq 12^2 - 11(12) + 1 = 13 > 10$, then we always can obtain a set Y' of internally-disjoint δ-$\eta\rho$-paths which satisfies $\cup_{P\in Y'} P = S'$ and:

$$|ext(Y')| \leq 3\left(\left\lceil\frac{6}{\delta}\right\rceil^2 - 11\left\lceil\frac{6}{\delta}\right\rceil + 1\right).$$

If S' is a minimal δ-$\eta\rho$-separator, then $Y = Y'$ is a δ-$\eta\rho$-spine. Otherwise, exists a minimal δ-$\eta\rho$-separator $S \subset S'$ and by deleting from each path in Y' the elements that are not contained in S we obtain a δ-$\eta\rho$-spine Y of T satisfying:

$$|ext(Y)| \leq 3\left(\left\lceil\frac{6}{\delta}\right\rceil^2 - 11\left\lceil\frac{6}{\delta}\right\rceil + 1\right).$$

□

4.4 PTAS

Using lemmata 2 and 3 we can state the following proposition:

Proposition 5. *Given* $0 < \delta \leq \frac{1}{2}$ *and a spanning tree* T *of* G, *there exists a* $\left(3\left(\left\lceil\frac{6}{\delta}\right\rceil^2 - 11\left\lceil\frac{6}{\delta}\right\rceil + 1\right)\right)$-*star* X *of* G, *such that* $C(X) \leq \frac{1}{1-\delta}C(T)$.

Let T^* be an optimal spanning tree for m-SROCT over G, by proposition 5 for any $0 < \delta \leq \frac{1}{2}$, there exists a $\left(3\left(\left\lceil\frac{6}{\delta}\right\rceil^2 - 11\left\lceil\frac{6}{\delta}\right\rceil + 1\right)\right)$-star X of G such that $C(X) \leq \frac{1}{1-\delta}C(T^*)$. Since an optimal $\left(3\left(\left\lceil\frac{6}{\delta}\right\rceil^2 - 11\left\lceil\frac{6}{\delta}\right\rceil + 1\right)\right)$-star X^* of G guarantees $C(X^*) \leq C(X)$, then $C(X^*) \leq \frac{1}{1-\delta}C(T^*)$.

Lemma 4. *Given* $0 < \delta \leq \frac{1}{2}$ *an optimal* $\left(3\left(\left\lceil\frac{6}{\delta}\right\rceil^2 - 11\left\lceil\frac{6}{\delta}\right\rceil + 1\right)\right)$-*star of* G *is a* $\frac{1}{1-\delta}$-*approximation for* m-SROCT.

The results of lemmata 1 and 4 complete the necessary tools for providing the PTAS:

Theorem 1. *There exists a PTAS for* m-SROCT, *such that a* $\left(1 + \frac{\delta}{1-\delta}\right)$-*approximation can be found in* $O\left(n^{6\left(\left\lceil\frac{6}{\delta}\right\rceil^2 - 11\left\lceil\frac{6}{\delta}\right\rceil + 1\right)+1}\log^2(n)\right)$ *time complexity where* $0 < \delta \leq \frac{1}{2}$.

5 Conclusions

In this work we present a PTAS for m-SROCT, a NP-hard particular case of OCT. The best previously known result for this problem was a 2-approximation algorithm due to [7]. Many questions remain open regarding OCT and related problems. One could improve the approximation ratio for SROCT or other particular case of OCT. In future works we will attempt to answer this question for some of these problems.

Acknowledgments. This research is supported by the following projects: FAPESP 2013/03447-6, CNPq 477203/2012-4 and CNPq 302736/2010-7.

References

1. Hu, T.C.: Optimum communication spanning trees. SIAM J. Comput. 3(3), 188–195 (1974)
2. Wu, B.Y., Chao, K.M.: Spanning Trees and Optimization Problems. Chapman & Hall / CRC (2004) ISBN: 1584884363
3. Johnson, D.S., Lenstra, J.K., Rinnooy Kan, A.H.G.: The complexity of the network design problem. Networks 8, 279–285 (1978)
4. Wu, B.Y., Lancia, G., Bafna, V., Chao, K.M., Ravi, R., Tang, C.Y.: A polynomial time approximation scheme for minimum routing cost spanning trees. SIAM J. on Computing 29(3), 761–778 (2000)
5. Bartal, Y.: Probabilistic approximation of metric spaces and its algorithmic applications. In: Proceedings of the 37th Annual IEEE Symposium on Foundations of Computer Science, pp. 184–1963 (1996)
6. Talwar, K., Fakcharoenphol, J., Rao, S.: A tight bound on approximating arbitrary metrics by tree metrics. In: Proceedings of the 35th Annual ACM Symposium on Theory of Computing, pp. 448–455 (2003)
7. Wu, B.Y., Chao, K.M., Tang, C.Y.: Approximation algorithms for some optimum communication spanning tree problems. Discrete and Applied Mathematics 102, 245–266 (2000)
8. Wu, B.Y., Chao, K.M., Tang, C.Y.: A polynomial time approximation scheme for optimal product-requirement communication spanning trees. J. Algorithms 36, 182–204 (2000)
9. Farley, A.M., Fragopoulou, P., Krumme, D., Proskurowski, A., Richards, D.: Multi-source spanning tree problems. Journal of Interconnection Networks 1(1), 61–71 (2000)
10. Wu, B.Y.: A polynomial time approximation scheme for the two-source minimum routing cost spanning trees. J. Algorithms 44, 359–378 (2002)
11. Orlin, B.J.: A faster strongly polynomial minimum cost flow algorithm. Operations Research 41(2), 338–350 (1993)

Constant Approximation for Broadcasting in k-cycle Graph

Puspal Bhabak and Hovhannes A. Harutyunyan

Department of Computer Science and Software Engineering
Concordia University
Montreal, QC, H3G 1M8, Canada

Abstract. *Broadcasting* is an information dissemination problem in a connected graph in which one vertex, called the *originator*, must distribute a message to all other vertices by placing a series of calls along the edges of the graph. Every time the informed vertices aid the originator in distributing the message. Finding the broadcast time of any vertex in an arbitrary graph is NP-complete. The problem is NP-Complete for even more restricted classes of graphs, such as for 3-regular planar graphs. The best approximation algorithm for broadcast problem is $O(\frac{\log(|V|)}{\log\log(|V|)} b(G))$. The polynomial time solvability is shown only for certain tree-like graphs; trees, unicyclic graphs, tree of cycles. The problem becomes very difficult when cycles intersect. In this paper we study the broadcast problem in a simple cactus graph called k-cycle graph. For any originator we present a $(2 - \epsilon)$-approximation algorithm in the arbitrary k-cycle graph. We also prove that our algorithm generates the optimal broadcast time for some subclasses of this graph.

1 Introduction

In today's world, due to massive parallel processing, as processors have become faster and more efficient, communication has become a larger concern for bottlenecks. In recent years, a lot of work has been dedicated in order to find the best communication structures for parallel and distributed computing. One of the main problems of information dissemination investigated in this research area is broadcasting. *Broadcasting* is the message dissemination problem in a connected network in which one informed node, called the *originator*, must distribute a message to all other nodes by placing a series of calls along the communication lines of the network. Every time the informed nodes aid the originator in distributing the message. This is assumed to take place in discrete time units. The broadcasting is to be completed as quickly as possible subject to the following constraints: (1) Each call requires one unit of time. (2) A vertex can participate in only one call per unit of time. (3) Each call involves only two adjacent vertices, a sender and a receiver.

Given a connected graph G and a message originator, vertex u, the *broadcast time* of vertex u, denoted $b(u, G)$ or $b(u)$ is the minimum number of time units required to complete broadcasting in graph G from vertex u. The broadcast time

S. Ganguly and R. Krishnamurti (Eds.): CALDAM 2015, LNCS 8959, pp. 21–32, 2015.

$b(G)$ of the graph G is defined as $\max\{b(u)|u \in V\}$. A broadcast scheme is the set of calls that accomplish broadcasting. These set of calls define a broadcast tree in the graph G. It is easy to see that for any vertex u in a connected graph G with n vertices, $b(u) \geq \lceil \log n \rceil$ (all log's in the paper are base 2), since during each time unit the number of informed vertices can at most double. In general, solving the broadcast problem for an arbitrary originator u in an arbitrary graph G has been proved to be NP-complete in [19]. The problem remains NP-Complete even for 3-regular planar graphs [15] and for a graph whose vertex set can be partitioned into a clique and an independent set [13]. The best theoretical upper bound is obtained by the approximation algorithm in [4] which produces a broadcast scheme with $O(\frac{\log(|V|)}{\log\log(|V|)}b(G))$ rounds. Research in [18] has showed that the broadcast time cannot be approximated within a factor $\frac{57}{56} - \epsilon$. However this result has been improved within a factor of $3 - \epsilon$ in [4]. As a result research has been made in the direction of finding approximation or heuristic algorithms to determine the broadcast time in arbitrary graphs (see [12], [1], [2], [4], [5], [6], [8], [14], [16], [17], [7], [11]). However, there is a linear algorithm to determine the broadcast time of any tree [19]. Recent research shows that there are polynomial time algorithms for the broadcast problem in tree-like graphs where two cycles do not intersect - unicyclic graphs, tree of cycles, or in graphs containing no intersecting cliques - fully connected trees and tree of cliques ([9], [10]). The broadcasting problem becomes very difficult when two cycles intersect. [3] presents a $(4 - \epsilon)$-approximation algorithm to find the broadcast time in simple graphs where intersection of two cycles is a path, called a k-path graph.

In this paper we consider broadcasting in simple graphs where cycles intersect at single vertex. The simplest such graph where several cycles have only one intersecting vertex is called a k-cycle graph. A k-cycle graph is a collection of k cycles of arbitrary lengths connected by a central vertex on one end. Note that k-cycle graph is a cactus graph. Broadcasting in the k-cycle graph is different from broadcasting in the k-path graph in a way that in k-path graph, after a certain time, broadcasting depends on the strategy how the two intersecting vertices select the paths to send the message. However, in k-cycle graph, the entire broadcast scheme is strongly dependent on the single central vertex.

In section 2 we give lower bounds on broadcast time. In section 3 we present a constant approximation algorithm to find the broadcast time of an arbitrary k-cycle graph. In section 4 we show the optimality of our algorithm for some subclasses of k-cycle graph.

2   Lower Bounds on Broadcast Time

Definition 1. *Let* $G_k = (V, E)$ *be a connected graph consisting of* k *cycles* $C_1, C_2, C_3, ..., C_k$ *and an intersecting vertex* u *connected on one end point of all cycles. Vertex* u *is called central vertex of* G_k *(see Fig 1).*

Let $l_1 \geq l_2 \geq ... \geq l_k \geq 2$, where l_i be the number of vertices in cycle C_i (excluding vertex u) for all $1 \leq i \leq k$.

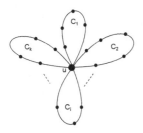

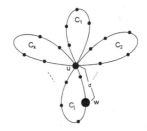

Fig. 1. k-cycle graph   **Fig. 2.** k-cycle graph with originator w

2.1   Lower Bounds When Originator is the Central Vertex

In this section we give lower bounds on the broadcast time of G_k from u.

Lemma 1. *Let G_k be a k-cycle graph where the originator is the central vertex u. Then*
(i) $b(u) \geq k + 1$. (ii) $b(u) \geq \left\lceil \frac{l_j + 2j - 1}{2} \right\rceil$ for any j, $1 \leq j \leq k$.
(iii) $b(u) \geq \left\lceil \frac{2k + l_j + 2j + 1}{4} \right\rceil$ for any j, $1 \leq j \leq k$.

Proof. (i): Under any minimum time broadcast scheme, k time units are necessary to inform at least one vertex in each of the k cycles from vetex u. Since $l_j \geq 2$ for any j, where $1 \leq j \leq k$, at least one more time unit is required to inform the second vertex on the cycle which initially receives the message from u at time unit k. So, $b(u) \geq k + 1$.

(ii): We consider any cycle C_j where $1 \leq j \leq k$. Under any minimum time broadcast scheme all vertices in C_j must be informed. u informs the k cycles in some order and assume it initially informs C_j at time unit j or later. Then u informs its second neighboring vertex in C_j no sooner than time unit $j + 1$. At time unit j there are at least $l_j - 1$ uninformed vertices in C_j. Starting at time $j + 1$ onwards, C_j receives the message from both directions from u. At each time unit 2 new vertices on C_j will get informed. So, $b(u) \geq j + \left\lceil \frac{l_j - 1}{2} \right\rceil$ $= \left\lceil \frac{l_j + 2j - 1}{2} \right\rceil$. Suppose, by contradiction u initially calls path C_j before time j. Then by pigeonhole principle there exists m, $1 \leq m \leq j - 1$ such that u initially calls C_m at time j. Similarly at time unit j there are at least $l_m - 1$ uninformed vertices in C_m. If, starting at time $j + 1$ onwards, C_m receives the message from both directions from u, then $b(u) \geq j + \left\lceil \frac{l_m - 1}{2} \right\rceil = \left\lceil \frac{l_m + 2j - 1}{2} \right\rceil \geq \left\lceil \frac{l_j + 2j - 1}{2} \right\rceil$ as $l_m \geq l_j$. Hence, $b(u) \geq \left\lceil \frac{l_j + 2j - 1}{2} \right\rceil$.

For the proof of (iii), we combine the inequalities in (i) and (ii). We get $2b(u) \geq k + 1 + \left\lceil \frac{l_j + 2j - 1}{2} \right\rceil \geq \left\lceil \frac{l_j + 2j + 2k + 1}{2} \right\rceil$. Hence, $b(u) \geq \left\lceil \frac{l_j + 2j + 2k + 1}{4} \right\rceil$ for any j, $1 \leq j \leq k$.   □

Lemma 2. *Let G_k be a k-cycle graph where the originator is the central vertex u and n is the total number of vertices in G_k. Then*
(i) $b(u) \geq \lceil \frac{n-1}{2k} + k - \frac{1}{2} \rceil$ if $b(u) \geq 2k$.
(ii) $b(u) \geq \lceil \sqrt{(2n - \frac{7}{4})} - \frac{1}{2} \rceil$ if $k + 1 \leq b(u) \leq 2k - 1$.

Proof. (i): Since $b(u) \geq 2k$, then u will be busy informing its adjacent vertices in k different cycles at time units $1, 2, ..., 2k$. By $b(u)$ time units, u can inform at most $b(u)$, $b(u) - 1,..., b(u) - (2k - 1)$ vertices in these k different cycles. So, $n \leq b(u) + b(u) - 1 +...+ b(u) - (2k - 1) + 1 \Rightarrow n \leq 2kb(u) - k(2k - 1) + 1$. Hence, $b(u) \geq \lceil \frac{n-1}{2k} + k - \frac{1}{2} \rceil$.

(ii): Since $k + 1 \leq b(u) \leq 2k - 1$, then u can inform its adjacent vertices in k different cycles at time units $1, 2, ..., b(u)$, where $b(u) \leq 2k - 1$. By $b(u)$ time units, u can inform at most $b(u)$, $b(u) - 1,..., 1$ vertices in these k different cycles. So, $n \leq b(u) + b(u) - 1 +...+ 1 + 1 \Rightarrow n \leq \frac{b(u)(b(u)+1)}{2} + 1 \Rightarrow b(u)^2 + b(u) - (2n - 2) \geq 0$. Roots of $b(u)$ are $\frac{-1 \pm \sqrt{8n-7}}{2}$. Considering the positive root of $b(u)$, we get $b(u) \geq \lceil \sqrt{(2n - \frac{7}{4})} - \frac{1}{2} \rceil$. □

Let us now consider the originator in G_k to be any vertex w on a cycle C_j, for some $1 \leq j \leq k$. Let us assume the length of the shorter path from w to the central vertex u be d. Then the length of the longer path from w to $u = l_j + 1 - d$ and $d \leq l_j + 1 - d$ (see Fig 2).

Lemma 3. *There is a minimum time broadcast scheme from w in G_k in which w first sends the information along the shortest path towards vertex u.*

Proof. Let S_1 be a minimum broadcast scheme, $b_{S_1}(w) = b(w, G_k)$ under which w first informs its adjacent vertex along the longer path towards vertex u. We will construct a new broadcast scheme S_2 under which w first sends information towards the shorter path. We will show that $b_{S_2}(w) \leq b_{S_1}(w) = b(w, G_k)$.

According to scheme S_1, w informs its adjacent vertex along the shorter path at time two. Now we construct a new broadcast scheme S_2 where w informs its adjacent vertex along the shorter path at time one. The order in which u broadcasts along the remaining $k-1$ cycles is the same in both schemes. However, under S_2, every vertex along the longer path towards vertex u from w will receive the message exactly one time unit later compared to S_1. To prove that $b_{S_2}(w) = b(w, G_k)$ we consider two cases:

Case 1: under S_1, u is informed along the shorter path at time $b_1 \leq b(w, G_k)$: Under S_2 all the vertices along the shorter path will be informed exactly one time unit earlier. So, u is informed at time $b_1 - 1$. u has exactly one free time unit immediately after $b_1 - 1$ to inform its adjacent vertex along the longer path towards w. Since the broadcast time in the remaining $k - 1$ paths remains the same, $b_{S_2}(w) \leq b_{S_1}(w)$.

Case 2: under S_1, u is informed along the longer path from w:

Recall the length of the shorter path is d and the length of the longer path is $l_j + 1 - d$. Under S_1, u is informed along the longer path from w when either $d = l_j + 1 - d$ or $d + 1 = l_j + 1 - d$.

When $d = l_j + 1 - d$, it is quite trivial that $b_{S_2}(w) \le b_{S_1}(w)$ since the broadcast time in the remaining $k - 1$ paths remains the same.

When $d + 1 = l_j + 1 - d$: Recall that under S_2 all the vertices along the shorter path will be informed exactly one time unit earlier. So u is informed at time unit d instead of time unit $l_j + 1 - d = d + 1$ under scheme S_1. u has exactly one free time unit immediately after d to inform its adjacent vertex along the longer path towards w. Since the broadcast time in the remaining $k - 1$ paths remains the same, $b_{S_2}(w) \le b_{S_1}(w)$.   □

2.2 Lower Bounds when Originator is Not the Central Vertex

In this section we give lower bounds on the broadcast time of G_k from w.

Lemma 4. *Let G_k be a k-cycle graph where the originator is any vertex w on a cycle C_m and the length of the shortest path from w to vertex u is d. Then*
(i) $b(w) \ge d + k$. (ii) $b(w) \ge d + \left\lceil \frac{l_j + 2j - 2}{2} \right\rceil$ for any j, $1 \le j \le k$.
(iii) $b(w) \ge d + \left\lceil \frac{2k + l_j + 2j - 2}{4} \right\rceil$ for any j, $1 \le j \le k$.

Proof. (i): By Lemma 3 there is a minimum time broadcast scheme from originator w in G_k in which w first sends the information along the shorter path towards vertex u. Considering this minimum broadcast scheme, u is informed no earlier than d time units. It takes another $k - 1$ time units to inform at least one vertex in each of the remaining $k - 1$ cycles from u. Recall that $l_j \ge 2$ for any j, where $1 \le j \le k$. So, at least one more time unit is required to inform the second vertex on the cycle which initially receives the message from u at time unit $d + k - 1$. So, $b(w) \ge d + k$.

(ii): Similarly, at least d time units are necessary for u to receive the message from w. Now, we consider any cycle C_j where $1 \le j \le k$ and $j \ne m$. Under any minimum time broadcast scheme all vertices in C_j must be informed. u informs the remaining $k - 1$ cycles in some order and assume it initially informs C_j at time unit $d + j$ or later. Then u informs C_j along the second branch no sooner than time unit $d + j + 1$. At time unit $d + j$ there are at least $l_j - 1$ uninformed vertices in C_j. Similar to the argument given in Lemma 1(ii), we can write $b(w) \ge d + j + \left\lceil \frac{l_j - 1}{2} \right\rceil = d + \left\lceil \frac{l_j + 2j - 1}{2} \right\rceil \ge d + \left\lceil \frac{l_j + 2j - 2}{2} \right\rceil$.

When $j = m$, the number of uninformed vertices in C_m at time d, denoted as $\tau(m) = l_m - (2d - 1)$. Considering $j = 1$ and $l_j = \tau(m)$ for the cycle C_m, we get $b(w) \ge d + \left\lceil \frac{l_j + 2j - 2}{2} \right\rceil$ for any j, $1 \le j \le k$ included m.

For the proof of (iii), we combine the inequalities in (i) and (ii). We get $2b(w) \ge d + k + d + \left\lceil \frac{l_j + 2j - 2}{2} \right\rceil \ge 2d + \left\lceil \frac{l_j + 2j + 2k - 2}{2} \right\rceil$. Hence, $b(w) \ge d + \left\lceil \frac{l_j + 2j + 2k - 2}{4} \right\rceil$ for any j, $1 \le j \le k$.   □

3   Approximation Algorithm

In this section we present broadcast algorithm S for graph G_k. We consider any vertex x to be the originator. When the originator is u then the algorithm S in G_k starts by informing the longest cycle C_1 in the first time unit.

When the originator is a non-central vertex w on cycle C_m then the scheme S in G_k starts by informing along the shorter path towards u. u is informed at time d. u informs the cycle with maximum number of uninformed vertices at time $d + 1$.

At time i, where $i \geq 1$ when $x = u$; else $i \geq d+1$, consider the following 3 sets of cycles: a) The set X_0 consists of the cycles where there are no informed vertices and let the cycle C_{10} has the maximum uninformed vertices of length l_{10}. b) The set X_1 consists of the cycles where at least one vertex has been informed along one branch from u and let the cycle C_{11} has the maximum uninformed vertices of length l_{11}. c) The set X_2 consists of the cycles which has been informed from u along both directions. Depending on the lengths of l_{10} and l_{11}, u decides either to inform C_{10} or C_{11} at time $i+1$. If there is no cycle in X_1 at time i, then u has no other choice but to inform C_{10} at time $i + 1$. Finally when there is no cycle in X_0 (at least one cycle will be present in X_1 at this moment), u broadcasts along the cycle having maximum number of uninformed vertices from the set X_1.

Broadcast Algorithm S_{cycle}:
INPUT: A k-cycle graph G_k where $l_1 \geq l_2 \geq ... \geq l_k \geq 2$ and any originator x.
OUTPUT: Broadcast time $b_{S_{cycle}}(x)$ and scheme of G_k.
BROADCAST-SCHEME-$S_{cycle}(G_k, l_1 \geq l_2 \geq ... \geq l_k \geq 2, x)$

1. when $x = u$, u broadcasts along C_1 at time unit 1.
2. when $x = w$, w first informs along the shorter path towards u.
    2.1 u is informed at time d.
    2.2 u informs the cycle with maximum number of uninformed vertices at time $d + 1$.
3. At time i, where $i \geq 1$ when $x = u$, else $i \geq d + 1$ consider the following 3 sets of cycles:
    3.1 X_0 : It consists of the cycles where there are no informed vertices. Let there are r cycles such that $l_{10} \geq l_{20} \geq ... \geq l_{r0}$, where l_{j0} is the length of the cycle C_{j0} in X_0 and $1 \leq j \leq r$. $C_{10}, C_{20}, ..., C_{r0}$ is a combination of r cycles from $C_1, ..., C_k$.
    3.2 X_1 : It consists of the cycles where at least one vertex has been informed along one branch from u. Let there are m cycles such that $l_{11} \geq l_{21} \geq ... \geq l_{m1}$, where l_{j1} is the number of uninformed vertices in the cycle C_{j1} in X_1 at time i and $1 \leq j \leq m$. $C_{11}, ..., C_{m1}$ is a combination of m cycles from $C_1, ..., C_k$ but not in set X_0.
    3.3 X_2 : It consists of the cycles which has been informed from u along both directions. Let there are p such cycles and $m + r + p = k$.
4. Starting at time $i + 1$ onwards until there is no cycle in X_0 do:
    4.1. If there is at least one cycle in X_1

4.1.1. Select l_{10} and l_{11}
4.1.2. If $l_{10} \geq l_{11} - 1$
     u broadcasts along C_{10} at time $i + 1$.
4.1.3. Else-If $l_{10} < l_{11} - 1$
     u broadcasts along C_{11} at time $i + 1$.
4.2. Else-If there is no cycle in X_1
     u broadcasts along C_{10} at time $i + 1$.
4.3. If u informs C_{10}
     update $X_1 = X_1 + \{C_{10}\}$ and $X_0 = X_0 - \{C_{10}\}$.
4.4. Else-If u informs C_{11}
     update $X_2 = X_2 + \{C_{11}\}$ and $X_1 = X_1 - \{C_{11}\}$.
4.5. For every cycle in X_1 do
     $l_{j1} = l_{j1} - 1$.
4.6. Arrange the cycles in X_1 in decreasing order of the number of uninformed vertices if u informs along C_{10}.
5. When there are cycles in X_1
   u informs the cycle having maximum number of uninformed vertices.

Complexity Analysis
Steps 3 and 4.6 take $O(k \log k)$ time to arrange the cycles in sorted order. Steps 4.3 and 4.4 take $O(k)$ time to update the information in X_0, X_1 and X_2. Broadcasting done in steps 1, 2, 4.1, 4.2, 4.5 and 5 will take another $O(|V|)$ time to finish. Thus, the complexity of the algorithm is $O(|V| + k \log k)$.

Theorem 1. *Algorithm S is a $(2 - \epsilon)$-approximation for any originator in the k-cycle graph G_k.*

Proof. 1) when originator is u: Under algorithm S, at any time unit u informs the cycle either in X_0 or in X_1 depending on the lengths of the cycles C_{10} and C_{11} where C_{10} and C_{11} are the cycles from $C_1, ..., C_k$ having maximum number of uninformed vertices in X_0 and X_1 respectively. Assume that under scheme S, C_j is one of the cycles where broadcasting finishes at time unit $b_S(u)$. In scheme S, C_j has been informed from u at time $2j-1$ or sooner along its first branch. Let u informs its second adjacent vertex in C_j at time t_j, where $2j - 1 < t_j \leq 2k$. At time $t_j - 1$ number of uninformed vertices in cycle C_j will be $l_j - (t_j - 2j + 1)$. Since starting at time t_j, C_j receives the message from both directions from u, then

$$b_S(u) = t_j - 1 + \left\lceil \frac{l_j - t_j + 2j - 1}{2} \right\rceil = \left\lceil \frac{l_j + t_j + 2j - 3}{2} \right\rceil \leq \left\lceil \frac{l_j + 2k + 2j - 3}{2} \right\rceil \leq \frac{l_j + 2k + 2j - 2}{2} \text{ as}$$

$t_j \leq 2k$. From Lemma 1(iii), we can write $\frac{b_S(u)}{b(u)} \leq 2\frac{l_j + 2k + 2j - 2}{l_j + 2j + 2k + 1} = 2 - \frac{6}{l_j + 2j + 2k + 1} < 2$.

2) when originator is w on cycle C_m: Under algorithm S, w first sends the information along the shorter path towards vertex u. So u gets informed at time unit d. Consider the cycle C_j, where $1 \leq j \leq k$ and $j \neq m$. Similar to the proof when originator is u, we consider in the worst case any cycle C_j which finishes broadcasting in the last time unit in S. Similarly, u calls its first adjacent vertex in C_j at time $d + 2j - 1$ or sooner and informs its second adjacent vertex in C_j at time $d + t_j$, where $2j - 1 < t_j \leq 2k - 1$. The number of uninformed vertices in

C_j before time $d+t_j$ will be $l_j - (t_j - 2j + 1)$. Now, let us consider the cycle C_m. Since starting at time two onwards w sends the information along the longer path towards vertex u, the number of informed vertices in C_m at time d will be $2d - 1$. Hence, the number of uninformed vertices in C_m before time $d + t_j$ will be $l_m - (2d - 1) - (t_j - 1) = l_m + 2 - 2d - t_j = \tau(m) - t_j + 1 < \tau(m) - (t_j - 2j + 1)$ since $j \geq 1$ and $\tau(m) = l_m + 1 - 2d$. So, for $1 \leq j \leq k$ (including $j = m$) and $l_m = \tau(m)$, C_j has at most $l_j - (t_j - 2j + 1)$ uninformed vertices. Since starting at time $d + t_j$ onwards, C_j receives the message from both directions, then $b_S(w) \leq d + t_j - 1 + \left\lceil \frac{l_j - t_j + 2j - 1}{2} \right\rceil = \left\lceil \frac{2d + l_j + t_j + 2j - 3}{2} \right\rceil \leq \left\lceil \frac{2d + l_j + 2k + 2j - 4}{2} \right\rceil$
$\leq \frac{2d + l_j + 2k + 2j - 3}{2}$ as $t_j \leq 2k - 1$. From Lemma 4(iii), we can write $\frac{b_S(w)}{b(w)} \leq$
$2 \frac{2d + l_j + 2k + 2j - 3}{4d + l_j + 2j + 2k - 2} = 2 - \frac{4d + 2}{4d + l_j + 2j + 2k - 2} < 2$. □

The above algorithm S is a $(2 - \epsilon)$-approximation algorithm in general, but it generates the exact broadcast time for some subclasses of k-cycle graph.

4   Optimality of Approximation Algorithm S for Some Subclasses of G_k

In this section we consider several cases depending on the length of C_j and for some cases we will present an optimal algorithm.

Theorem 2. *If $l_j \geq l_{j+1} + 4$ for all $1 \leq j \leq k - 1$, then algorithm S generates the optimal broadcast time in the k-cycle graph from originator u.*

Proof. In scheme S, u first informs one of its adjacent vertices along the cycle C_1 and C_1 is placed in the set X_1. Since $l_j \geq l_{j+1} + 4$, at time one, the number of uninformed vertices in C_1 is at least three more than the number of uninformed vertices in C_2 (C_2 is a cycle in X_0 having the maximum number of vertices among all cycles in X_0). So, according to scheme S, u informs C_1 at time two. In other words, u informs the two adjacent vertices along cycle C_j at times $2j - 1$ and $2j$ for $1 \leq j \leq k$. Since $l_j \geq l_{j+1} + 4$, when C_{j+1} gets informed from u at time $2(j + 1) - 1$, the number of uninformed vertices, call it l'_j in C_j are in the order $l'_j \geq l'_{j+1}$ for all $1 \leq j \leq k - 1$. Starting at time $2(j + 1)$ onwards, C_{j+1} also receives the message from both directions from u similar to C_j. As a result all the vertices in C_1 will be informed no sooner than the vertices of any other cycle in G_k. So in the worst case, we consider the time taken to inform all the vertices in C_1. In scheme S, starting at time 2 onwards, C_1 is informed from both directions from u. Thus, $b_S(u) \leq 1 + \left\lceil \frac{l_1 - 1}{2} \right\rceil = \left\lceil \frac{l_1 + 1}{2} \right\rceil$.

Using Lemma 1(ii), for $j = 1$ we get $b(u) \geq \left\lceil \frac{l_1 + 1}{2} \right\rceil \geq b_S(u)$. □

Theorem 3. *If $l_{j+1} + 4 \geq l_j \geq l_{j+1} + 3$ for all $1 \leq j \leq k - 1$, then algorithm S is a 1.2-approximation in the k-cycle graph G_k from originator u, for $k \geq 3$.*

Proof. As a result $l_k + 4(k - j) \geq l_j \geq l_k + 3(k - j)$ and the total number of vertices in G_k, denoted as $n \geq l_k + (l_k + 3) + ... + (l_k + 3(k - 1)) + 1 = kl_k + \frac{3}{2}k(k - 1) + 1$.

In scheme S, u first informs one of its adjacent vertices along the cycle C_1 and C_1 is placed in the set X_1. Since $l_1 \geq l_2 + 3$, at time one, the number of uninformed vertices in C_1 is at least two more than the number of uninformed vertices in C_2 (C_2 is a cycle in X_0 having the maximum number of vertices among all cycles in X_0). So, according to scheme S, u informs C_1 at time two. In other words, u informs the two adjacent vertices along cycle C_j at times $2j - 1$ and $2j$ for $1 \leq j \leq k$. At time $2j - 1$, number of uninformed vertices in C_j will be $l_j - 1$. Now, $l_k + 4(k - j) \geq l_j \geq l_k + 3(k - j) \Rightarrow l_k + 4(k - j) - 1 \geq l_j - 1 \geq l_k + 3(k - j) - 1 > 0$ for $j \leq k$ and $l_k \geq 2$. As a result, all the cycles will receive the message twice from u. Since starting at time $2j$ onwards, C_j receives the message from both directions from u, then $b_S(u) \leq 2j - 1 + \left\lceil \frac{l_j - 1}{2} \right\rceil \leq 2j - 1$

$+ \left\lceil \frac{l_k + 4k - 4j - 1}{2} \right\rceil = \left\lceil \frac{l_k + 4k - 3}{2} \right\rceil \leq \frac{l_k + 4k - 2}{2}$ as $l_k + 4(k - j) - 1 \geq l_j - 1$.

Now using Lemma 2(i), $b(u) \geq \left\lceil \frac{n-1}{2k} + k - \frac{1}{2} \right\rceil \geq \left\lceil \frac{kl_k + \frac{3}{2}k(k-1)}{2k} + k - \frac{1}{2} \right\rceil = \left\lceil \frac{2l_k + 7k - 5}{4} \right\rceil$.

Hence, $\frac{b_S(u)}{b(u)} \leq 2\frac{l_k + 4k - 2}{2l_k + 7k - 5} = \frac{2l_k + 8k - 4}{2l_k + 7k - 5} = 1 + \frac{k+1}{2l_k + 7k - 5} \leq 1 + \frac{k+1}{7k - 1} = 1 + \frac{k+1}{5k + 5 + 2k - 6} \leq 1 + \frac{k+1}{5k + 5} = 1.2$, for $k \geq 3$ and $l_k \geq 2$. □

Observation: Note that algorithm S gives $\frac{7}{6}$-approximation for $k \geq 10$ for the case $l_{j+1} + 4 \geq l_j \geq l_{j+1} + 3$. Moreover if k is large enough then the approximation ratio of algorithm S approaches $\frac{8}{7}$.

Theorem 4. If $l_j = l_{j+1} + 2$ for all $1 \leq j \leq k - 1$, then algorithm S generates the optimal broadcast time in the k-cycle graph from originator u.

Proof. As a result $l_j = l_k + 2(k - j)$ and total number of vertices in G_k, denoted as $n = l_k + (l_k + 2) + \ldots + (l_k + 2(k - 1)) + 1 = kl_k + k(k - 1) + 1$.

In scheme S, u first informs one of its adjacent vertices along the cycle C_1 and C_1 is placed in the set X_1. Since $l_1 = l_2 + 2$, at time one, the number of uninformed vertices in C_1 is one more than the number of uninformed vertices in C_2 (C_2 is a cycle in X_0 having the maximum number of vertices among all cycles in X_0). So, according to scheme S, u informs C_2 at time two. In general, during the first k time units, u informs C_j at time j for $1 \leq j \leq k$. At time k, number of uninformed vertices in C_j will be $l_j - (k - (j - 1)) = l_k + 2k - 2j - k + j - 1 = l_k + k - j - 1$. In other words, at time k, the number of uninformed vertices in the cycles $C_1, \ldots, C_k$ forms an arithmetic series with difference 1 starting from $l_k - 1$ up to $l_k + k - 2$. Starting at time $k + 1$ onwards u informs the cycle with maximum number of uninformed vertices. Now we are going to consider two cases:

a) $k < l_k$: This ensures that all the cycles will get informed twice from u. So in general, u informs the second vertex in cycle C_i at time $k + i$ ($1 \leq i \leq k$). Thus,

$b_S(u) = \max \left\{ k + \left\lceil \frac{l_k + k - 2}{2} \right\rceil, k + 1 + \left\lceil \frac{l_k + k - 3 - 1}{2} \right\rceil, \ldots, k + i - 1 + \left\lceil \frac{l_k + k - i - 1 - (i-1)}{2} \right\rceil \right.$

$= k + \left\lceil \frac{l_k + k - 2}{2} \right\rceil, \ldots, k + k - 1 + \left\lceil \frac{l_k - 1 - (k-1)}{2} \right\rceil = k + \left\lceil \frac{l_k + k - 2}{2} \right\rceil \right\} = \left\lceil \frac{l_k + 3k - 2}{2} \right\rceil$.

b) $k \geq l_k$: Since, at time k, the number of uninformed vertices in the cycles $C_1, ..., C_k$ forms an arithmetic series with difference 1 starting from $l_k - 1$ and $k \geq l_k$, some of the cycles will not receive the message from u twice. Assume there are p cycles $C'_1, ..., C'_p$ which will receive the information along its second branch from u starting at time $k + 1$ onwards, where $C'_1, ..., C'_p$ is a combination of p cycles from $C_1, ..., C_k$. u finishes broadcasting all its adjacent vertices along these p cycles by time $k + p$. All the vertices in the remaining $k - p$ cycles must have been informed within $k + p$ time units. From the proof in part a) it is clear that the time taken to inform any of the p cycles will be $\lceil \frac{l_k+3k-2}{2} \rceil = k + p - 1 + \lceil \frac{l_k+k-2p}{2} \rceil$. Recall that at time k, number of uninformed vertices in C'_p is $l_k + k - p - 1$. As a result, $l_k + k - 2p$ is the number of uninformed vertices in C'_p before time unit $k + p$. Since u informs C'_p at time $k + p$, then $l_k + k - 2p \geq 1$. Thus, $k + p - 1 + \lceil \frac{l_k+k-2p}{2} \rceil \geq k + p$. Hence, $b_S(u) \leq \lceil \frac{l_k+3k-2}{2} \rceil$ as in a).

Now using Lemma 2(i), $b(u) \geq \lceil \frac{n-1}{2k} + k - \frac{1}{2} \rceil = \lceil \frac{kl_k+k(k-1)}{2k} + k - \frac{1}{2} \rceil = \lceil \frac{l_k-2+3k}{2} \rceil \geq b_S(u)$. □

Theorem 5. *If $l_j = l_{j+1}$ for all $1 \leq j \leq k - 1$, then algorithm S generates the optimal broadcast time in the k-cycle graph G_k from originator u.*

Proof. As a result $l_j = l_k$.

The order in which u initially informs the cycles is similar to the proof of Theorem 4. In scheme S, u first informs one of its adjacent vertices along the cycle C_1 and C_1 is placed in the set X_1. At time one, the number of uninformed vertices in C_1 is one less than the number of uninformed vertices in C_2 (C_2 is a cycle in X_0 having the maximum number of vertices among all cycles in X_0). So, u informs C_2 at time two. In general, during the first k time units, u informs C_j at time j for $1 \leq j \leq k$. At time k, number of uninformed vertices in C_j will be $l_j - (k - (j - 1)) = l_k - k + j - 1$. According to scheme S, starting from time $k + 1$ onwards, u informs C_k, C_{k-1},..., C_j,..., C_1 at times $k + 1$, $k + 2$,..., $2k + 1 - j$,..., $2k$ respectively. In general, C_j will have $l_k - k + j - 1 - (k - j) = l_k - 1 - 2(k - j) \geq l_k - (2k - 1)$ uninformed vertices before $2k + 1 - j$ time units as $j \geq 1$. There are two cases to consider.

a) $l_k \geq 2k$: This guarantees that u has enough time to inform all its adjacent vertices in k cycles. Starting at time $2k + 1 - j$ onwards, C_j receives the message from both directions from u. Thus $b_S(u) \leq 2k - j + \lceil \frac{l_k-1-2k+2j}{2} \rceil = \lceil \frac{l_k-1+2k}{2} \rceil$.

b) $l_k < 2k$: As a result, some of the cycles will not receive the message from u twice. Let us assume there are $k - p + 1$ such cycles $C'_k, ..., C'_p$ which will receive the information along its second branch from u starting at time $k + 1$ onwards, where $C'_k, ..., C'_p$ is a combination of $k - p + 1$ cycles from $C_1, ..., C_k$. u finishes broadcasting all its adjacent vertices along these $k - p + 1$ cycles by time $2k + 1 - p$. All the vertices in the remaining $p - 1$ cycles must have been informed within $2k + 1 - p$ time units. From the proof in part a) it is clear that the time taken to inform any of the $k - p + 1$ cycles will be $\lceil \frac{l_k+2k-1}{2} \rceil = 2k - p + \lceil \frac{l_k-1-2k+2p}{2} \rceil$. Now, $l_k - 1 - 2k + 2p$ is the number of uninformed vertices in C'_p before time unit $2k+1-p$. Since u informs C'_p at time $2k+1-p$, then $l_k - 1 - 2k + 2p \geq 1$. Thus, $2k - p + \left\lceil \frac{l_k - 1 - 2k + 2p}{2} \right\rceil \geq 2k + 1 - p$. Hence, $b_S(u) \leq \left\lceil \frac{l_k + 2k - 1}{2} \right\rceil$ as in a).

Now using Lemma 2(i), $b(u) \geq \left\lceil \frac{n-1}{2k} + k - \frac{1}{2} \right\rceil$ (n is the total number of vertices in G_k) $= \left\lceil \frac{kl_k}{2k} + k - \frac{1}{2} \right\rceil = \left\lceil \frac{l_k - 1 + 2k}{2} \right\rceil \geq b_S(u)$ as $n - 1 = kl_k$. □

Theorem 6. *If $l_j \leq l_{j+1} + 1$ for all $1 \leq j \leq k - 1$, then algorithm S is a $(1.5 - \epsilon)$-approximation in the k-cycle graph G_k from originator u.*

Proof. As a result $l_j \leq l_k + k - j$ and total number of vertices in G_k, denoted as $n \leq l_k + (l_k + 1) + \dots + (l_k + k - 1) + 1 = kl_k + \frac{k(k-1)}{2} + 1$.

In scheme S, u first informs one of its adjacent vertices along the cycle C_1 and C_1 is placed in the set X_1. Since $l_1 \leq l_2 + 1$, at time one, the number of uninformed vertices in C_1 is either exactly the same or one less than the number of uninformed vertices in C_2 (C_2 is a cycle in X_0 having the maximum number of vertices among all cycles in X_0). So, according to scheme S, u informs C_2 at time two. In general, during the first k time units, u informs C_j at time j for $1 \leq j \leq k$. At time k, number of uninformed vertices in C_j will be $l_j - (k - (j - 1)) \leq l_k + k - j - k + j - 1 = l_k - 1$. Starting from time $k + 1$ onwards, u informs along the cycle having maximum number of uninformed vertices. Now, we are going to consider two cases:

a) $k < l_k$: This ensures that all the cycles receive the message twice from u. Thus, $b_S(u) \leq \max\{k + \lceil \frac{l_k - 1}{2} \rceil, k + 1 + \lceil \frac{l_k - 2}{2} \rceil, \dots, k + (i - 1) + \lceil \frac{l_k - 1 - i + 1}{2} \rceil = k + \lceil \frac{l_k + i - 2}{2} \rceil, \dots, k + k - 1 + \lceil \frac{l_k - 1 - (k-1)}{2} \rceil\} = k + \lceil \frac{l_k + k - 2}{2} \rceil \leq \frac{l_k + 3k - 1}{2}$

b) $k \geq l_k$: By time k, u has informed at least one vertex in each cycle and each cycle has at most $l_k - 1$ uninformed vertices at time unit k. As a result, it will take at most another $l_k - 1$ time units to inform all the vertices in G_k. Thus, $b_S(u) \leq k + l_k - 1 = k + \frac{2l_k - 2}{2} < k + \frac{2l_k - 1}{2} < k + \frac{l_k + k - 1}{2} = \frac{l_k + 3k - 1}{2}$ as $k > l_k$.

Thus, for both cases, we get $b_S(u) \leq \frac{l_k + 3k - 1}{2}$.

Now using Lemma 1(ii) for $j = k$ we get, $\frac{b_S(u)}{b(u)} \leq \frac{l_k + 3k - 1}{l_k + 2k - 1} = 1 + \frac{k}{l_k + 2k - 1}$ $\leq 1 + \frac{k}{2k+1} < 1.5$ as $l_k \geq 2$.

□

5   Conclusion and Future Work

In this paper we considered a simple graph where cycles intersect at single vertex, namely k-cycle graph. As it turned out finding the exact broadcast time in this graph is not very simple. We give an approximation algorithm for the k-cycle graph, with the approximation ratio 2. We also show that our algorithm generates the optimum broadcast time when the difference of cycle lengths between each pair is at least 4. Minimum time broadcasting in k-cycle graph is difficult when the lengths of the paths form an arithmetic series with difference either 1 or 3. The future work in this area of course will be to design a polynomial algorithm to find the exact broadcast time or to prove that the broadcast problem in arbitrary k-cycle graph is NP-hard.

References

1. Bar-Noy, A., Guha, S., Naor, J., Schieber, B.: Multicasting in heterogeneous networks. In: Proceedings of the Thirtieth Annual ACM Symposium on Theory of Computing (STOC 1998), pp. 448–453 (1998)
2. Beier, R., Sibeyn, J.F.: A powerful heuristic for telephone gossiping. In: Proceedings of the 7th International Colloquium on Structural Information Communication Complexity (SIROCCO 2000), pp. 17–36 (2000)
3. Bhabak, P., Harutyunyan, H.A.: Approximation algorithm for the broadcast time in k-path graph (abstract only). In: 3rd International Symposium on Combinatorial Optimization (ISCO 2014), p. 83 (2014)
4. Elkin, M., Kortsarz, G.: Combinatorial logarithmic approximation algorithm for directed telephone broadcast problem. In: Proceedings of the thirty-fourth Annual ACM Symposium on Theory of Computing (STOC 2002), pp. 438–447 (2002)
5. Elkin, M., Kortsarz, G.: Sublogarithmic approximation for telephone multicast: path out of jungle (extended abstract). In: Proceedings of the Fourteenth Annual ACM-SIAM Symposium on Discrete Algorithms (SODA 2003), pp. 76–85 (2003)
6. Fraigniaud, P., Vial, S.: Approximation algorithms for broadcasting and gossiping. J. Parallel and Distrib. Comput. 43(1), 47–55 (1997)
7. Fraigniaud, P., Vial, S.: Heuristic algorithms for personalized communication problems in point-to-point networks. In: Proceedings of the 4th Colloquium on Structural Information Communication Complexity (SIROCCO 1997), pp. 240–252 (1997)
8. Fraigniaud, P., Vial, S.: Comparison of heuristics for one-to-all and all-to-all communication in partial meshes. Parallel Processing Letters 9, 9–20 (1999)
9. Harutyunyan, H.A., Maraachlian, E.: On broadcasting in unicyclic graphs. J. Comb. Optim. 16(3), 307–322 (2008)
10. Harutyunyan, H.A., Maraachlian, E.: Broadcasting in fully connected trees. In: Proceedings of the 2009 15th International Conference on Parallel and Distributed Systems (ICPADS 2009), pp. 740–745 (2009)
11. Harutyunyan, H.A., Shao, B.: An efficient heuristic for broadcasting in networks. J. Parallel Distrib. Comput. 66(1), 68–76 (2006)
12. Harutyunyan, H.A., Wang, W.: Broadcasting algorithm via shortest paths. In: Proceedings of the 2010 IEEE 16th International Conference on Parallel and Distributed Systems (ICPADS 2010), pp. 299–305 (2010)
13. Jansen, K., Muller, H.: The minimum broadcast time problem for several processor networks. Theoretical Computer Science 147, 69–85 (1995)
14. Kortsarz, G., Peleg, D.: Approximation algorithms for minimum time broadcast. SIAM J. Discrete Math. 8, 401–427 (1995)
15. Middendorf, M.: Minimum broadcast time is np-complete for 3-regular planar graphs and deadline 2. Inf. Proc. Lett. 46, 281–287 (1993)
16. Ravi, R.: Rapid rumor ramification: approximating the minimum broadcast time. In: Proceedings of the 35th Annual Symposium on Foundations of Computer Science (FOCS 1994), pp. 202–213 (1994)
17. Scheuermann, P., Wu, G.: Heuristic algorithms for broadcasting in point-to-point computer networks. IEEE Trans. Comput. 33(9), 804–811 (1984)
18. Schindelhauer, C.: On the inapproximability of broadcasting time. In: Jansen, K., Khuller, S. (eds.) APPROX 2000. LNCS, vol. 1913, pp. 226–237. Springer, Heidelberg (2000)
19. Slater, P.J., Cockayne, E.J., Hedetniemi, S.T.: Information dissemination in trees. SIAM J. Comput. 10(4), 692–701 (1981)

Three Paths to Point Placement

Md. Shafiul Alam and Asish Mukhopadhyay[*]

School of Computer Science, University of Windsor,
401 Sunset Avenue, Windsor, ON, N9B 3P4, Canada

Abstract. The point placement problem is to determine the positions
of a set of n *distinct* points, $P = \{p_1, p_2, p_3, \ldots, p_n\}$, on a line uniquely,
up to translation and reflection, from the fewest possible distance queries
between pairs of points. Each distance query corresponds to an edge in
a graph, called point placement graph (*ppg*), whose vertex set is P. The
uniqueness requirement of the placement translates to line rigidity of the
ppg. In this paper, we show how to construct in 2 rounds a line rigid *ppg*
of size $9n/7 + O(1)$. This improves the best known result of $4n/3 + O(1)$.
We also improve the lower bound on 2-round algorithms from $14n/13$ to
$9n/8$.

1 Introduction

The Problem. Let $P = \{p_1, p_2, \ldots, p_n\}$ be a set of n distinct points on a line
L. In this paper, we address the problem of determining a unique placement
(up to translation and reflection) of the p_i's on L, by making the minimum
number of adversarial distance queries between some pairs of points p_i and p_j,
$1 \leq i, j \leq n$, spread over one or more rounds. It is assumed that the distances
returned by the adversary are valid, which means that there exists a placement
consistent with these distances. The queries are presented to the adversary as
a graph, called a point placement graph (*ppg*) or query graph, whose edges
connnect pairs of points whose distances are being queried. Depending upon
the adversary's answers, the *ppg* is adaptively modified by inserting additional
edges and querying these. This process is repeated over the successive rounds.
A *ppg* G is said to be line-rigid if (or simply rigid) if its vertices have a unique
placement on L for a valid assignment of lengths to its edges. The length of an
edge is the same as the distance bewteen its end-points. The *ppg* of Fig. 1 is line-
rigid because each of its constituent triangle,with a common base, is line-rigid
and leads to a 1-round point-placement algorithm that makes $2n - 3$ distance
queries.

Is it possible that the query complexity can be lowered if the *ppg* consisted of
quadrilaterals hung from a common base ? Unfortunately, quadrilaterals, unlike
triangles, are not rigid as there is a valid assignment of lengths that makes a
quadrilateral a parallelogram whose vertices have two distinct layouts (see Fig.2).

---

[*] Research supported by an NSERC discovery grant awarded to this author.

S. Ganguly and R. Krishnamurti (Eds.): CALDAM 2015, LNCS 8959, pp. 33–44, 2015.

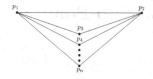

Fig. 1. Query graph using triangles

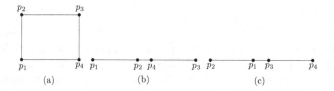

Fig. 2. Point placement graph in the shape of a quadrilateral (a) with opposite edges being equal have 2 placements as shown in (b) and (c)

Though a quadrilateral is not line-rigid, it becomes so if we require a pair of opposite sides to be unequal; for example, $|p_2p_3| \neq |p_1p_4|$. Such a rigidity condition can be met by a 2-round algorithm that makes use of the following useful observation and the *ppg* of Fig. 3, presented to an adversary in the first round.

Observation 1. *At most two equal length edges that are collinear with a line L can be incident to a point p on L.*

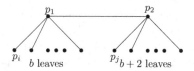

Fig. 3. Query graph for first round in a 2-round algorithm using quadrilaterals

In the second round, we form rigid quadrilaterals $p_1p_ip_jp_2$ by querying edges joining pairs of points p_i and p_j, $3 \leq i,j \leq n$, that satisfy the rigidity condition $|p_1p_i| \neq |p_2p_j|$. In view of the Observation 1, this is ensured by having 2 extra edges at p_2. This algorithm makes a total of $3n/2 - 1$ distance queries.

Motivation. The origins of this problem can be traced back to areas as diverse as computational biology, learning theory, computational geometry, etc. For example, in learning theory [5] this problem is one of learning a set of points on a line non-adaptively, when learning has to proceed based on a fixed set of given distances, or adaptively when learning proceeds in rounds, with the edges queried in one round depending on those queried in the previous rounds.

Prior Work. Early research on this problem was reported in [8,7]. In [5] it was shown that the jewel (see Fig. 4)and $K_{2,3}$ are both rigid, as also how to build large rigid graphs of density 8/5 (this is an asymptotic measure of the number of edges per vertex as the number of vertices go to infinity) out of the jewel. In a subsequent paper, Damaschke [6] proposed a randomized 2-round strategy that makes $(1 + o(1))n$ distance queries with high probability and also showed that this is not possible with 2-round deterministic strategies. The computational complexity of this algorithm is exponential, making it a completely impractical point placement algorithm (our implementation coinfirms this, too). Chin et al. [3] improved many of the results of [5]. Their principal contributions are a 3-round construction of rigid graphs of density 5/4 from 6-cycles and a lower bound of $17n/16$ for any 2-round algorithm. They also introduced the following concept of a *layer graph*, useful for finding conditions that make a *ppg* rigid.

Definition 1. *We first choose two orthogonal directions* **x** *and* **y** *(actually, any 2 non-parallel directions will do). A graph G admits a layer graph drawing if the following 4 properties are satisfied:*

P1 *Each edge e of G is parallel to one of the two orthogonal directions* **x** *and* **y**.
P2 *The length of an edge e is the distance between the corresponding points on L.*
P3 *Not all edges are along the same direction (thus a layer graph has a two-dimensional extent).*
P4 *When the layer graph is folded onto a line, by a rotation either to the left or to the right about an edge of the layer graph lying on this line, no two vertices coincide.*

Chin *et al.* [3] proved the following result.

Theorem 1. *A ppg G is rigid iff it cannot be drawn as a layer graph.*

In [1] we proposed a 2-round algorithm that makes $4n/3 + O(1)$ distance queries to construct rigid *ppg* on n points using a 6:6 jewel as the basic component.

Our Contribution. In this paper, we propose a 2-round algorithm that queries $9n/7 + O(1)$ edges to construct a rigid *ppg* on n points, using a 3-path graph as the basic component, bettering the previous best known result $4n/3 + O(1)$ of [1]. We also improve the best previously known lower bound on any 2-round algorithm from $14n/13$ [1] to $9n/8$.

2 A New 2-Round Algorithm

The graph, G_{3p} of Fig. 5 is our basic building block. As it can be drawn as a layer graph by Theorem 1 it is not rigid.

To find a set of conditions that make G_{3p} rigid, first we fix the placements of p_1, p_2 and p_3. Next, we find conditions that make the 7-cycle $(p_1, q_1, r_1, s, r_2, q_2, p_2)$ rigid. Relative to the fixed placement of p_3 and s, we have a 4-cycle (p_3, q_3, r_3, s) with a virtual edge between s and p_3. A condition that makes this 4-cycle rigid, plus the set of rigidity conditions of the 7-cycle, gives us the required set.

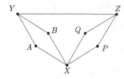

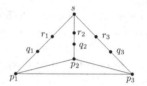

Fig. 4. The jewel graph   **Fig. 5.** The 3-path graph

By Theorem 2, a 7-cycle has 42 different layer graph representations. We find conditions that make these impossible. By Theorem 1, these constitute the set of conditions for the line rigidity of the 7-cycle. The 42 layer graphs for a 7-cycle form 6 groups, based on the number of edges on each side. For the 7-cycle $(p_1, q_1, r_1, s, r_2, q_2, p_2)$, Fig. 6 shows a representative layer graph for each group.

Theorem 2. *There are $2^{n-1} - \frac{n^2-n+2}{2}$ different layer graph representations of an n-vertex cycle [2].*

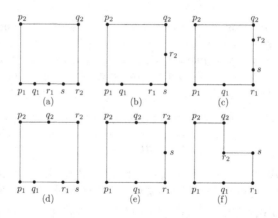

Fig. 6. An example of each group of layer graph for the 7-cycle $(p_1, q_1, r_1, s, r_2, q_2, p_2)$

From these layer graphs we derive the conditions for rigidity of this cycle. Thus, for the group represented by Fig. 6(a), the 7 layer graphs are shown in Fig. 7. These yield the rigidity conditions: $|p_1q_1| \neq |r_1s|, |q_1r_1| \neq |sr_2|, |r_1s| \neq |r_2q_2|, |sr_2| \neq |q_2p_2|, |r_2q_2| \neq |p_2p_1|, |q_2p_2| \neq |p_1q_1|$ and $|p_2p_1| \neq |q_1r_1|$. The conditions from all the groups are given in Lemma 1.

Lemma 1. *A 7-cycle $(p_1, q_1, r_1, s, r_2, q_2, p_2)$ is rigid if*

1. $|p_1p_2| \neq |q_2r_2|, |p_1p_2| \neq |q_1r_1|, |p_2q_2| \neq |r_2s|, |p_1q_1| \neq |r_1s|, |q_2r_2| \neq |r_1s|,$
   $|q_1r_1| \neq |r_2s|, |p_1q_1| \neq |p_2q_2|.$
2. $||p_1p_2| \pm |p_2q_2|| \neq |r_2s|, ||p_2q_2| \pm |q_2r_2|| \neq |r_1s|, ||p_1p_2| \pm |p_1q_1| \pm |p_2q_2|| \neq |r_1s|,$
   $|p_1q_1| \neq ||r_1s| \pm |r_2s||, |p_1p_2| \neq ||q_1r_1| \pm |r_1s||, ||p_1q_1| \pm |q_1r_1|| \neq |p_2q_2|,$
   $||p_1q_1| \pm |p_1p_2|| \neq |q_2r_2|.$

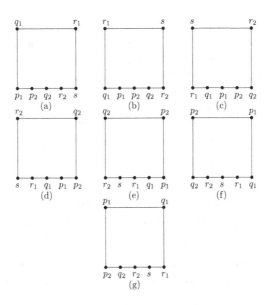

Fig. 7. A group of layer graphs for the 7-cycle $(p_1, q_1, r_1, s, r_2, q_2, p_2)$ where 4 edges lie on one edge of the layer graph

3. $||p_1p_2|\pm|p_1q_1|| \neq |r_1s|, ||p_1q_1|\pm|q_1r_1|| \neq |r_2s|, ||p_1p_2|\pm|p_1q_1|\pm|p_2q_2|| \neq |r_2s|,$
$|p_2q_2| \neq ||r_1s| \pm |r_2s||, |p_1p_2| \neq ||q_2r_2| \pm |r_2s||, ||p_2q_2| \pm |q_2r_2|| \neq |p_2q_2|,$
$||p_2q_2| \pm |p_1p_2|| \neq |q_1r_1|.$

4. $|p_1p_2| \neq |r_2s|, |p_1p_2| \neq |r_1s|, |p_2q_2| \neq |r_1s|, |p_1q_1| \neq |r_2s|, ||p_1q_1| \pm |p_2q_2| \pm |p_1p_2|| \neq ||r_1s| \pm |r_2s||, |p_2q_2| \neq |q_1r_1|, |p_1q_1| \neq |q_2r_2|.$

5. $|p_2q_2| \neq ||p_1p_2|\pm|r_1s||, |p_1q_1| \neq ||p_1p_2|\pm|r_2s||, |p_1q_1| \neq ||p_1p_2|\pm|r_1s|\pm|r_2s||,$
$|p_2q_2| \neq ||p_1p_2|\pm|r_1s|\pm|r_2s||, |p_1q_1| \neq ||q_2r_2|\pm|r_2s||, |p_2q_2| \neq ||q_1r_1|\pm|r_1s||,$
$|p_1p_2| \neq ||r_1s| \pm |r_2s||.$

6. $||p_1q_1| \pm |q_1r_1|| \neq ||p_2q_2| \pm |r_2s||, ||p_2q_2| \pm |q_2r_2|| \neq ||p_1q_1| \pm |r_1s||, |q_1r_1| \neq$
$||p_2q_2| \pm |r_2s||, |q_2r_2| \neq ||p_1q_1| \pm |r_1s||, |p_2q_2| \neq ||p_1q_1| \pm |r_1s||, |p_1q_1| \neq$
$||p_2q_2| \pm |r_2s||, |p_2q_2| \neq ||p_1q_1| \pm |r_1s| \pm |r_2s||.$

Proof. Omitted for lack of space. □

As the above conditions involve all the edges of the 7-cycle, edge lengths returned in the first query round may fail to satisfy all the rigidity conditions. Thus we re-engineer these so as not to involve q_1r_1 and q_2r_2, whose lengths are queried only in the second round; similarly, we can defer querying the length of edge q_3r_3 of the four cycle to the second round as its rigidity condition is captured by:

$$|p_3q_3| \neq |r_3s|. \tag{1}$$

The trade-off is that we will have to provide a sufficient number of choices for the edges p_1q_1, p_2q_2, p_3q_3, so that the rigidity conditions on the lengths of these edges, expressed as a function of the lengths of other edges, can be met no matter what the lengths of these other edges are.

Among the 42 rigidity conditions in Lemma 1 for the 7-cycle, 20 involve $q_1 r_1$ or $q_2 r_2$. We next show how to replace these 20 conditions with ones sans these edges. To this end, for each of these conditions, first we try to use other edges of the cycle in the layer graph representation corresponding to that condition. If this fails, then we embed the layer graph representation, corresponding to that condition, into all possible layer graph representations of the entire G_{3p}, and derive a rigidity condition from each such embedding.

2.1　Replacing Conditions

Replacing $|p_1 p_2| \neq |q_2 r_2|$

The rigidity condition $|p_1 p_2| \neq |q_2 r_2|$ corresponds to the layer graph of Fig. 6(a). To replace this condition we find a set of conditions that prevent the drawing of layer graph of the 7-cycle $(p_1, q_1, r_1, s, r_2, q_2, p_2)$ in the configuration of Fig. 6(a).

Lemma 2. *The 7-cycle* $(p_1, q_1, r_1, s, r_2, q_2, p_2)$ *of* G_{3p} *cannot be drawn as the layer graph of Fig. 6(a) if the edges of* G_{3p} *satisfy the following set of conditions:*
$\{|p_1 p_3| \neq ||p_3 q_3| \pm |r_3 s||, |p_1 p_3| \neq |r_3 s||, ||p_3 q_3| \pm |sr_2|| \neq |p_2 q_2|, ||p_3 q_3| \pm |sr_2| \pm |sr_3|| \neq |p_2 q_2|\}.$

Proof. Omitted for lack of space

Similarly, we can replace the remaining 7 conditions corresponding to the layer graphs in groups 1 and 2, involving the edges $q_1 r_1$ and $q_2 r_2$. The result is summarized in the next two lemmas:

Lemma 3. *The 7-cycle* $(p_1, q_1, r_1, s, r_2, q_2, p_2)$ *of* G_{3p} *cannot be drawn as the layer graphs corresponding to the conditions* $|p_1 p_2| \neq |q_1 r_1|$, $|q_2 r_2| \neq |r_1 s|$ *and* $|q_1 r_1| \neq |r_2 s|$ *if the edges of* G_{3p} *satisfy the following conditions:*

1. $||p_1 p_3| \pm |r_3 s|| \neq |p_3 q_3|$, $|r_3 s| \neq ||p_2 p_3| \pm |r_2 s| \pm |p_2 q_2||$, $|p_3 q_3| \neq |r_1 s|$ *and* $|p_3 q_3| \neq ||r_1 s| \pm |r_3 s||$.
2. $||p_2 p_3| \pm |r_3 s|| \neq |p_3 q_3|$, $|r_3 s| \neq ||p_1 p_3| \pm |r_1 s| \pm |p_1 q_1||$, $|p_3 q_3| \neq |r_2 s|$ *and* $|p_3 q_3| \neq ||r_2 s| \pm |r_3 s||$.
3. $||p_3 q_3| \pm |r_3 s|| \neq |p_2 p_3|$, $|p_2 p_3| \neq |r_2 s|$, $|p_3 q_3| \neq ||p_1 q_1| \pm |r_1 s||$ *and* $||p_3 q_3| \pm |r_3 s|| \neq ||p_1 q_1| \pm |r_1 s||$.

Proof. Omitted for lack of space.　　　　　　　　　　　　　　　　　□

Lemma 4. *The 7-cycle* $(p_1, q_1, r_1, s, r_2, q_2, p_2)$ *of* G_{3p} *cannot be drawn as the layer graphs corresponding to the conditions* $||p_2 q_2| \pm |q_2 r_2|| \neq |r_1 s|$, $|p_1 p_2| \neq ||q_1 r_1| \pm |r_1 s||$, $||p_1 q_1| \pm |q_1 r_1|| \neq |p_2 q_2|$ *and* $||p_1 q_1| \pm |p_1 p_2|| \neq |q_2 r_2|$ *if the edges of* G_{3p} *satisfy the following conditions:*

1. $||p_3 q_3| \pm |r_3 s|| \neq ||p_2 p_3| \pm |r_2 s||$, $||p_2 p_3| \pm |r_2 s|| \neq |r_3 s|$, $|p_3 q_3| \neq |r_1 s|$ *and* $||p_3 q_3| \neq |r_1 s| \pm |r_3 s||$.
2. $||p_3 q_3| \pm |r_3 s|| \neq |p_2 p_3|$, $|p_2 p_3| \neq |r_3 s|$, $|p_1 q_1| \neq |r_1 s|$, $|p_2 q_2| \neq |r_2 s|$, $|p_3 q_3| \neq |p_1 q_1|$ *and* $||p_3 q_3| \pm |p_1 q_1|| \neq |r_3 s|$.

3. $||p_3q_3| \pm |r_3s| \pm |r_1s|| \neq |p_1p_3|$, $|p_2q_2| \neq |r_2s|$, $|p_1p_3| \neq |r_1s| \pm |r_3s|$, $|p_2q_2| \neq |p_3q_3|$, $|p_1q_1| \neq |r_1s|$ and $|p_2q_2| \neq ||p_3q_3| \pm |r_3s||$.

4. $|p_3q_3| \pm |r_3s| \pm |p_1q_1| \neq |p_1p_3|$, $|p_2q_2| \neq |r_2s|$, $|p_1q_1| \pm |p_1p_3| \neq |r_3s|$, $|p_2q_2| \pm |r_2s| \neq |p_3q_3|$, $|p_1q_1| \neq |r_1s|$ and $||p_2q_2| \pm |r_2s|| \neq ||p_3q_3| \pm |r_3s||$.

Proof. Omitted for lack of space.   □

By similar considerations we can replace all the conditions in groups 3-6 that involve q_1r_1 and/or q_2r_2 by another set of rigidity conditions.

2.2   Rigidity Conditions

From Eq. 1 and the results of the previous section, we have the following lemma for the rigidity of the 7-cycle $(p_1, q_1, r_1, s, r_2, q_2, p_2)$ of G_{3p}.

Lemma 5. *The 7-cycle $(p_1, q_1, r_1, s, r_2, q_2, p_2)$ of G_{3p} is rigid if the edges of G_{3p} satisfy the following conditions.*

1. $|p_1p_2| \notin \{|r_1s|, |r_2s|, ||r_1s| \pm |r_2s||\}$,
2. $|p_2p_3| \notin \{|r_2s|, |r_3s|, ||r_2s| \pm |r_3s||\}$,
3. $|p_3p_1| \notin \{|r_3s|, |r_1s|, ||r_3s| \pm |r_1s||\}$,
4. $|p_1q_1| \notin \{|r_1s|, |r_2s|, ||r_1s| \pm |r_2s||, ||p_1p_2| \pm |r_1s||, ||p_1p_2| \pm |r_2s||, ||p_1p_3| \pm |r_1s||, ||p_1p_3| \pm |r_3s||, ||p_1p_2| \pm |r_1s| \pm |r_2s||, ||p_1p_3| \pm |r_1s| \pm |r_3s||\}$,
5. $|p_2q_2| \notin \{|r_1s|, |r_2s|, |p_1q_1|, ||r_1s| \pm |r_2s||, ||p_1p_2| \pm |r_1s||, ||p_1p_2| \pm |r_2s||, ||p_2p_3| \pm |r_2s||, ||p_2p_3| \pm |r_3s||, ||p_1q_1| \pm |r_1s||, ||p_1q_1| \pm |r_2s||, ||p_1p_2| \pm |r_1s| \pm |r_2s||, ||p_2p_3| \pm |r_2s| \pm |r_3s||, ||p_1q_1| \pm |r_1s| \pm |r_2s||, ||p_1q_1| \pm |p_1p_2| \pm |r_1s||, ||p_1q_1| \pm |p_1p_2| \pm |r_2s||, ||p_1q_1| \pm |p_1p_2| \pm |r_1s| \pm |r_2s||\}$,
6. $|p_3q_3| \notin \{|r_1s|, |r_2s|, |r_3s|, |p_1q_1|, |p_2q_2|, ||r_2s| \pm |r_3s||, ||r_3s| \pm |r_1s||, ||p_1p_3| \pm |r_3s||, ||p_2p_3| \pm |r_3s||, ||p_1q_1| \pm |r_1s||, ||p_1q_1| \pm |r_3s||, ||p_2q_2| \pm |r_2s||, ||p_2q_2| \pm |r_3s||, ||p_1p_3| \pm |r_1s| \pm |r_3s||, ||p_2p_3| \pm |r_2s| \pm |r_3s||, ||p_1q_1| \pm |r_1s| \pm |r_3s||, ||p_2q_2| \pm |r_2s| \pm |r_3s||, ||p_1q_1| \pm |p_1p_3| \pm |r_3s||, ||p_2q_2| \pm |p_2p_3| \pm |r_3s||, ||p_1q_1| \pm |p_1p_3| \pm |r_1s| \pm |r_2s||, ||p_2q_2| \pm |p_2p_3| \pm |r_2s| \pm |r_3s||\}$.

The union of the two sets of conditions in Eq. 1 and Lemma 5 constitutes a set of sufficient conditions for the rigidity of G_{3p}. Eliminating overlapping conditions between the two sets, we have 55 distinct conditions for the rigidity of G_{3p} and hence the following lemma:

Lemma 6. *The 3-path ppg G_{3p} having the vertices p_1, p_2 and p_3 rigid in the first round, is rigid if its edges satisfy the conditions mentioned in Lemma 5.*

2.3   Algorithm

To fix the placement of the vertices (p_1, p_2, p_3) of each 3-path component in the first round we have to satisfy the rigidity conditions on the edges p_1p_2, p_2p_3 and p_3p_1 (Conditions 1-3 of Lemma 5). This is done by picking (p_1, p_2, p_3) from a sufficiently large pool of vertices, S, whose layout is fixed in the first round.

We fix the placement of the remaining 7 vertices of each 3-path component in the second round.

In addition, we need to attach the 3-path components to the vertices of S in such a way that the number of components attached to any two vertices differ by at most a constant number. The number of basic components attached to a vertex in S is called its *valence* and the subset of vertices with valence d is named S_d.

The next result specifies how big S has to be to satisfy the above two constraints.

Lemma 7. *A set S of 35 vertices is sufficient to ensure that the valences of any two vertices in S differ by at most 2.*

Proof. Omitted for lack of space.

We fix the placments of the vertices in S by using 6 different jewel graphs [5] on 32 vertices and a triangle graph on the remaining 3. This accounts for a total of 55 edges queries in the first round.

Query Graph Construction. Let the total number of vertices be $n = 245b + 4,419$, where b is a positive integer. We attach at least $3b$ and at most $3b + 2$ 3-path components (Fig. 5) to each of the 35 rigid vertices in S, with the total number of such components being at most $35b + 11$.

The first-round distance queries correspond to the edges of the *ppg* of Fig. 8. The box encloses the vertices of S and edges connecting these (not shown). Each of the vertices p_i, p_j, or p_k $(i, j, k = 1, ..., 35)$ of S has $b + 124$ leaves to attach $3b$, $3b + 1$ or $3b + 2$ 3-path components (Fig. 5). As there there are $35b + 11$ 3-path components, we make $35b + 11$ groups of 4 vertices $(r_{il}, r_{jl}, r_{kl}, s_l)$, $(l = 1, ..., 35b + 11)$. We query the distances $|r_{il}s_l|$, $|r_{jl}s_l|$ and $|r_{kj}s_l|$, $(l = 1, ..., 35b + 11)$ in the first round. This makes up a total of $210b + 4,428$ pairwise distance queries in the first round for the placement of $n = 245b + 4,419$ vertices.

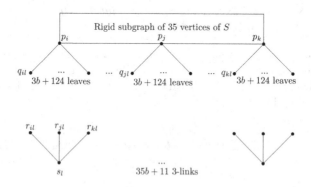

Fig. 8. Queries in the first round

In the second round, for each 3-link $(r_{il}, r_{jl}, r_{kl}, s_l), l = 1, ..., 35b + 11$, we construct a 3-path component (Fig. 5), satisfying all its rigidity conditions as in

Lemma 5. For each such 3-link we select a vertex p_i from S of minimum valence. Since the placements of the vertices of S are fixed in the first round, for any fixed pair vertices $\{p_i, p_j\}(i, j = 1, \dots 35; i \neq j)$ the distance $|p_i p_j|$ is known. So, for each pair of vertices $\{p_i, p_j\}(i, j = 1, \dots, 35; i \neq j)$, we shall use (p_i, p_j) as an edge in the construction of the 3-path component of Fig. 5.

From the remaining set of vertices of S we select another vertex $p_j (j \neq i)$ such that $|p_i p_j|$ satisfies all four rigidity conditions as in serial number 1 of Lemma 5, and, in addition, has the lowest valence of all qualifying vertices. We can always find p_j, as there are at most 8 edges $(p_i p_j)$ whose lengths do not satisfy the rigidity conditions on it (Lemma 5) whereas we have 34 vertices from which to choose the vertex p_j. Similar arguments show that the size of S is large enough to choose a $p_k (k \neq i, k \neq j)$ of minimum valence to satisfy the rigidity condition on the lengths $|p_j p_k|$ and $|p_k p_i|$ as stated in items 2 and 3 of Lemma 5.

Next we find an edge $p_i q_{il}$ rooted at p_i, satisfying the 20 rigidity conditions as in item 4 of Lemma 5, and again another edge $p_j q_{jl}$, rooted at p_j and satisfying the 45 conditions on it as in item no. 5 of Lemma 5; finally, we select $p_k q_{kl}$ rooted at p_k, satisfying the 61 conditions on it as in item no. 6 of Lemma 5.

Then for each $l, (l = 1, \dots, 35b + 11)$, we query the distances $|q_{il} r_{il}|$, $|q_{jl} r_{jl}|$ and $|q_{kl} r_{kl}|$ to form a 3-path component with vertices $p_i, p_j, p_k, q_{il}, q_{jl}, q_{kl}, r_{il}, r_{jl}, r_{kl}$ and s_l. Its edges will satisfy all the rigidity conditions of Lemma 5. Thus, all the $35b + 11$ 3-links will be used up to construct $35b + 11$ 3-path components. This accounts for $105b + 33$ distances queried in the second round.

This leaves us with 4,307 unused leaves $q_{il}/q_{jl}/q_{kl}$; the placements of 4306 can be fixed by the 4-cycle algorithm discussed in the Introduction, and that of the residual vertex by using a triangle as the ppg. A total of 2,153 distance queries are made to complete the 4-cycles and one more to construct the triangle. Thus the total number of distance queries made in the second round is $105b + 2,187$.

Theorem 3. *The ppg constructed by Algorithm is rigid.*

Proof. Omitted for lack of space.   □

The number of queries in the first and second rounds are $210b + 4,428$ and $105b + 2,187$ respectively. Thus, in 2 rounds a total of $315b + 6,615$ pairwise distances are to be queried for the placement of $245b + 4,419$ points. Now, $315b + 6,615 = (315/245) * (245b + 4419) - (9/7) * 4419 + 6615 = 9n/7 + (46305 - 39771)/7 = 9n/7 + 6534/7$. The following theorem summarizes the discussion above.

Theorem 4. $9n/7 + 6534/7$ *queries are sufficient to place n distinct points on a line in two rounds.*

3   An Improved Lower Bound for Two Rounds

Let $G_1 = (V, E_1)$ and $G_2 = (V, E_1 \cup E_2)$ be the query graphs for the first and second round respectively.

We use an adversarial argument as in [6,3] to improve the lower bound for a 2-round algorithm to $\frac{9}{8}$. In each of the 2 rounds, the edge lengths of the query graph returned by the adversary $\mathcal{B}$ are designed to keep the placement of the vertices ambiguous, thereby forcing the algorithm to add more edges to disambiguate the placement. This increases the density ($\rho = |E|/|V|$) of the graph.

Definition 2. *A vertex of G_i, $i = 1, 2$ is heavy if its degree is 3 or more.*

Definition 3. $P_k =< p_1, p_2, ..., p_k >$ *denotes a path of distinct degree 2 vertices in G_i of length k. It is maximal if the vertices p_0 and p_{k+1} adjoining p_1 and p_k are not of degree 2. If p_0 and p_{k+1} are both heavy then P_k is a class A path.*

Definition 4. *A connected subgraph H of G_i ($i = 1, 2$) is called a handle [6] in G_i if the layout of H is ambiguous in the i-th round, though the layout of all the remaining vertices of G_i are fixed in round i.*

For example, the subgraphs $(\{p_1\}, \phi), (\{p_1, p_2\}, \{p_1 p_2\})$ and $(\{p_1, p_2, p_3\}, \{p_1 p_2, p_2 p_3\})$ are handles in the the graphs whose layer graphs are shown in Figs. 9(a), 9(b) and 9(c) respectively[1]. Handles capture the ambiguity of a placement.

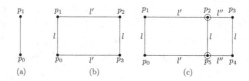

Fig. 9. Three different handles

Over the two rounds, $\mathcal{B}$ assigns lengths to the edges according to the following strategies. The algorithm is oblivious of these strategies.

S1. *The adversary fixes the layout of all heavy vertices except the following 3 types of degree 3 vertices. Let p_0 be a vertex of degree 3. The exceptions are:*
(1) The length of each of the 3 degree 2 maximal paths in G_1 connected to p_0 is at most 1 and the other terminals of the path are not heavy. (Fig. 10)
(2) The vertex p_0 is connected to exactly one heavy vertex by a degree 2 maximal path of length 1 in G_1 and the length of each of the remaining 2 degree 2 maximal paths in G_1 connected to p_0 is at most 1 and remaining two terminals not heavy.(Fig. 11)
(3) The vertex p_0 is adjacent either to exactly one heavy vertex in G_1, or to the start or end vertex of a class A path; and the length of each of the remaining 2 degree 2 maximal paths in G_1 connected to p_0 is exactly 1. (Fig. 12)
We call the vertex p_0 of exception (3) as *specialOne* vertex and its adjacent vertex in G_1 as *specialTwo* vertex if it is heavy in G_1.

---

[1] Heavy vertices are circled in the figures of this section.

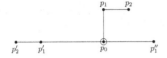

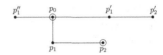

Fig. 10. The layout of heavy vertex p_0 is not fixed in round 1

Fig. 11. The layout of heavy vertex p_0 is not fixed in round 1

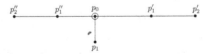

Fig. 12. The layout of heavy vertex p_0 is not fixed in round 1

$S2.$ *For all degree 2 vertices, if one of the incident edges is also incident on a degree 1 vertex, the adversary sets the length of one of the two incident edges to be the same, say c, over all these degree 2 vertices.*

$S3.$ *For each degree 2 maximal path $\mathcal{P}_k$ of length at least 2 in G_1, say $< p_1, p_2, ...,$ $p_k >$, $k \geq 2$, let p_0 and p_{k+1} be non-degree 2 vertices in G_1 adjacent to p_1 and p_k respectively in G_1. (a) The adversary sets $|p_{i-1}p_i| = |p_{i+1}p_{i+2}|$ for $i = 1$ (mod 3). In addition, (b) if $\mathcal{P}_k$ is a class A path then $\mathcal{B}$ fixes the layout of p_i, $i = 0$ (mod 3) and sets $|p_ip_{i+1}| = |p_{i-1}p_{i+2}|$ for $i = 1$ (mod 3), with the exception that if p_0 is a specialOne vertex then $\mathcal{B}$ keeps option for potentially fixing p_0 such that $|p_1p_2| = |p_0p_3|$, and that if $pk + 1$ is a specialOne vertex and $k + 1 = 0$ (mod 3) then $\mathcal{B}$ keeps option for potentially fixing p_{k+1} such that $|p_{k-1}p_k| = |p_{k-2}p_{k+1}|$; finally, (c) if at least one of them, say p_{k+1}, is of degree one the adversary sets the lengths of alternate edges equal.*

The following lemma is a consequence of parts (a) and (b) of strategy $S3$.

Lemma 8. *[3] For each class A path, say $< p_1, p_2, ..., p_k >$, $k \geq 2$, there exists at least one edge in E_2 incident to either p_i or p_{i+1} for $i = 1$ (mod 3) in G_2.*

We can also prove the following result.

Lemma 9. *Strategies $S2$ and $S3$ of $\mathcal{B}$ are mutually consistent.*

$S4.$ *(1) If a vertex, say p_0, of degree 3 has 2 degree 2 maximal paths the other ends of which are not attached to any heavy vertex, and if p_0 is incident on only one degree 2 maximal path of length 1 of which the other end is incident on a heavy vertex, then set the length of one of the edges of this third path as c.*

(2) If 2 specialOne vertices p_0 and p_0' are adjacent in G_1 then set $|p_0p_0'| = c$. If a specialTwo vertex p_0' of degree 3 in G_1 has exactly 2 adjacent vertices of type specialOne then $\mathcal{B}$ sets the length of the edge incident to p_0' and one of

the *specialOne* vertices adjacent to p'_0 as c. Let p_0 be any *specialOne* vertex and the 2 degree 2 paths of length 1 attached to it be $< p_0, p_1, p_2 >$ and $< p_0, p''_1, p''_2 >$. Then $\mathcal{B}$ sets $|p_1 p_2| = |p''_1 p''_2| = c$.

Lemma 10. *In a degree 2 maximal path in G_2 that contains at least one edge from E_2, there can be at most 2 consecutive edges from E_1.*

The above results together with $S2$ and $S3$ implies the following characterization [1].

Lemma 11. *The length of any maximal path of degree 2 vertices in G_2 is at most 3.*

Theorem 5. *Any deterministic 2-round algorithm for solving the 1-dimensional point placement problem requires at least $9n/8$ queries in the worst case.*

Proof. Omitted for lack of space.

4   Conclusions

We implemented the above algorithm. It runs much faster than the randomized algorithm, as it maintains 2^r different signed sums for r randomly added points. It is challenging to close the gap between the upper and lower bounds for two rounds. Improving the upper bound of $5n/4$ for three rounds [3] is another challenge. A further line of work is to generalize this problem to higher dimensions. In three dimensions this problem has strong ties with the molecular conformation problem [4].

References

1. Alam, M.S., Mukhopadhyay, A.: A new algorithm and improved lower bound for point placement on a line in two rounds. In: CCCG 2010: Proceedings of the 22nd Canadian Conference on Computational Geometry, pp. 229–232 (2010)
2. Alam, M.S., Mukhopadhyay, A., Sarker, A.: Generalized jewels and the point placement problem. In: CCCG 2009: Proceedings of the 21st Canadian Conference on Computational Geometry, pp. 45–48 (2009)
3. Chin, F.Y.L., Leung, H.C.M., Sung, W.K., Yiu, S.M.: The point placement problem on a line – improved bounds for pairwise distance queries. In: Giancarlo, R., Hannenhalli, S. (eds.) WABI 2007. LNCS (LNBI), vol. 4645, pp. 372–382. Springer, Heidelberg (2007)
4. Crippen, G., Havel, T.: Distance geometry and molecular conformation, vol. 15. Research Studies Press Taunton Somerset England (1988)
5. Damaschke, P.: Point placement on the line by distance data. Discrete Applied Mathematics 127(1), 53–62 (2003)
6. Damaschke, P.: Randomized vs. deterministic distance query strategies for point location on the line. Discrete Applied Mathematics 154(3), 478–484 (2006)
7. Mumey, B.: Probe location in the presence of errors: a problem from DNA mapping. Discrete Applied Mathematics 104(1-3), 187–201 (2000)
8. Redstone, J., Ruzzo, W.L.: Algorithms for a simple point placement problem. In: Bongiovanni, G., Petreschi, R., Gambosi, G. (eds.) CIAC 2000. LNCS, vol. 1767, pp. 32–43. Springer, Heidelberg (2000)

Vertex Guarding in Weak Visibility Polygons

Pritam Bhattacharya, Subir Kumar Ghosh, and Bodhayan Roy

Tata Institute of Fundamental Research, Mumbai-400005, India
pritamb@tcs.tifr.res.in, {ghosh,bodhayan}@tifr.res.in

Abstract. The art gallery problem enquires about the least number of guards that are sufficient to ensure that an art gallery, represented by a polygon P, is fully guarded. In 1998, the problems of finding the minimum number of point guards, vertex guards, and edge guards required to guard P were shown to be APX-hard by Eidenbenz, Widmayer and Stamm. In 1987, Ghosh presented approximation algorithms for vertex guards and edge guards that achieved a ratio of $\mathcal{O}(\log n)$, which was improved upto $\mathcal{O}(\log \log OPT)$ by King and Kirkpatrick in 2011. It has been conjectured that constant-factor approximation algorithms exist for these problems. We settle the conjecture for the special class of polygons that are weakly visible from an edge and contain no holes by presenting a 6-approximation algorithm for finding the minimum number of vertex guards that runs in $\mathcal{O}(n^2)$ time. On the other hand, for weak visibility polygons with holes, we present a reduction from the Set Cover problem to show that there cannot exist a polynomial time algorithm for the vertex guard problem with an approximation ratio better than $((1 - \epsilon)/12) \ln n$ for any $\epsilon > 0$, unless NP = P.

1 Introduction

1.1 The Art Gallery Problem and Its Variants

The art gallery problem enquires about the least number of guards that are sufficient to ensure that an art gallery (represented by a polygon P) is fully guarded, assuming that a guard's field of view covers 360° as well as an unbounded distance. This problem was first posed by Victor Klee in a conference in 1973, and in the course of time, it has turned into one of the most investigated problems in computational geometry.

A *polygon* P is defined to be a closed region in the plane bounded by a finite set of line segments, called edges of P, such that, between any two points of P, there exists a path which does not intersect any edge of P. If the boundary of a polygon P consists of two or more cycles, then P is called a *polygon with holes*. Otherwise, P is called a *simple polygon* or a *polygon without holes*. An art gallery can be viewed as an n-sided polygon P (with or without holes) and guards as points inside P. Any point $z \in P$ is said to be *visible* from a guard g if the line segment zg does not intersect the exterior of P. In general, guards may be placed anywhere inside P. In 1975, Chvátal [4] showed that $\lfloor \frac{n}{3} \rfloor$ stationary guards are sufficient and sometimes necessary for guarding a simple polygon. In 1978, Fisk [9] presented a simpler and more elegant proof of this result.

S. Ganguly and R. Krishnamurti (Eds.): CALDAM 2015, LNCS 8959, pp. 45–57, 2015.
© Springer International Publishing Switzerland 2015

1.2   Related Hardness and Approximation Results

The decision version of the art gallery problem is to determine, given a polygon P and a number k as input, whether the polygon P can be guarded with k or fewer guards. The problem was first proved to be NP-complete for polygons with holes by O'Rourke and Supowit [19]. For guarding simple polygons, it was proved to be NP-complete for vertex guards by Lee and Lin [18], and their proof was generalized to work for point guards by Aggarwal [1]. The problem is NP-hard even for simple orthogonal polygons as shown by Katz and Roisman [16] and Schuchardt and Hecker [20]. Each one of these hardness results hold irrespective of whether we are dealing with vertex guards, edge guards, or point guards.

In 1987, Ghosh [10,12] provided an $\mathcal{O}(\log n)$-approximation algorithm for the case of vertex and edge guards by discretizing the input polygon and treating it as an instance of the Set Cover problem. In fact, applying methods for the Set Cover problem developed after Ghosh's algorithm, it is easy to obtain an approximation factor of $\mathcal{O}(\log OPT)$ for vertex guarding simple polygons or $\mathcal{O}(\log h \log OPT)$ for vertex guarding a polygon with h holes. Deshpande et al. [5] obtained an approximation factor of $\mathcal{O}(\log OPT)$ for point guards or perimeter guards by developing a sophisticated discretization method that runs in pseudopolynomial time. Efrat and Har-Peled [6] provided a randomized algorithm with the same approximation ratio that runs in fully polynomial expected time. For guarding simple polygons using vertex guards and perimeter guards, King and Kirkpatrick [17] obtained $\mathcal{O}(\log \log OPT)$ approximation ratio in 2011.

In 1998, Eidenbenz, Stamm and Widmayer [7,8] proved that the problem is APX-complete, implying that an approximation ratio better than a fixed constant cannot be achieved unless P=NP. They also proved that if the input polygon is allowed to contain holes, then there cannot exist a polynomial time algorithm for the problem with an approximation ratio better than $((1-\epsilon)/12) \ln n$ for any $\epsilon > 0$, unless NP $\subseteq$ TIME$(n^{\mathcal{O}(\log \log n)})$. Contrastingly, in the case of simple polygons without holes, the existence of a constant-factor approximation algorithm for vertex guards and edge guards has been conjectured by Ghosh [10,13] since 1987. However, this conjecture has not yet been settled even for special classes of polygons such as weak visibility polygons, monotone polygons, orthogonal polygons etc.

1.3   Our Contributions

A polygon P is said to be a *weak visibility polygon* if every point in P is visible from some point of an edge [11]. In Section 2, we present a 6-approximation algorithm, which has running time $\mathcal{O}(n^2)$, for vertex guarding polygons that are weakly visible from an edge and contain no holes. This result can be viewed as a step forward towards solving Ghosh's conjecture for a special class of polygons. Following the construction of Eidenbenz, Stamm and Widmayer [7], we establish a reduction from Set Cover and show that, for the special class of polygons containing holes that are weakly visible from an edge, there cannot exist a polynomial time algorithm for the vertex guard problem with an approximation ratio better than $((1 - \epsilon)/12) \ln n$ for any $\epsilon > 0$, unless NP = P. For details of this reduction, refer to the full version of the paper [3].

2 Placement of Guards in Weak Visibility Polygons

Let P be a simple polygon. If there exists an edge uv in P (where u is the next clockwise vertex of v) such that P is weakly visible from uv, then it can be located in $\mathcal{O}(n^2)$ time [2,14]. Henceforth, we assume that such an edge uv has been located. Let $bd_c(p, q)$ (or, $bd_{cc}(p, q)$) denote the clockwise (respectively, counterclockwise) boundary of P from a vertex p to another vertex q. Note that, by definition, $bd_c(p, q) = bd_{cc}(q, p)$. The *visibility polygon* of P from a point z, denoted by $VP(z)$, is defined to be the set of all points in P that are visible from z. In other words, $VP(z) = \{q \in P : q \text{ is visible from } z\}$.

The *shortest path tree* of P rooted at a vertex r of P, denoted by $SPT(r)$, is the union of Euclidean shortest paths from r to all the vertices of P. This union of paths is a planar tree, rooted at r, which has n nodes, namely the vertices of P. For every vertex x of P, let $p_u(x)$ and $p_v(x)$ denote the parent of x in $SPT(u)$ and $SPT(v)$ respectively. In the same way, for every interior point y of P, let $p_u(y)$ and $p_v(y)$ denote the vertex of P next to y in the Euclidean shortest path to y from u and v respectively.

2.1 Guarding All Vertices of a Polygon

Suppose a guard is placed on each non-leaf vertex of $SPT(u)$ and $SPT(v)$. It is obvious that these guards see all points of P. However, the number of guards required may be very large compared to the size of an optimal guarding set. In order to reduce the number of guards, placing guards on every non-leaf vertex should be avoided. Let A be a subset of vertices of P. Let S_A denote the set which consists of the parents $p_u(z)$ and $p_v(z)$ of every vertex $z \in A$. Then, A should be chosen such that all vertices of P are visible from guards placed at vertices of S_A. We present a method for choosing A and S_A as follows:-

---

Algorithm 2.1. An $\mathcal{O}(n^2)$-algorithm for computing a guard set S_A for all vertices of P

Compute $SPT(u)$ and $SPT(v)$
Initialize all the vertices of P as unmarked
Initialize $A \leftarrow \emptyset$, $S_A \leftarrow \emptyset$ and $z \leftarrow u$
while $z \neq v$ **do**
    $z \leftarrow$ the vertex next to z in clockwise order on $bd_c(u, v)$
    if z is unmarked **then**
        $A \leftarrow A \cup \{z\}$ and $S_A \leftarrow S_A \cup \{p_u(z), p_v(z)\}$
        Place guards on $p_u(z)$ and $p_v(z)$
        Mark all vertices of P that become visible from $p_u(z)$ or $p_v(z)$
    end if
end while
return the guard set S_A

Now, assume a special condition such that for every vertex $z \in A$, all vertices of $bd_c(p_u(z), p_v(z))$ are visible from $p_u(z)$ or $p_v(z)$. We prove that, in such a situation, $|S_A| \leq 2.|S_{opt}|$, where S_{opt} denotes an optimal vertex guard set.

Lemma 1. *Any guard $g \in S_{opt}$ that sees vertex z of P must lie on $bd_c(p_u(z), p_v(z))$.*

Proof. Since $p_u(z)$ is the parent of z in $SPT(u)$, z cannot be visible from any vertex of $bd_c(u, p_u(z))$. Similarly, since $p_v(z)$ is the parent of z in $SPT(v)$, z cannot be visible from any vertex of $bd_{cc}(v, p_v(z))$. Hence, any guard $g \in S_{opt}$ that sees z must lie on $bd_c(p_u(z), p_v(z))$.

Lemma 2. *Let z be a vertex of P such that all vertices of $bd_c(p_u(z), p_v(z))$ are visible from $p_u(z)$ or $p_v(z)$. For every vertex x lying on $bd_c(p_u(z), p_v(z))$, if x sees a vertex q of P, then q must also be visible from $p_u(z)$ or $p_v(z)$.*

Proof. If q lies on $bd_c(p_u(z), p_v(z))$, then it is visible from from $p_u(z)$ or $p_v(z)$ by assumption. So, consider the case where q lies on $bd_{cc}(p_u(z), p_v(z))$. Now, either q lies on $bd_c(u, p_u(z))$ or q lies on $bd_{cc}(v, p_v(z))$. In the former case, if $bd_{cc}(q, p_v(z))$ intersects the segment $qp_v(z)$, then q or $p_v(z)$ is not weakly visible from uv (see Fig. 1). Moreover, no other portion of the boundary can intersect $qp_v(z)$ since qx and $zp_v(z)$ are internal segments. Hence, q must be visible from $p_v(z)$. Analogously, if q lies on $bd_{cc}(v, p_v(z))$, q must be visible from $p_u(z)$.

Lemma 3. *Assume that every vertex $z \in A$ is such that every vertex of $bd_c(p_u(z), p_v(z))$ is visible from $p_u(z)$ or $p_v(z)$. Then, $|A| \leq |S_{opt}|$.*

Proof. Assume on the contrary that $|A| > |S_{opt}|$. This implies that Algorithm 2.1 includes two distinct vertices z_1 and z_2 belonging to A which are both visible from a single guard $g \in S_{opt}$. Moreover, it follows from Lemma 1 that g must lie on $bd_c(p_u(z_1), p_v(z_1))$. Without loss of generality, let us assume that vertex z_1 is added to A before z_2 by Algorithm 2.1. In that case, Algorithm 2.1 places guards at $p_u(z_1)$ and $p_v(z_1)$. Now, as vertex z_2 is visible from g, it follows from Lemma 2 that z_2 is also visible from $p_u(z_1)$ or $p_v(z_1)$. Therefore, z_2 is already marked, and hence, Algorithm 2.1 does not include z_2 in A, which is a contradiction.

Lemma 4. $|S_A| = 2|A|$.

Proof. For every $z \in A$, since Algorithm 2.1 includes both the parents $p_u(z)$ and $p_v(z)$ of z in S_A, it is clear that $|S_A| \leq 2|A|$. If both the parents of every $z \in A$ are distinct, then $|S_A| = 2|A|$. Otherwise, there exists two distinct vertices z_1 and z_2 in A that share a common parent, say p. Without loss of generality, let us assume that vertex z_1 is added to A before z_2 by Algorithm 2.1. In that case, Algorithm 2.1 places a guard at p, which results in z_2 getting marked. Thus, Algorithm 2.1 cannot include z_2 in A, which is a contradiction. Hence, $|S_A| = 2|A|$ must be true.

Theorem 1. *If every vertex $z \in A$ is such that all vertices of $bd_c(p_u(z), p_v(z))$ are visible from $p_u(z)$ or $p_v(z)$, then $|S_A| \leq 2|S_{opt}|$.*

Proof. By Lemma 4, $|S_A| = 2|A|$. Also, by Lemma 3, $|A| \leq |S_{opt}|$. So, $|S_A| = 2|A| \leq 2|S_{opt}|$.

The above bound does not hold if there exists $z \in A$ such that some vertices of $bd_c(p_u(z), p_v(z))$ are not visible from $p_u(z)$ or $p_v(z)$. Now, consider Fig. 2. For each $i \in \{1, 2, \ldots, k - 1\}$, z_{i+1} is not visible from $p_u(z_i)$ or $p_v(z_i)$, which forces Algorithm 2.1 to place guards at $p_u(z_{i+1})$ and $p_v(z_{i+1})$. Therefore, Algorithm 2.1 includes $z_1, z_2, z_3, \ldots, z_k$ in A and places a total of $2k$ guards at vertices $u, p_{v1}, p_{u2}, p_{v2}, \ldots, p_{uk}, p_{vk}$. However, all vertices of P are visible from just two guards placed at u and g. Hence, $|S_A| = 2k$ whereas $|S_{opt}| = 2$. Since the construction in Fig. 2 can be extended for any arbitrary integer k, $|S_A|$ can be arbitrarily large compared to $|S_{opt}|$. So, we present a new and better algorithm which gives us a 4-approximation of $|S_{opt}|$.

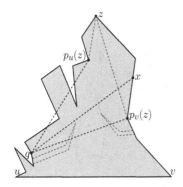

Fig. 1. Case in Lemma 2 where the segment $qp_v(z)$ is intersected by $bd_c(u, p_u(z))$

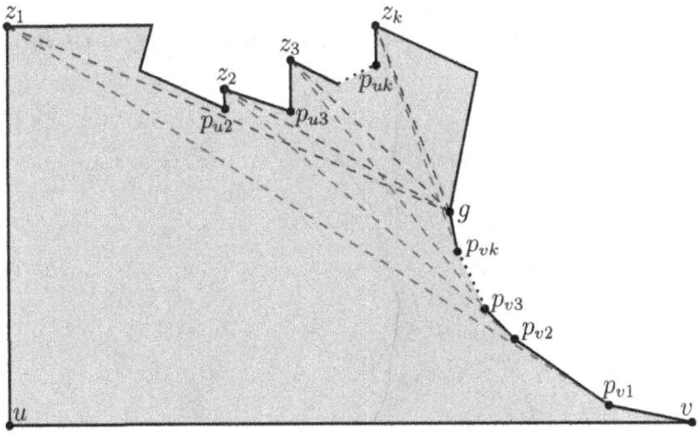

Fig. 2. An instance where the guard set S_A computed by Algorithm 2.1 is arbitrarily large compared to an optimal guard set S_{opt}

In the new algorithm, $bd_c(u, v)$ is scanned to identify a set of unmarked vertices, denoted as B, such that all vertices of P are visible from guards in $S_B = \{p_u(z) | z \in B\} \cup \{p_v(z) | z \in B\}$. During the scan, let z denote the current unmarked vertex under consideration. At every step, the algorithm maintains the invariance that, for every unmarked vertex y of $bd_c(u, z)$ (excluding z), $p_u(y)$ and $p_v(y)$ see all unmarked vertices of $bd_c(p_u(y), y)$. Let z' denote the next unmarked vertex of $bd_c(z, p_v(z))$ in clockwise order from z such that z' is not visible from either $p_u(z)$ or $p_v(z)$. Depending on whether z' exists, the current vertex z must satisfy one of the following properties.

(A) All vertices of $bd_c(z, p_v(z))$ are already marked due to the guards currently included in S_B (see Fig. 3).

(B) Every unmarked vertex of $bd_c(z, p_v(z))$ is visible from $p_u(z)$ or $p_v(z)$ (see Fig. 4).

(C) Not all unmarked vertices of $bd_c(p_u(z'), z')$ are visible from $p_u(z')$ or $p_v(z')$ (see Fig. 5).

(D) Every unmarked vertex of $bd_c(p_u(z'), z')$ is visible from $p_u(z')$ or $p_v(z')$ (see Fig. 6).

If z satisfies property (A) or (B), then z is included in B and the first unmarked vertex of $bd_c(p_v(z), v)$ in clockwise order from $p_v(z)$ becomes the new z. If z satisfies property (C), then z is included in B and z' becomes the new z. If z satisfies property (D), then z' becomes the new z. Whenever z is included in B, $p_u(z)$ and $p_v(z)$ are included in S_B and all unmarked vertices that become visible from $p_u(z)$ or $p_v(z)$ are marked. After doing so, if there remain unmarked vertices on $bd_{cc}(z, u)$, then $bd_{cc}(z, u)$ is scanned from z in counterclockwise order and more guards are included in S_B according to the following strategy:-

(i) $y \leftarrow p_u(z)$
(ii) Scan $bd_{cc}(y, u)$ from y in counterclockwise till an unmarked vertex x is located.
(iii) Add x to B. Add $p_u(x)$ and $p_v(x)$ to S_B.
(iv) Mark every vertex visible from $p_u(x)$ or $p_v(x)$.
(v) $y \leftarrow p_u(x)$
(vi) Repeat steps (ii)-(v) until all vertices of $bd_{cc}(z, u)$ are marked.

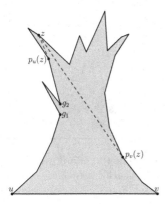

Fig. 3. All vertices of $bd_c(z, p_v(z))$ are already marked due to guards at g_1 and g_2

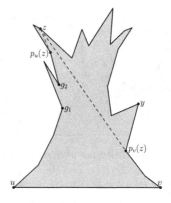

Fig. 4. The only unmarked vertex y of $bd_c(z, p_v(z))$ is visible from $p_v(z)$

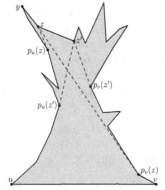

Fig. 5. Guards at $p_u(z')$ and $p_v(z')$ do not see the unmarked vertex y of $bd_c(p_u(z'), z')$

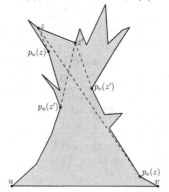

Fig. 6. Guards at $p_u(z')$ and $p_v(z')$ see all unmarked vertices of $bd_c(p_u(z'), z')$

Initially, z is chosen to be the first unmarked vertex of $bd_c(u, v)$ in clockwise order from u. Then, for each z under consideration along the clockwise scan of $bd_c(u, v)$, the appropriate action is performed corresponding to the property of z. Then, z is updated and the process is repeated till v is reached. The set of vertices S_B is returned by the algorithm as a guard set. The entire process is described in pseudocode as Algorithm 2.2.

Algorithm 2.2. An $\mathcal{O}(n^2)$-algorithm for computing a guard set S for all vertices of P

Compute $SPT(u)$ and $SPT(v)$
Initialize all the vertices of P as unmarked
Initialize $B \leftarrow \emptyset$, $S_B \leftarrow \emptyset$ and $z \leftarrow u$
while there exists an unmarked vertex in P **do**
    $z \leftarrow$ the first unmarked vertex on $bd_c(u, v)$ in clockwise order from z
    if every unmarked vertex of $bd_c(z, p_v(z))$ is visible from $p_u(z)$ or $p_v(z)$ **then**
        $B \leftarrow B \cup \{z\}$ and $S_B \leftarrow S_B \cup \{p_u(z), p_v(z)\}$
        Mark all vertices of P that become visible from $p_u(z)$ or $p_v(z)$
        $z \leftarrow p_v(z)$
    else
        $z' \leftarrow$ the first unmarked vertex on $bd_c(z, v)$ in clockwise order
        while every unmarked vertex of $bd_c(p_u(z'), z')$ is visible from $p_u(z')$ or $p_v(z')$
do
            $z \leftarrow z'$ and $z' \leftarrow$ the first unmarked vertex on $bd_c(z', v)$ in clockwise order
        end while
        $w \leftarrow z$
        while there exists an unmarked vertex on $bd_c(u, z)$ **do**
            $B \leftarrow B \cup \{w\}$ and $S_B \leftarrow S_B \cup \{p_u(w), p_v(w)\}$
            Mark all vertices of P that become visible from $p_u(w)$ or $p_v(w)$
            $w \leftarrow$ the first unmarked vertex on $bd_{cc}(w, u)$ in counterclockwise order
        end while
    end if
end while
return the guard set $S = S_B$

For showing an upper bound on S, a bipartite graph $G = (B \cup S_{opt}, E)$ is constructed such that the degree of each vertex in B is exactly 1 and the degree of each vertex in S_{opt} is at most 2.

The graph G is constructed as follows. For every vertex $b_i \in B$, there exists a $g \in S_{opt}$ such that g sees b_i. By Lemma 1, g must lie on $bd_c(p_u(b_i), p_v(b_i))$. If g lies on $bd_c(p_u(b_i), b_i)$, then the edge gb_i is added to E. Observe that any vertex q lying on $bd_{cc}(p_u(b_i), b_i)$ that is visible from g is also visible from $p_u(b_i)$ or $p_v(b_i)$ (see the proof of Lemma 2). So, q is marked on inclusion of b_i in B, and therefore q cannot be included in B. Hence, for $k > i$, no vertex $b_k \in B$ exists that can add an edge gb_k.

If g lies on $bd_c(b_i, b_i')$ and sees another $b_j \in B$, then the edges gb_i and gb_j are added to G. Similar arguments as above show that, for $k > j$, no vertex $b_k \in B$ exists that can add an edge gb_k.

If g lies on $bd_c(b_i', p_v(b_i'))$ (see Fig. 7), there exists a vertex x_i on $bd_c(p_u(b_i), b_i)$ such that x_i is visible from $p_u(b_i)$ or $p_v(b_i)$, but not from $p_u(b_i')$ or $p_v(b_i')$. So, in order to see x_i, there must exist another guard $g' \in S_{opt}$ lying on $bd_c(p_v(b_i'), p_v(x_i))$. The edge $g'b_i$ is added to G. Let $V_{g'}$ denote the set of vertices of P lying on $bd_c(p_v(b_i'), p_v(x_i))$ that are visible from g'. If $V_{g'}$ does not contain any vertex of B, then the degree of g' is 1 in G. Otherwise, the first vertex $b_j \in B$ of $V_{g'}$ in clockwise order from $p_v(b_i')$ is located and the edge $g'b_j$ is added to G. Now, every vertex belonging to $V_{g'}$ must be visible from $p_u(b_j)$ or $p_v(b_j)$, which means that no other vertex of $V_{g'}$ can be included in B. Hence, for $k > j$, no vertex $b_k \in B$ exists which can have an edge $g'b_k$ incident on g', ensuring that the degree of g' is at most 2 in G.

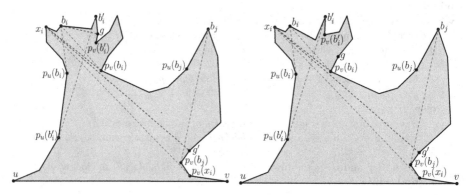

Fig. 7. The guard $g \in S_{opt}$ is located on $bd_c(b_i', p_v(b_i'))$

Fig. 8. The guard $g \in S_{opt}$ is located on $bd_c(p_v(b_i'), p_v(b_i))$

If g lies on $bd_c(p_v(b_i'), p_v(b_i))$ (see Fig. 8), then add the edge gb_i to G. Observe that no vertex lying on $bd_c(b_i', p_v(b_i'))$ can be visible from g. Moreover, at most one other vertex $b_j \in B$ lying on $bd_c(p_v(b_i'), p_v(b_i))$ is visible from g, as explained earlier for the case of $g' \in S_{opt}$ seeing x_i. If b_j exists, then the edge gb_j is added to G, ensuring the degree of g is at most 2 in G. As a direct consequence of the above construction, we have the following results.

Lemma 5. *In the bipartite graph G, the degree of each vertex in B is exactly 1 and degree of each vertex in S_{opt} is at most 2.*

Corollary 1. $|B| \leq 2|S_{opt}|$.

Theorem 2. $|S| \leq 4|S_{opt}|$.

Proof. By arguments similar to those in the proof of Lemma 4, it can be shown that $|S_B| = 2|B|$. Also, by Corollary 1, $|B| \leq 2|S_{opt}|$. Therefore, $|S| = |S_B| = 2|B| \leq 4|S_{opt}|$.

2.2   Guarding All Interior Points of a Polygon

In the previous subsection, we presented an algorithm (see Algorithm 2.2) which returns a guard set S such that all vertices of P are visible from guards in S. However, it may not always be true that all interior points of P are also visible from guards in S. Consider the polygon shown in Fig. 9. While scanning $bd_c(u, v)$, our algorithm places guards at $p_u(z)$ and $p_v(z)$ as all vertices of $bd_c(p_u(z), p_v(z))$ become visible from $p_u(z)$ or $p_v(z)$. Observe that in fact all vertices of P become visible from these two guards. However, the triangular region $P \setminus (VP(p_u(z)) \cup VP(p_v(z)))$, bounded by the segments x_1x_2, x_2x_3 and x_3x_1, is not visible from $p_u(z)$ or $p_v(z)$. Also, one of the sides x_1x_2 of the triangle $x_1x_2x_3$ is a part of the polygonal edge a_1a_2. In fact, for any such region invisible from guards in S, one of the sides must always be a part of a polygonal edge. Otherwise, there should exist another guard g (see Fig. 9) from which the entire polygonal side (x_1x_2) of the region is visible and yet some portion of the region (including x_3) is not visible. However, such a vertex g cannot be weakly visible from the edge uv, which is a contradiction. Henceforth, any such region invisible from guards in S is referred to as an *invisible cell*, and the polygonal edge which contributes as a side to the invisible cell is referred to as its corresponding *partially invisible edge*. One additional guard is required in order to see each invisible cell entirely. For example, in Fig. 9, an extra guard is required at a vertex of $bd_c(z, w)$, since none of the vertices outside this boundary can see all points of the triangular invisible cell $x_1x_2x_3$.

The boundary of the visibility polygon $VP(s)$ of any vertex s consists of polygonal edges and constructed edges. A *constructed edge* yx is an edge formed by extending the segment sy (which could be either an edge of P or an internal segment), where y is some other vertex of P, till it touches the boundary of P at a point x. If y lies on $bd_c(s, x)$, the region of P bounded by $bd_c(y, x)$ and xy is referred to as the *left pocket* of $VP(z)$. Similarly, if y lies on $bd_{cc}(s, x)$, then the region of P bounded by $bd_{cc}(y, x)$ and xy is referred to as the *right pocket* of $VP(z)$. In both these cases, we refer to the vertex y as the *lid vertex* and the point x as the *lid point* of the corresponding left or right pocket.

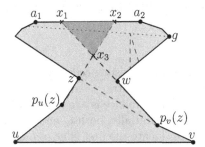

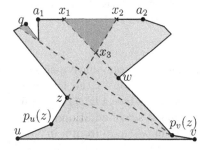

Fig. 9. All vertices are visible from $p_u(z)$ or $p_v(z)$, but triangle $x_1x_2x_3$ is invisible

Fig. 10. The left pocket of $VP(p_u(z))$ can contain only one invisible cell

Observe that each invisible cell must be wholly contained within the intersection region (which is a triangle) of a left pocket and a right pocket. For example, in Fig. 9, the invisible cell $x_1 x_2 x_3$ is actually the entire intersection region of the left pocket of $VP(p_u(z))$ and the right pocket of $VP(p_v(z))$. Also, z is the lid vertex and x_2 is the lid point of the left pocket of $VP(p_u(z))$. Similarly, w is the lid vertex and x_1 is the lid point of the right pocket of $VP(p_v(z))$.

Suppose $bd_c(z, x_2)$ contains reflex vertices (see Fig. 10). In that case, in addition to the invisible cell $x_1 x_2 x_3$, the left pocket of $VP(p_u(z))$ may contain several regions that are not visible from $p_v(z)$. However, in each such region there exists a vertex, say q, that is not visible from $p_v(z)$, which contradicts the fact that all vertices of $bd_c(p_u(z), p_v(z))$ are visible from $p_u(z)$ or $p_v(z)$. So, the left pocket of $VP(p_u(z))$ can contain only one invisible cell. Analogously, the right pocket of $VP(p_v(z))$ can contain only one invisible cell.

Now consider the situation (as shown in Fig. 11) where $VP(p_u(z))$ has several left pockets and $VP(p_v(z))$ has several right pockets which intersect pairwise to create multiple invisible cells. In order to guard these invisible cells, additional guards are placed as follows. Let c_1 be the lid point of the left pocket containing the first invisible cell in clockwise order. Then, guards are placed at $p_u(c_1)$ and $p_v(c_1)$. Now, for every invisible cell T, the portions of T are removed that are visible from $p_u(c_1)$ or $p_v(c_1)$. Note that some of these cells may turn out to be totally visible and hence may be eliminated altogether. This process is repeated until all invisible cells become totally visible.

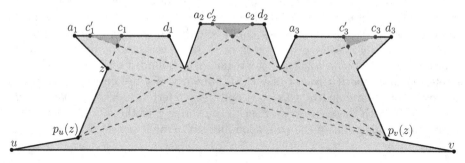

Fig. 11. Multiple invisible cells exist within the polygon that are not visible from the guards placed at $p_u(z)$ and $p_v(z)$

In general, we may have a situation where multiple invisible cells are created by the intersection of the left and right pockets of arbitrary pairs of guards belonging to S (see Fig. 12). In this scenario, all invisible cells are guarded by introducing a set of additional guards S' as follows. Initially, both C and S' are empty. Scan $bd_c(u, v)$ from u in clockwise order to locate the first edge $a_i d_i$ that is not totally visible from guards in $S \cup S'$, where d_i is the next clockwise vertex of a_i. Let $c'_i c_i$ be the portion of $a_i d_i$ that is not visible from guards in $S \cup S'$, where $c'_i \in bd_c(a_i, c_i)$ and $c_i \in bd_c(c'_i, d_i)$. In other words, $c'_i c_i$ is the polygonal side of the first invisible cell. Add $p_u(c_i)$ and $p_v(c_i)$ to S'. Also, add c_i to C. Repeat this process until all the edges of P are totally visible from guards in $S \cup S'$. At its termination, let us assume that $C = \{c_1, c_2, \ldots, c_k\}$. The entire procedure is described in pseudocode as follows.

Algorithm 2.3. An $\mathcal{O}(n^2)$-algorithm for computing a guard set $S \cup S'$ for guarding P entirely

---

Compute $SPT(u)$ and $SPT(v)$
Compute the set of guards S using Algorithm 2.2.
Initialize $C \leftarrow \emptyset$, $S' \leftarrow \emptyset$ and $z \leftarrow u$
while there exists an edge in P that is partially visible from guards in $S \cup S'$ **do**
    $z' \leftarrow$ the vertex next to z in clockwise order on on $bd_c(u, v)$
    if if the edge zz' is partially visible from guards in $S \cup S'$ **then**
        $c_i \leftarrow$ the lid point of the left pocket on zz'
        $C \leftarrow C \cup \{c_i\}$ and $S' \leftarrow S' \cup \{p_u(c_i), p_v(c_i)\}$
    end if
    $z \leftarrow z'$
end while
return the guard set $S \cup S'$

Theorem 3. *The running time of Algorithm 2.3 is $\mathcal{O}(n^2)$.*

Proof. $SPT(u)$ and $SPT(v)$ can be computed in $\mathcal{O}(n)$ time [15]. Then, the computation of the guard set S takes $\mathcal{O}(n^2)$ time, since it involves scanning the boundary of P and identifying vertices to be marked whenever new guards are placed. The number of lid points on an edge can be at most $\mathcal{O}(n)$. Therefore, each time a new vertex is added to S', the invisible portion of the first partially visible edge in clockwise order can be determined in $\mathcal{O}(n)$ time. Hence, the overall running time of Algorithm 2.3 is $\mathcal{O}(n^2)$.

Theorem 4. $2|C| = |S'| \leq 2|S_{opt}|$.

Proof. For every $c_i \in C$, there exists an invisible cell T_i. For every such invisible cell T_i, let l_i and r_i respectively denote the lid vertices of the left and right

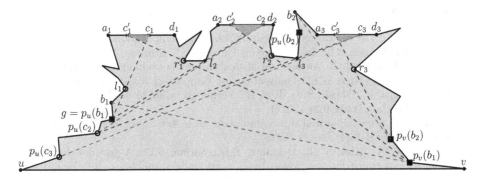

Fig. 12. Placement of guards to in order to see all invisible cells

pockets intersecting to form T_i (see Fig. 12). Let $g \in S$ be the guard such that l_i is the lid vertex of a left pocket of $VP(g)$. Similarly, let $g' \in S$ be the guard such that r_i is the lid vertex of a right pocket of $VP(g')$.

Assume that, for every T_i, there exists at least one guard in S_{opt} that sees all points of T_i. Now, consider any guard $g_{opt} \in S_{opt}$ that sees all points of T_i. Then, g_{opt} can lie on $bd_c(l_i, r_i)$. Also, g_{opt} can lie on $bd_c(p_u(c_i), g)$, but only when $p_u(c_i) \neq l_i$ and $p_u(c_i)$ lies on $bd_c(u, g)$. Now, let z be the vertex such that $p_v(z) = g'$. Then, no vertex of $bd_c(z, g')$ is visible from any vertex of $bd_c(g', v)$. Further, if z is such that $p_u(z) = g$, then z has to lie on $bd_c(g, l_i)$. Otherwise, z has to lie on $bd_c(l_i, c'_i)$. In either case, g_{opt} cannot lie on $bd_c(g', v)$ since c'_i lies on $bd_c(z, g')$.

Since the guard set S' includes $p_u(x)$ and $p_v(x)$ for every $z \in C$, clearly $|S'| = 2|C|$. If for every i, there exists an unique vertex belonging to S_{opt} that sees all points of T_i, then obviously $|S'| \leq 2|S_{opt}|$. Consider the special situation where $l_{i+1} = r_i$ for some i (see Fig. 11) so that both T_i and T_{i+1} are totally visible from r_i. Since all points of T_i are visible from r_i, it must be the case that $p_v(c_i) = r_i$. Moreover, r_i can be a vertex of S_{opt}. Therefore, no additional guards are chosen for T_{i+1} because all points of T_{i+1} become visible from the guard already placed at r_i.

If no vertex of $bd_c(l_i, r_i)$ belongs to S_{opt}, then there must be a vertex of S_{opt} lying on $bd_c(p_u(c_i), g)$ and $p_u(c_i)$ must belong to $bd_c(u, g)$. If $p_u(c_{i-1})$ also belongs to $bd_c(u, g)$, then S_{opt} must have a vertex on the boundary $bd_c(p_u(c_i), p_v(c_{i-1}))$ in order to see T_{i-1} because l_{i-1} is the lid vertex of a left pocket of $VP(p_u(c_{i-1}))$. Hence, $2|C| = |S'| \leq 2|S_{opt}|$.

Finally, if we remove the assumption that there exists at least one guard in S_{opt} that sees all points of T_i, then the size of S_{opt} increases but the size of our guard set S' remains the same. Therefore, the bound $|S'| \leq 2|S_{opt}|$ is still preserved.

Theorem 5. $|S \cup S'| \leq 6|S_{opt}|$.

Proof. By Theorem 2, $|S| \leq 4|S_{opt}|$ and by Theorem 4, $|S'| \leq 2|S_{opt}|$. Therefore, $|S \cup S'| \leq |S| + |S'| \leq 4|S_{opt}| + 2|S_{opt}| \leq 6|S_{opt}|$.

References

1. Aggarwal, A.: The art gallery theorem: its variations, applications and algorithmic aspects. PhD thesis, The Johns Hopkins University, Baltimore, MD (1984)
2. Avis, D., Toussaint, G.: An optimal algorithm for determining the visibility of a polygon from an edge. IEEE Transactions on Computers 30, 910–914 (1981)
3. Bhattacharya, P., Ghosh, S.K., Roy, B.: Vertex guarding in weak visibility polygons. CoRR, abs/1409.4621 (2014)
4. Chvatal, V.: A combinatorial theorem in plane geometry. Journal of Combinatorial Theory, Series B 18(1), 39–41 (1975)
5. Deshpande, A., Kim, T.-J., Demaine, E.D., Sarma, S.E.: A pseudopolynomial time $o(\log n)$-approximation algorithm for art gallery problems. In: Dehne, F., Sack, J.-R., Zeh, N. (eds.) WADS 2007. LNCS, vol. 4619, pp. 163–174. Springer, Heidelberg (2007)

6. Efrat, A., Har-Peled, S.: Guarding galleries and terrains. Information Processing Letters 100(6), 238–245 (2006)
7. Eidenbenz, S., Stamm, C., Widmayer, P.: Inapproximability of some art gallery problems. In: Canadian Conference on Computational Geometry, pp. 1–11 (1998)
8. Eidenbenz, S., Stamm, C., Widmayer, P.: Inapproximability results for guarding polygons and terrains. Algorithmica 31(1), 79–113 (2001)
9. Fisk, S.: A short proof of Chvátal's watchman theorem. Journal of Combinatorial Theory, Series B 24(3), 374 (1978)
10. Ghosh, S.K.: Approximation algorithms for art gallery problems. In: Proc. of Canadian Information Processing Society Congress, pp. 429–434 (1987)
11. Ghosh, S.K.: Visibility Algorithms in the Plane. Cambridge University Press (2007)
12. Ghosh, S.K.: Approximation algorithms for art gallery problems in polygons. Discrete Applied Mathematics 158(6), 718–722 (2010)
13. Ghosh, S.K.: Approximation algorithms for art gallery problems in polygons and terrains. In: Rahman, M. S., Fujita, S. (eds.) WALCOM 2010. LNCS, vol. 5942, pp. 21–34. Springer, Heidelberg (2010)
14. Ghosh, S.K., Maheshwari, A., Pal, S., Saluja, S., Veni Madhavan, C.E.: Characterizing and recognizing weak visibility polygons. Computational Geometry 3(4), 213–233 (1993)
15. Guibas, L.J., Hershberger, J., Leven, D., Sharir, M., Tarjan, R.E.: Linear-time algorithms for visibility and shortest path problems inside triangulated simple polygons. Algorithmica 2, 209–233 (1987)
16. Katz, M.J., Roisman, G.S.: On guarding the vertices of rectilinear domains. Computational Geometry 39(3), 219–228 (2008)
17. King, J., Kirkpatrick, D.G.: Improved approximation for guarding simple galleries from the perimeter. Discrete & Computational Geometry 46(2), 252–269 (2011)
18. Lee, D.T., Lin, A.: Computational complexity of art gallery problems. IEEE Transactions on Information Theory 32(2), 276–282 (1986)
19. O'Rourke, J., Supowit, K.J.: Some NP-hard polygon decomposition problems. IEEE Transactions on Information Theory 29(2), 181–189 (1983)
20. Schuchardt, D., Hecker, H.-D.: Two NP-Hard Art-Gallery Problems for Ortho-Polygons. Mathematical Logic Quarterly 41, 261–267 (1995)

On Collections of Polygons Cuttable
with a Segment Saw

Adrian Dumitrescu[1,*], Anirban Ghosh[1,**], and Masud Hasan[2]

[1] Department of Computer Science, University of Wisconsin-Milwaukee
Milwaukee, WI 53201-0784, USA
{dumitres,anirban}@uwm.edu
[2] Department of Computer Science, Taibah University
Madina Munawwara, Saudi Arabia
mhares@taibahu.edu.sa

Abstract. (I) Given a cuttable polygon P drawn on a piece of planar material Q, we show how to cut P out of Q by a (small) segment saw with a total length no more than 2.5 times the optimal. We revise the algorithm of Demaine et al. (2001) so as to achieve this ratio.

(II) We prove that any collection $\mathcal{R}$ of n disjoint axis-parallel rectangles[1] is cuttable by at most $4n$ rays and present an algorithm that runs in $O(n \log n)$ time for computing a suitable cutting sequence. In particular the same result holds for cutting with an arbitrary segment saw (of any length).

(III) In contrast, we show that there exist collections of disjoint rectangles (in arbitrary orientations) that are uncuttable by a segment saw. We also present various uncuttable collections of disjoint polygons, including triangles.

Keywords: cuttable polygon, cuttable collection, separability, line cut, ray cut, segment cut, approximation algorithm.

1 Introduction

The problem of cutting out a simple polygon P drawn on a planar material Q was introduced by Overmars and Welzl in their seminal paper [20] from 1985. Since then, the problem has attracted the interest of many computational geometers.

In a geometric setting, a *saw* is a moving line segment. A *saw cut* may split (divide) Q into a number of pieces—those that lie left of the cut and pieces that lie right of the cut, or stop short of splitting Q, case in which the material remains one solid piece; however, we do not allow a cut to run through the interior of P. Several variants have been studied, primarily depending on the cutting tool used [1, 3, 6, 7, 10–14, 20, 23]. The cutting tools that have been

* Supported in part by NSF grant DMS-1001667.
** Supported by NSF grant DMS-1001667.
[1] For brevity, a collection of convex bodies with pairwise disjoint interiors is referred to as a collection of disjoint convex bodies.

S. Ganguly and R. Krishnamurti (Eds.): CALDAM 2015, LNCS 8959, pp. 58–68, 2015.

studied are line cuts, ray cuts and segment cuts; they are described subsequently. The type of tool used in cutting determines the class of polygons that can be cut within that model.

The measures of efficiency commonly considered in polygon cutting are the total length of the cuts and the total number of cuts. Polygon cutting problems are useful in industry applications such as metal sheet cutting, paper cutting, furniture manufacturing and numerous other areas of engineering, where smart cutting techniques with high efficiency may result in the reduction of production costs. For instance, reducing the total length of the cuts may result in lesser power requirement and also increase the life of the cutting tool. Similarly, reducing the total number of cuts may save cutting time and extend the life of the cutting tool.

A *line cut* (also called a *guillotine cut*) is a line that does not go through P and divides Q or the current piece of material containing P into two pieces. For cutting P out of Q by line cuts, P must be convex. The most studied efficiency measure for line cutting is the total length of the cuts and several approximation algorithms have been obtained [1, 3, 6, 7, 10–13, 20, 23], including a PTAS proposed by Bereg et al. [3].

A *ray cut* comes from infinity and can stop at any point outside P, again, not necessarily splitting the piece of material into pieces. Ray cuts are usually used to cut out non-convex polygons; however, not all non-convex polygons can be cut by ray cuts. Minimizing the total length of cuts is again the most studied efficiency measure in this model. Algorithms for deciding whether a polygon is ray cuttable and several approximation algorithms can be found in [7, 10, 11, 23].

A *segment cut* (by slightly abusing notation, it may be also called *saw cut*), is similar to a ray cut, but is not required to start at infinity, it may start at some finite point. The saw (also referred to as *circular saw* in [11]) is abstracted as a line segment, which cuts through Q when moved along its supporting line. A small example of a cutting sequence appears in Figure 1. In saw cuts, turns are impossible essentially because the projection of the blade is a positive-length line segment, and the width of the cut is equal to the width of the blade. If a small free space within Q is available, a saw cut can be initiated there by maneuvering the saw. The space required for maneuvering the saw is proportional to the length of the saw. The problem of cutting a polygon by a segment saw was introduced by Demaine et al. [11] in 2001. The authors gave a characterization of the class of polygons cuttable by a (possibly small) segment saw: a polygon is cuttable in this model, i.e., by a sufficiently small segment saw, if and only if it does not have two adjacent reflex vertices (vertices with interior angle $> \pi$).

For ease of analysis the length of the saw is assumed to be arbitrarily small (i.e., the segment abstracting the saw is as short as needed). Consequently, a saw cut can be initiated from an arbitrarily small available free space. It is also assumed that a segment saw cannot ever make turns even if the blade is small. In this model, several parts may result after a cut is made, and any of them can be removed (lifted) from the original plane. Moreover, free space may appear within the pieces of material from where future saw cuts can be initiated. The cutting

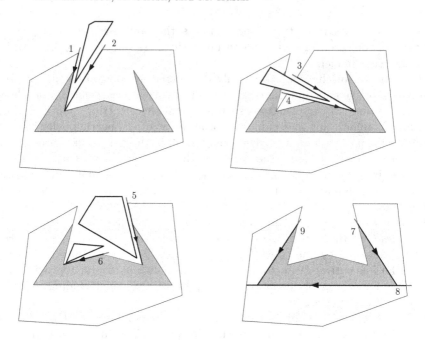

Fig. 1. A cutting sequence with a segment saw (in red) consisting of 9 cuts

process may continue independently on any of the separated pieces of material, for instance if the resulting parts contain subcollections of a larger collection to be cut out.

When the piece of material is the convex hull of the polygon to be cut out, Demaine et al. [11] presented an algorithm for cutting P out of its convex hull, with a total number of cuts and total length of the cuts within constant factors of the respective optima. With regard to the total number of cuts, this number is within 2.5 times the respective optimum. Dumitrescu and Hasan [14] improved the above approximation ratio from 2.5 to 2 on the total number of cuts. More-over, the new approximation guarantee is in a stronger sense than that offered by Demaine et al. [11]. While theirs achieves ratio 2.5 for cutting out P from its convex hull, the new algorithm achieves ratio 2 for cutting out P from any enclosing polygon Q.

With regard to the total length measure of the cuts, Demaine et al. [11] reported that their algorithm achieves the same ratio, 2.5, as for the number of cuts. It turns out that the approximation factor on the total length of the cuts for their algorithm (as implied by their proof) is in fact 3. In Section 2 we show that with a little care, one can recover the claimed ratio 2.5; it requires a small change in their algorithm to obtain the following.

Theorem 1. *Given a cuttable polygon P, drawn on a piece of planar material Q, P can be cut out of Q by an arbitrarily small segment saw with the total length of the cuts within 2.5 times the optimum.*

We now proceed to our main results regarding collections of polygons. In the conclusion of their paper, Demaine et al. [11] left several open questions. Here we focus on one of them: "What collections of nonoverlapping polygons in the plane can be simultaneously cut out by a circular saw?"[2] The authors pointed out that the problem is nontrivial when some of the polygons share edges. We study the case of axis-parallel rectangles in Section 3, where it is shown that the answer is always positive. Moreover, Section 4 gives evidence that the problem is nontrivial even if the polygons do not share edges.

It is easy to draw collections of disjoint convex polygons (even with axis-parallel rectangles) that are uncuttable by line cuts. Motivated by this state of affairs, Pach and Tardos have studied the problem of separating a large subfamily from a given family of pairwise disjoint compact convex sets on a sheet of glass, using line cuts [21]. For the case of axis-parallel rectangles, the authors show how to separate a subcollection with $\Omega(n/\log n)$ members out of given n. From the other direction, there exist instances of n such rectangles such that at most cn of them can be separated in this model, where $c < 1$ is a positive constant. Far weaker guarantees, sublinear in n, can be made for arbitrary convex polygons. For some other related results see [2, 8, 9, 18, 22]. In Section 3 we prove the following.

Theorem 2. *Given a collection $\mathcal{R}$ of n interior-disjoint axis-parallel rectangles drawn on a planar piece of material, $\mathcal{R}$ is cuttable by rays, so in particular by a segment saw of any size. The cutting sequence can be computed in $O(n \log n)$ time and uses at most $4n$ ray cuts, which is optimal in the worst case.*

In Section 4 we show some uncuttable collections of interior-disjoint polygons.

Theorem 3. *There exist collections of interior-disjoint polygons that are uncuttable by any segment saw. Such collections can be realized with convex or not necessarily convex polygons, and even with rectangles, or triangles.*

We conclude in Section 5 with an open problem.

Preliminaries. We distinguish between various cutting tools, such as: cuttable by a segment saw, cuttable by an arbitrarily small segment saw, ray cuttable, etc. For instance: A collection $\mathcal{P}$ of disjoint polygons drawn on a planar piece of material is *ray cuttable* (or cuttable by rays) if there exists a sequence of ray cuts after which every polygon in the collection is cut out (along its sides) as a separate piece. Otherwise, we say that the collection is *uncuttable* by rays. Cuttability with other tools (such as segment saw, or arbitrarily small segment saw) are defined similarly.

The following observations are in order.

1. If a collection of polygons is cuttable by rays, then it is also cuttable by a segment saw of any size.
2. If a collection of polygons is cuttable by a segment saw (segment) s of length $|s|$, then it is also cuttable by any segment saw of smaller length.

---

[2] Recall, this means cuttable by some (possibly small) segment saw.

2 Cutting Out a Single Polygon

In this section we prove Theorem 1. Let OPT and ALG denote the lengths of an optimal cutting sequence and that of a given algorithm being analyzed. We first show that the approximation algorithm of Demaine et al. [11] for cutting out a polygon achieves ratio 3 in the length measure. Moreover, this ratio cannot be improved as long as one uses the trivial lower bound on OPT given by the perimeter of P.

The algorithm cuts the polygon P out of its convex hull, conv(P), by removing the material in the pockets of P (the maximal connected components of conv(P)\ P). Each pocket K of P is cut out sequentially in an arbitrary direction, clockwise or counterclockwise. We refer the reader to their paper for details. Note that reflex vertices in K correspond to convex vertices of the target polygon P, and vice versa. By the characterization of Demaine et al. [11], P is cuttable by a (small) segment saw if and only if no two reflex vertices of P are consecutive; equivalently, no two convex vertices of any pocket K are consecutive.

Let a_i and b_i (in this order) be the two edges incident to a convex vertex of K, along the chosen direction. By the above characterization, any two terms a_i and b_j are disjoint, i.e., they denote distinct edges of K. Let $i = 1, \ldots, k$ be the sequence of convex vertices along the same direction. Write $A = \sum_{i=1}^{k} |a_i|$ and $B = \sum_{i=1}^{k} |b_i|$. The cutting algorithm (illustrated in [11, Fig. 5, page 74]) gives a cost arbitrarily close to $2A + 3B$, while obviously OPT $\geq A + B$ (the trivial lower bound on OPT). If $A \to 0$ and $A \ll B$ then the ratio can be arbitrarily close to 3:

$$\frac{\text{ALG}}{\text{OPT}} \leq \frac{2A + 3B}{A + B} \to 3.$$

Moreover, such a polygon (with $A \to 0$ and $A \ll B$) is easy to construct by

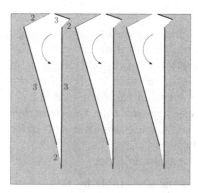

Fig. 2. A polygon with many thin pockets for which cutting length is close to 3 times the perimeter. The B-parts (which generate cuts of length close to $3B$) are drawn in bold lines; the pockets are processed in the order indicated by the arrow. The number of cuts (2 or 3) corresponding to each edge is specified.

choosing small $|a_i|$ and large $|b_i|$ in the pockets and minimizing other parts on the perimeter in comparison with pockets. See Figure 2.

The revised algorithm chooses the best direction for cutting each pocket K from the two possible, clockwise or counterclockwise, namely the direction for which $A \geq B$. Recall that the cutting algorithm gives a cost arbitrarily close to $2A + 3B$, while OPT $\geq A + B$. Hence the ratio is arbitrarily close to

$$\frac{\text{ALG}}{\text{OPT}} \leq \frac{2A + 3B}{A + B} = 2 + \frac{B}{A + B} \leq 2.5,$$

as desired.

3   Cutting Out a Collection of Axis-Parallel Rectangles

In this section we prove Theorem 2. As it turns out, the problem of cutting out a collection of axis-parallel rectangles by a segment saw is very much related to the problem of separating such a collection by moving the rectangles, one at at time, to infinity, using translations. We start by recalling the following classical result of Guibas and Yao [17] concerning translations of rectangles in a common direction.

Lemma 1 (Guibas and Yao [17]). *Let $\mathcal{R}$ be any set of n disjoint axis-parallel rectangles in the plane, and θ be any direction. Then there is an ordering $R_1, \ldots, R_n$ of the rectangles such that R_i can be moved continuously to infinity in direction θ without colliding with the rectangles R_j, $1 \leq i < j \leq n$. Such an ordering can be computed in $O(n \log n)$ time.*

The reader can verify that the proof of Lemma 1 in [17] implies the following stronger result; see also [15, Lemma 3].

Lemma 2 (Guibas and Yao [17]). *For any set of n disjoint axis-parallel rectangles in the plane, there is an ordering $R_1, \ldots, R_n$ of the rectangles such that R_i can be moved continuously to infinity in any direction between 0 and $\pi/2$ without colliding with the rectangles R_j, $1 \leq i < j \leq n$. Such an ordering can be computed in $O(n \log n)$ time.*

In Lemma 2 the set of directions under discussion make the closed interval $[0, \pi/2]$. It is now convenient to reformulate this lemma in terms of our interest.

Notation. Given two points $p, q \in \mathbb{R}^2$, p dominates q if the inequalities $x(p) > x(q)$ and $y(p) > y(q)$ hold among their x- and y-coordinates. Given two axis-parallel rectangles R', R'', write $R'' >_x R'$ if there exists a vertical line that separates R' and R'', so that R' lies in the left (closed) halfplane and R'' lies in the right (closed) halfplane determined by the line.

Lemma 3. *For any set of n disjoint axis-parallel rectangles in the plane, there is an ordering $R_1, \ldots, R_n$ of the rectangles such that R_i is unblocked in any direction between 0 and $\pi/2$ by any of the rectangles R_j, $1 \leq i < j \leq n$. Such an ordering can be computed in $O(n \log n)$ time.*

To be precise, R_i is *unblocked* in any direction between 0 and $\pi/2$ by any of the rectangles R_j, $i < j \leq n$, if and only if no vertex of such a rectangle dominates the lower left corner of R_i; see Figure 3.

Fig. 3. R is unblocked hence it is is cuttable

The two-step algorithm from [17] for computing the order is as follows.

STEP 1. Sort the rectangles by decreasing order of their y-coordinate of the top side; let $R_1, \ldots, R_n$ be the resulting order.

STEP 2. Start with an empty list L and add the rectangles R_i, for $i = 1, \ldots, n$, in this order. Place each new rectangle R in the first (i.e., leftmost) position in L consistent to the constraint that $R >_x S$ for every rectangle S following R in the list.

An illustration of the ordering produced appears in Figure 4. An obvious implementation of STEP 2 takes $O(n^2)$ time, however, Guibas and Yao [17] showed that it is possible to use balanced trees in a non-trivial way to reduce the time to $O(n \log n)$. Consequently, the two-step algorithm runs in $O(n \log n)$ time. Using the ordering guaranteed by Lemma 3, we get our desired result.

Proof of Theorem 2. We use the ordering provided by Lemma 3 and cut out rectangles one by one in this order; there are n iterations, $i = 1, \ldots, n$. The following invariant is maintained: in iteration i, rectangles $R_1, \ldots, R_{i-1}$ have been cut out (i.e., each has detached on a separate piece of material), and the current piece of material contains the subcollection $R_i, \ldots, R_n$.

Observe that in iteration i, rectangle R_i is unblocked in any direction between 0 and $\pi/2$, hence it is cuttable as follows; refer again to Figure 4.

Make two ray cuts from infinity (or from the boundary of the material): one vertical (going down) and one horizontal (going left); the two cuts meet at the lower left corner of R_i. The effect is detaching R_i from the piece of material containing the remaining rectangles R_j, $i < j \leq n$. The piece containing R_i contains no other rectangles, so two more cuts along two sides of R_i suffice to completely separate R_i completely, for a total of at most 4 cuts per rectangle.

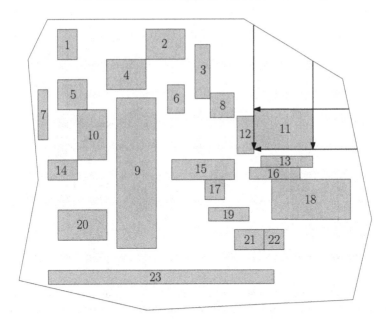

Fig. 4. STEP 2 of the algorithm. After $R_1, \ldots, R_{23}$ are successively inserted into L, the list is $11, 13, 12, 16, 18, 22, 8, 3, 2, 6, 15, 17, 19, 21, 4, 9, 1, 5, 10, 14, 20, 7, 23$. The first rectangle to be cut out is 11.

The process is continued until all rectangles are cut out in this way, with at most $4n$ cuts overall.

Clearly some collections require at least 4 cuts for each rectangle, e.g., if no two rectangle sides are collinear, for a total of at least $4n$ cuts. Hence the number of cuts executed in the algorithm is worst-case optimal.   □

Remark. There are cases when as few as $\Theta(\sqrt{n})$ cuts suffice, for instance when the rectangles are arranged in a square grid formation with their sides aligned.

4 Uncuttable Collections

It is easy to exhibit collections of disjoint polygons that are uncuttable by a segment saw, especially if one uses concave polygons or convex polygons with many sides. The problem becomes more interesting when one restricts the number of sides of the polygons or their shape. In this section we prove Theorem 3.

An uncuttable collection of rectangles (in arbitrary orientations) is shown in Figure 5 (left); this example has $n = 12$ rectangles, 6 drawn on the sides of a regular hexagon in its exterior, and 6 drawn in its interior. A similar pattern can be realized for every $n \geq 10$ (with $\lfloor n/2 \rfloor$ rectangles on the outer boundary of the union and the remaining $\lceil n/2 \rceil$ inside). Observe that every possible cut that can be initiated from outside is blocked by one of the small rectangles. Hence

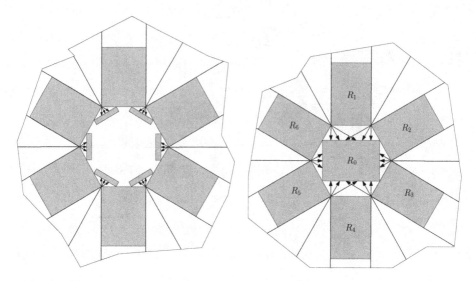

Fig. 5. Left: an uncuttable collection of $n = 12$ rectangles; arrows represent saw cuts that fail in cutting out any rectangle. Right: a cuttable collection of $n = 7$ rectangles.

none of the rectangles can be separated from the uncut material (shown in light yellow).

In contrast, the similar looking construction with $n = 7$ rectangles shown in Figure 5 (right) is cuttable by rays, hence in particular, by a segment saw of any size. This particular example has $n = 7$ rectangles, 6 drawn on the sides of a regular hexagon in its exterior, and one drawn in its interior. Start by cutting the rectangle on the top side of the hexagon by using two rays aligned with the long sides of its left and right neighbor rectangles. Observe that these two rays cross each other, so the piece of material containing the top rectangle can be detached. Once one of the rectangles has been cut out separately, the rest can be easily cut out one by one, and so the entire collection.

An uncuttable collection of triangles is shown in Figure 6 (left). Similarly, this small collection can be easily realized with larger numbers of triangles. It is straightforward to draw uncuttable k-gon collections for any $k \geq 4$ using $k + 1$ k-gons $P_0, P_1, ..., P_k$. Constructions for $k = 4$ and $k = 6$ are shown in Figure 6 (middle and right).

Remark. It is interesting to observe that separability by translations in a single direction holds for any collection of disjoint convex bodies; see also [4, Theorem 1], [15, Lemma 1], [19, Theorem 8.7.2]. Theorem 4 below appears in the work of Fejes Tóth and Heppes [16], but it can be traced back to de Bruijn [5]; the algorithmic aspects of the problem have been studied by Guibas and Yao [17].

Theorem 4. [5, 16, 17] *Any set of n convex objects in the plane can be separated via translations all parallel to any given fixed direction, with each object moving once only. If the topmost and bottommost points of each object are given (or can*

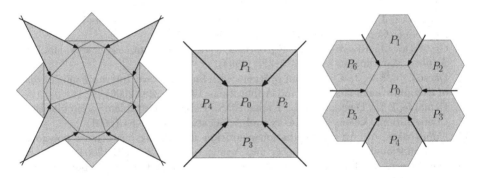

Fig. 6. Uncuttable k-gon collections. Left: construction with triangles ($k = 3$). Middle: $k = 4$. Right: $k = 6$. Arrows represent the only useful possible cuts in the collection.

be computed in $O(n \log n)$ time), an ordering of the moves can be computed in $O(n \log n)$ time.

However, as we have seen in Figures 5 and 6, this variant of separability does not guarantee the cuttability of arbitrary collections of convex bodies. The broader variant of separability present in Lemma 2 (for axis-aligned rectangles) makes the difference.

5   Conclusion

The obvious remaining open problem is devising an algorithm, which, given a collection of disjoint polygons in the plane determines whether it is cuttable by a segment saw, and computes a suitable cutting sequence if it is so. We conjecture that the problem admits a polynomial time algorithm.

References

1. Ahmed, S.I., Islam, M.A., Hasan, M.: Cutting a cornered convex polygon out of a circle. Journal of Computers 5(1), 4–11 (2010)
2. Alon, N., Katchalski, M., Pulleyblank, W.R.: Cutting disjoint disks by straight lines. Discrete & Computational Geometry 4, 239–243 (1989)
3. Bereg, S., Dăescu, O., Jiang, M.: A PTAS for cutting out polygons with lines. Algorithmica 53, 157–171 (2009)
4. Bereg, S., Dumitrescu, A., Pach, J.: Sliding disks in the plane. International Journal of Computational Geometry & Applications 15(8), 373–387 (2008)
5. de Bruijn, N.G.: Aufgaben 17 and 18 (in Dutch). Nieuw Archief voor Wiskunde 2, 67 (1954)
6. Bhadury, J., Chandrasekaran, R.: Stock cutting to minimize cutting length. European Journal of Operations Research 88, 69–87 (1996)
7. Chandrasekaran, R., Dăescu, O., Luo, J.: Cutting out polygons, In. In: Proceedings of the 17th Canadian Conference on Computational Geometry, pp. 183–186 (2005)

8. Czyzowicz, J., Rivera-Campo, E., Urrutia, J.: Separating convex sets in the plane. Discrete & Computational Geometry 7, 189–195 (1992)
9. Czyzowicz, J., Rivera-Campo, E., Urrutia, J.: Separation of convex sets. Discrete Applied Mathematics 51(3), 325–328 (1994)
10. Dăescu, O., Luo, J.: Cutting out polygons with lines and rays. International Journal of Computational Geometry and Applications 16, 227–248 (2006)
11. Demaine, E.D., Demaine, M.L., Kaplan, C.S.: Polygons cuttable by a circular saw. Computational Geometry: Theory and Applications 20, 69–84 (2001)
12. Dumitrescu, A.: An approximation algorithm for cutting out convex polygons. Computational Geometry: Theory and Applications 29, 223–231 (2004)
13. Dumitrescu, A.: The cost of cutting out convex n-gons. Discrete Applied Mathematics 143, 353–358 (2004)
14. Dumitrescu, A., Hasan, M.: Cutting out polygons with a circular saw. International Journal of Computational Geometry and Applications 23(2), 127–139 (2013)
15. Dumitrescu, A., Jiang, M.: On reconfiguration of disks in the plane and other related problems. Computational Geometry: Theory and Applications 46(3), 191–202 (2013)
16. Fejes Tóth, L., Heppes, A.: Über stabile Körpersysteme (in German). Compositio Mathematica 15, 119–126 (1963)
17. Guibas, L., Yao, F.F.: On translating a set of rectangles. In: Computational Geometry. Advances in Computing Research, vol. 1, pp. 61–67. JAI Press, London (1983)
18. Hope, K., Katchalski, M.: Separating plane convex sets. Math. Scand. 66, 44–46 (1990)
19. O'Rourke, J.: Computational Geometry in C, 2nd edn. Cambridge University Press (1998)
20. Overmars, M.H., Welzl, E.: The complexity of cutting paper. In: Proceedings of the 1st Annual ACM Symposium on Computational Geometry, pp. 316–321 (1985)
21. Pach, J., Tardos, G.: Cutting glass. Discrete & Computational Geometry 24, 481–495 (2000)
22. Pach, J., Tardos, G.: Separating convex sets by straight lines. Discrete Mathematics 241(1-3), 427–433 (2001)
23. Tan, X.: Approximation Algorithms for Cutting Out Polygons with Lines and Rays. In: Wang, L. (ed.) COCOON 2005. LNCS, vol. 3595, pp. 534–543. Springer, Heidelberg (2005)

Rectilinear Path Problems in Restricted Memory Setup

Binay K. Bhattacharya[1], Minati De[2], Anil Maheshwari[3,*],
Subhas C. Nandy[4,**], and Sasanka Roy[5,***]

[1] School of Computing Science, Simon Fraser University, Burnaby, Canada
[2] The Technion – Israel Institute of Technology, Haifa , Israel
[3] School of Computer Sciences, Carleton University, Ottawa, Canada
[4] Indian Statistical Institute, Kolkata, India
[5] Chennai Mathematical Institute, Chennai, India

Abstract. We study the *rectilinear path problem* in the presence of disjoint axis parallel rectangular obstacles in the *in-place* and *read-only* setup. The input to the problem is a set $\mathcal{R}$ of n axis-parallel rectangular obstacles in $\mathbb{R}^2$. We need to preprocess the members in $\mathcal{R}$ such that the following query can be answered efficiently.

Path-Query(p, q): Given a pair of points p and q, report an axis-parallel path from p to q avoiding the obstacles in $\mathcal{R}$.

In the *read-only* setup, we consider a restricted version of the *Path-Query*(p, q) problem, where the objective is to check the existence of an xy-monotone path between the given pair of points p and q avoiding the obstacles, and report it if such a path exists. Given $O(s)$ extra space, the problem can be solved in $O(\frac{n^2}{s} + n \log s + M_s \log n)$ time, where M_s is the time complexity for computing the median of n elements in read-only setup using $O(s)$ extra space.

In the *in-place* setup, we preprocess the input rectangles in a data structure such that for any pair of query points p and q, the problem *Path-Query*(p, q) can be solved efficiently. The time complexities for the preprocessing and query are $O(n \log n)$ and $O(n^{3/4} + \chi)$ respectively, where χ is the number of links (bends) in the path. The extra space requirement for both preprocessing and query answering are $O(1)$. We also show that among a set of unit square obstacles, there always exists a path of $O(\log n)$ links between a pair of query points. Here, we use a different in-place data structure with same preprocessing time complexity to answer the query in $O(\log n)$ time.

Keywords: Rectilinear path problem, in-place and read-only model of computation, constant work-space.

---

* Supported by NSERC.
** Part of work was done when this author was visiting Carleton University.
*** Supported by Infosys Foundation.

S. Ganguly and R. Krishnamurti (Eds.): CALDAM 2015, LNCS 8959, pp. 69–80, 2015.

1 Introduction

We study some new variations of *rectilinear path problem* in the restricted memory setup in a given closed planar environment containing a set $\mathcal{R}$ of n disjoint axis-parallel rectangular obstacles. The work is motivated from the *single point query problem* of Rezende et al. [17], and the two point query problem of Bint et al. [5]. Given a set $\mathcal{R}$ of rectangles and a fixed point p, Rezende et al. [17] preprocesses the input in a data structure so that given any arbitrary query point q, they can report the shortest path from p to q efficiently. Bint et al. [5] preprocesses the obstacles in $\mathcal{R}$ such that given a pair of query points (p, q), they can report the existence of an xy-monotone path from p to q efficiently. Both these algorithms use $O(n)$ extra work-space in addition to the space allocated for storing the input.

Recently, space-efficient algorithms are becoming popular to the algorithm researchers since it has wide applications in an environment where we need to handle very large data sets, or where we need to work on a input data sharing environment. As the low memory algorithms need only a small amount of extra memory space compared to the input size, a larger part of the data can be kept in a faster memory. This makes the algorithm to run faster. Here two different models are very popular, namely the *read-only model* and the *in-place model*. In the *in-place model*, the elements in the input location are allowed to permute during the execution. However, at any time during the execution, all the input elements should be available in the input locations [6, 7]. In the *read-only model*, the input data is not allowed to be permuted. Here one can only read the data and no write instruction can be executed in the input locations [1–3, 9, 10, 15]. Geometric shortest path problems in a restricted memory setup is studied in the literature. Asano and Doerr [2] showed that if the input is a weighted directed *grid graph* given in a read-only memory, then the shortest path between a pair of nodes u and v can be computed in $O(n^{\frac{1}{\epsilon}})$ time using $O(n^{\frac{1}{2}+\epsilon})$ extra space. Asano et al. [3] considered the geometric shortest path problem inside a simple polygon P given in a read-only array. For a pair of points $p, q \in P$, their proposed algorithm runs in $O(n^2)$ time using $O(1)$ extra space. Thus, the algorithm in [3] works in a single connected planar region. Our study in this paper is identifying axis-parallel paths in a planar region with rectangular holes.

In read-only setup, we study the problem of finding an xy-monotone path between a pair of points p and q in the presence of rectangular obstacles. We show that an xy-monotone path between a pair of points in such an environment can be computed in $O(\frac{n^2}{s} + n \log s)$ time using $O(s)$ extra space. We also consider the case where no xy-monotone path exists between the points p and q. Here we report an x-monotone (or a y-monotone) path between p and q, which always exists. We provide an algorithm for this version of the problem that also runs in $O(\frac{n^2}{s} + n \log s)$ time using $O(s)$ extra space.

Next, we consider the query version of the problem in the in-place setup. Here, we can prepare a data structure by permuting the obstacles during the preprocessing. The objective is, given a pair of query points p and q, to report a path between p and q efficiently. Our proposed algorithm needs $O(n \log n)$

preprocessing time, and can answer the query in $O(n^{\frac{3}{4}} + \chi)$ time, where χ is the number of links (bends) in the path from p to q. Both preprocessing and query need $O(1)$ extra space. Next, we consider a constrained version of this problem, where the input is a set of unit squares. We show that in this case, there always exists a path between any pair of points consisting of $O(\log n)$ links. Such a path can be reported in $O(\log n)$ time using a data structure which can be built using the in-place priority search tree [10], in $O(n \log n)$ time and $O(1)$ extra space.

2 Read-Only Setup

The proposed algorithms in this paper extensively use the computation of k-th order statistics among a set of elements given in a read-only array. The following table gives a comprehensive survey of results available on selection algorithms.

Table 1. Complexity results for selection algorithms when input is in a read-only array

Time $(T(n))$	Space $(S(n))$	Reference
$O(n^{1+\epsilon})$	$O(\frac{1}{\epsilon})$, where $2\sqrt{\frac{\log \log n}{\log n}} \leq \epsilon < 1$	[15]
$O(n \log^2 n)$	$O(\log n)$	[18]
$O(n \log_s n + n \log s)$	s, where $s \geq \log^2 n$	[12, 14]
$O(n)$	s, where $s \geq n/\log n$	[11]
$O(sn^{1+1/s} \log n)$	s, where $s \leq \log n$	[18]
$O(n \log \log_s n)$ (randomized)	s (any value)	[8, 15]

2.1 Reporting a Path - A Simple Algorithm

Here the input is a set of disjoint rectangles $\mathcal{R} = \{R_1, R_2 \ldots, R_n\}$, in $\mathbb{R}^2$ in a read-only array, and a pair of points $p, q \in \mathbb{R}^2 \setminus \mathcal{R}$. The objective is to compute an axis-parallel path from p to q avoiding the obstacles using $O(1)$ extra space.

Connect p and q by an "L-path" as shown in Figure 1(a). If the bend point r of this L-path lies inside a rectangle $R_i \in \mathcal{R}$, and the horizontal and vertical edges of this L-path intersect the rectangle R_i at the points a and b respectively, then update the path as $p \to a \to x \to b \to q$, where x is the appropriate corner of R_i (see Figure 1(b)). Now, try to get a path from p to a and a path from b to q avoiding the obstacles in $\mathcal{R}$. Otherwise, if the bend-point r does not lie inside any rectangle in $\mathcal{R}$, then the desired path is the concatenation of two obstacle avoiding paths, from p to r and from r to q, respectively. The identification of R_i (if any) can be done in $O(n)$ time by inspecting all the members in $\mathcal{R}$.

To Obtain a Path from p **to** a**.** Inspect all the rectangles in $\mathcal{R}$ to find a rectangle (if any) that intersects the line segment $[p, a]$ and is closest to p. Let it be R_1. Now, update the path $p \to a$ as the U-path $p \to a_1 \to x_1 \to y_1 \to b_1 \to a$, where a_1, x_1, y_1 and b_1 are as illustrated in Figure 1(c). Again find another rectangle (if any) that intersects $[b_1, a]$, and update the path accordingly. Proceed similarly, until the point a or the point r is reached. The same method is adopted to get a path from b (or r) to q.

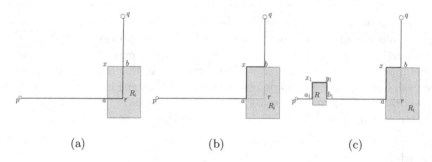

Fig. 1. Demonstration of different steps for drawing a path

Since the rectangles are non-overlapping, this method finds an obstacle avoiding path from p to q in $O(n^2)$ time using $O(1)$ extra space. If $O(s)$ extra workspace is available, then the time complexity can be improved to $O(\frac{n^2}{s})$ as follows. Let us consider the method of obtaining a path from p to a.

> Split the input array into s parts each containing at most $\lceil \frac{n}{s} \rceil$ rectangles. Initialize $\tau = p$.
> For each $i = 1, 2, \ldots, s$ do
> - Inspect all the rectangles in the i-th block to find the rectangle closest to τ that intersects the line segment $[\tau, a]$. Let it be R_i^*.
> - Create a priority queue Q as a heap with $R_i^*, i = 1, 2, \ldots, s$, where the key is the x-coordinate of the left-boundary of $R_i^*, i = 1, 2, \ldots, s$.
> - Repeat until Q becomes empty
>   * choose the element R from the root of Q, and update the path $\tau \rightarrow a$ as the U-path $\tau \rightarrow a_1 \rightarrow x_1 \rightarrow y_1 \rightarrow b_1 \rightarrow a$.
>   * update $\tau = b_1$.
>   * if the top element of Q is from the i-th block of the input array, then inspect the elements of the i-th block to obtain a rectangle R' which intersects the line segment $[\tau, a]$, and is closest to τ in comparison with the other rectangles in the i-th block.
>   * insert R' in Q.

We can implement the priority queue using a binary heap data structure in the array S. The initialization of the priority queue takes $O(n)$ time. Each step needs $O(\frac{n}{s} + \log s)$ time in the worst case. Thus, we have the following result.

Theorem 1. *Given a set $\mathcal{R}$ of n disjoint rectangular obstacles in $\mathbb{R}^2$ and a pair of points $p, q \in \mathbb{R}^2 \setminus \mathcal{R}$, a rectilinear path from p to q avoiding the obstacles can be computed in $O(\frac{n^2}{s} + n \log s)$ time using $O(s)$ extra space in the* read-only *setup.*

3 xy-Monotone Path

We borrow definitions of monotone and preferred paths from [5]. Let $p, q \in \mathbb{R}^2 \setminus \mathcal{R}$ be two points such that p is to the south-west of q.

Definition 1. *[5]* **(Monotone Path)** *An xy-monotone path from p to q is a path which follows the +x direction for the horizontal line segments and +y direction for the vertical line segments. Similarly we can define $(-x)y$, $x(-y)$ and $(-x)(-y)$-monotone paths.*

Definition 2. *[5]* **(Preferred Path)** *A y-preferred xy-path from p to q is an xy-monotone path which follows the +y direction whenever possible. If it encounters an obstacle it follows the +x-direction until the obstacle is there. When it reaches the end of the obstacle it follows the +y direction again.*

Checking the Existence of an xy-Monotone Path

Let H_p and H_q (resp. V_p and V_q) be the horizontal (resp. vertical) lines through the points p and q, respectively. We compute two unbounded paths P_1 and P_2 from p, where P_1 is the x-preferred xy-path and P_2 is the y-preferred xy-path (see Figure 2). The path P_1 crosses V_q and P_2 crosses H_q. If q lies inside the region bounded by these two paths, H_q and V_q, then there exists an xy-monotone path from p to q [17]. An $O(n^2)$ time

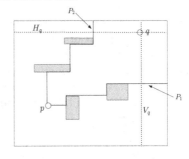

Fig. 2. Existence of an xy-monotone path

and $O(1)$ extra-space algorithm for computing P_1 (resp. P_2) is as follows.

Perform linear search in the input array to get the first bend on the path P_1 starting from p. This is equivalent to finding a rectangle R that is closest to p and is intersected by the horizontal line H_p. Suppose it hits R at a point a. Here two cases can arise depending on whether or not R intersects the vertical line V_q. If so, then the point of intersection of P_1 with V_q is the intersection of V_q and the top boundary of R. Otherwise, shoot a horizontal ray from the top-right corner of R to identify a rectangle $R' \in \mathcal{R}$ that it hits first. This can be done by inspecting all the rectangles in $\mathcal{R}$. Proceed similarly until the line V_q is reached.

As in the earlier section, here also it can be shown that with an extra space of size s ($s > \log n$), the time complexity can be improved to $O(\frac{n^2}{s} + n \log s)$. Thus, we have the following result.

Lemma 1. *Given a set $\mathcal{R}$ of n disjoint rectangular obstacles in $\mathbb{R}^2$ and a pair of points $p, q \in \mathbb{R}^2 \setminus \mathcal{R}$, the existence of an xy-monotone path from p to q avoiding the obstacles in $\mathcal{R}$ can be checked in $O(\frac{n^2}{s} + n \log s)$ time using $O(s)$ extra space.*

Reporting the Path. Suppose the above confirms the existence of an xy-monotone path from p to q. We, now, consider the problem of reporting such a path. We compute (i) the x-preferred xy-path Π_1 from p up to V_q, and (ii) the $(-y)$-preferred $(-x)(-y)$-path Π_2 from q up to H_p. Let these intersect at a point r. We report an xy-monotone path from p to q as the concatenation of two xy-monotone path-segments from p to r along Π_1 and from q to r along Π_2.

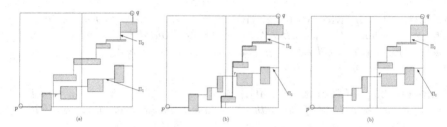

Fig. 3. Computing the point of intersection r of two paths Π_1 and Π_2

Computing the Point r**.** We use two scalar variables τ_1 and τ_2 to indicate the set S of *useful rectangles*, whose left boundaries have the x-coordinate in the interval $[\tau_1, \tau_2]$. We perform a binary search in this interval to find a vertical line on which the point of intersection of the paths Π_1 and Π_2 lies. Initially, we set $\tau_1 = x(p)$ and $\tau_2 = x(q)$. In each iteration of the binary search, we compute the median μ of the left boundaries of the rectangles in the set S, and consider the vertical line $W : x = \mu$. We compute the x-preferred xy-path and $(-y)$-preferred $(-x)(-y)$-path from p and q respectively. Let these meet W at θ_1 and θ_2 respectively. Here the following two situations may arise:

1. $y(\theta_1) < y(\theta_2)$ (Figure 3(a)). Here, point r is to the left of W. We set $\tau_2 = \mu$.
2. $y(\theta_1) > y(\theta_2)$ (Figure 3(b)). Here, point r is to the right of W. We set $\tau_1 = \mu$.
3. Π_2 intersects H_p to the right of W (see Figure 3(c)). Here also the point r is to the right of W. We set $\tau_1 = \mu$.

Iteration terminates when τ_1 and τ_2 correspond to the two consecutive left-boundaries. Here, the point r lies between the vertical lines $W : x = \tau_1$ and $W' : x = \tau_2$. We compute r by consulting the edges of Π_1 and Π_2 that intersect the vertical lines W and W'.

Complexity Analysis. The number of levels of recursion required for computing the point r may be $O(\log n)$ in the worst case. In each level, getting the line W needs computation of median in a read-only array. This needs $O(M_s \log n)$ time in total, where M_s is the time complexity of median finding algorithm using $O(s)$ extra work-space (see Table 1).

In each level of recursion, we can use the points of intersection of Π_1 and Π_2 with the vertical lines defined by τ_1 and τ_2, and compute the portion of Π_1 and Π_2 between two vertical lines defined by τ_1 and τ_2. As the number of steps in the staircase paths Π_1 and Π_2 in the i-th level of recursion is at most $\frac{n}{2^i}$, so the total number of steps of Π_1 and Π_2 to be computed in the entire process is $O(n)$. Since computation of each step of Π_1 and Π_2 needs $O(\frac{n}{s} + \log s)$ time using $O(s)$ space, we have the following result.

Theorem 2. *Given a set of n disjoint rectangular obstacles and a pair of points $p, q \in \mathbb{R}^2 \setminus \mathcal{R}$, reporting an xy-monotone path from p to q needs $O(\frac{n^2}{s} + n \log s + M_s \log n)$ time using $O(s)$ extra space in the* read-only *setup, where M_s is the time complexity of computing median of n elements using $O(s)$ space in the* read-only *setup.*

Note: Here, we want to mention an important issue. After checking the existence of an xy-monotone path, we have reported it as a concatenation of two paths of different directions: (i) from p to r and (ii) from q to r, for at a specific point r on the plane. It needs to study whether this path can be reported in one direction from p to q using the $O(\log n)$ size compressed stack data structure of [4].

4 x-Monotone Path

It is well-known that the shortest path problem in a graph is NL-complete [13]. However, the shortest rectilinear path between a pair of points p and q among a set of rectangular obstacles is either x- or y-monotone [17]. Here, our objective is to compute an x-monotone path from p to q using sub-linear extra space.

Without loss of generality, we assume that p is to the south-west of q. We first check for the existence of an xy-monotone path from p to q using the algorithm in Section 3. If it exists, we report it. Otherwise, we proceed as follows. Let A be the upper boundary of the rectangle bounding all the members in $\mathcal{R}$. We compute y-preferred xy-monotone path from p, and y-preferred $(-x)y$-monotone path from q. Suppose these two paths intersect the horizontal line A at α and β respectively. If $x(\alpha) < x(\beta)$, then the x-monotone path is obtained by concatenating the path from $p \rightsquigarrow \alpha$, $\alpha \to \beta$ and $\beta \rightsquigarrow q$. However, if $x(\alpha) > x(\beta)$ then these two paths intersect at a point or at an edge of a rectangle which is common to both the paths. The point of intersection r of these two paths can be obtained as was done earlier for the xy-monotone path. Thus, we have the following result.

Theorem 3. *Given a set of rectangular obstacles $\mathcal{R}$ and a pair of points $p, q \in \mathbb{R}^2 \setminus \mathcal{R}$, an x-monotone path between p and q avoiding the obstacles can be computed in $O(\frac{n^2}{s} + n \log s)$ time using $O(s)$ extra space in the read-only setup.*

5   In-Place Setup

In the in-place model, we are allowed to permute the elements in the array $\mathcal{R}$. Here, we consider the query version of the problem, where the obstacles in $\mathcal{R}$ is to be preprocessed in the same array, such that given a pair of points in $\mathbb{R}^2 \setminus \mathcal{R}$, a path from p to q avoiding the obstacles can be reported in an output sensitive manner with a sub-linear extra time. Next, we show that if the obstacles are unit squares, then there exists a path of length $O(\log n)$, and it can be computed in polylogarithmic time. For both the algorithms, the construction of necessary data structures and query answering take $O(1)$ extra space.

5.1   Path Query Among Arbitrary Obstacles

We adopt the same strategy of reporting a path between two query points as we did in Section 2.1. This needs maintaining of a data structure in the input array such that given any arbitrary horizontal query line segment ℓ, we can answer the set of rectangles that are stabbed by ℓ.

Here the input is a set $\mathcal{R} = \{R_i, i = 1, 2, \ldots, n\}$ of rectangles, where each rectangle R_i is specified by a pair of coordinates $[(a_i, b_i), (c_i, d_i)]$ indicating the bottom-left and top-right corners of R_i respectively. The query input is a horizontal line segment $\ell = [u, v]$, where the coordinates of u and v are (α, y) and (β, y), respectively.

Brönnimann et al. [7] proposed an *in-place k-D tree* data structure for a set of n points in $\mathbb{R}^2$, which can be constructed in $O(n \log n)$ time, and any k dimensional rectangular range query can be answered in $O(n^{1-\frac{1}{k}} + \chi)$ time, where χ is the size of the reported answers. We consider n points $\{(a_i, c_i, b_i, d_i), i = 1, 2, \ldots, n\}$ in $\mathbb{R}^4$ corresponding to the n rectangles in $\mathcal{R}$, and construct an *in-place k-D tree* $\mathcal{T}$ with $k = 4$.

Let us name the coordinate axes as A, C, B and D, respectively. We assume that none of the two end-points of the query line segment ℓ lies inside any rectangle in $\mathcal{R}$. This can be tested in $O(\log n)$ time using the data structure $\mathcal{T}$. If one of the end-points, say v, lies in a rectangle R_i, we need to shorten it up to the point of intersection (v') of ℓ and R_i (as we considered line segment $[p, a]$ in Figure 1(b)), so that both the end-points of the line segment $[u, v']$ is outside the rectangles of $\mathcal{R}$.

We follow a *divide-and-conquer strategy* to answer this query. In order to intersect a rectangle $R_i = [(a_i, b_i), (c_i, d_i)]$ by the line ℓ, we must have (i) $\alpha < a_i$, (ii) $\beta > c_i$, and (iii) $b_i < y < d_i$. We search $\mathcal{T}$ to identify those rectangles satisfying the constraints (i), (ii) and (iii).

We start searching the tree $\mathcal{T}$ from its root. If the partition line with respect to A at the root node is to the left of α, then the elements (rectangles) in the left subtree can not span over ℓ (since u is outside of any rectangle). Thus, the left subtree can be ignored; otherwise both the subtrees need to be checked. Similarly, in the next level, in the relevant partition, if the partition line with respect to C is to the right of β, then the elements in the right subtree can be ignored; otherwise both the subtrees need to be checked. Similarly, in the next level of the relevant partition, the partition line with respect to B is above y, then the elements in the right subtree can be ignored; otherwise both the subtrees need to be checked, and finally in the next level of the relevant partition, if the partition line with respect to D is below y, then the elements in the left subtree can be ignored; otherwise both the subtrees need to be checked. It needs to mention that, if at any level both the subtrees needs to be processed, then the left subtree is processed first, and then the right subtree is processed. Lemma 2 says that the rectangles intersecting ℓ will be reported in an ordered manner, so that the corresponding axis-parallel path can be reported as in Section 2.1.

Lemma 2. *The rectangles intersected by the line ℓ will be generated in an ordered manner from left to right.*

Proof. Consider the first 4 consecutive levels of the tree $\mathcal{T}$. We need to analyze the situation when the search spans both the sides in each level. Surely, with respect to the axis A, the elements that are intersecting ℓ in the left subtree (having left boundary to the left of the partition line) will be reported first, and then the elements that are intersecting ℓ in the right subtree (having left boundary to the right of the partition line) will be reported. Now, we will show that in the first level if more than one element of its left subtree intersect ℓ, then they will be reported in the ordered manner.

Consider the elements in the left subtree of the first level. The rectangles intersected by ℓ are ordered from left to right with respect to their right boundaries. Among these rectangles, if one of them (say R_1) has right boundary to the left the right boundary of the other (say R_2), then the left boundary of R_1 is also to the left of the left boundary of R_2. Thus, if the rectangles of both the side of the partition line in the next level (with respect to axis C) are candidate for reporting, then the order is maintained since those in the left partition line are reported first, and then those in the right partition are reported. Note that, there may exist rectangles R_1 and R_2 with right boundary of R_1 to the left of that of R_2 and left boundary of R_1 is to the right of that of R_2. Here, both R_1 and R_2 can not be intersected by ℓ, and they will be discriminated in the next two levels of comparison with respect to the axes B and D. If still there are more than one element in the same partition that are candidate of reporting, they will be split in the subsequent levels of division with respect to axes A and C. Thus, the resulting elements are printed at the leaf level of the tree in order. □

In order to report a path from $(x(u), y(u))$ to $(x(v), y(u))$, we draw the concatenation of "U-paths" corresponding to the intersected rectangles in order, as in Section 2.1. The path in the vertical direction from $(x(v), y(u))$ to $(x(v), y(v))$ will be reported similarly. Thus, we have the following result.

Theorem 4. *Given a set of n disjoint rectangles in an array of size n, we can preprocess them in the same array in $O(n \log n)$ time, so that for any pair of query points $p, q \in \mathbb{R}^2 \setminus \mathcal{R}$, a path from p to q can be reported in $O(n^{3/4} + \chi)$ time, where χ is the number of links in the reported path.*

5.2   Path Query among Unit Square Obstacles

Here, a set $\mathcal{R}$ of non-overlapping *unit squares* in $\mathbb{R}^2$ is given in an input array $\mathcal{R}$. The objective is to show that one can maintain a data structure in the array $\mathcal{R}$ itself such that given any arbitray pair of query points $p, q \in \mathbb{R}^2 \setminus \mathcal{R}$, an axis-parallel path between p and q avoiding the obstacles in $\mathcal{R}$ can be reported in poly-logarithmic time.

Definition 3. *For a unit square $R \in \mathcal{R}$,*

- *its* center-point *is the point of intersection of its two diagonals,*
- *its* left-point *is its top-left corner,*
- *its* right-point *is its top-right corner.*

We arrange the members of $\mathcal{R}$ in the input array in an in-place manner such that their *center-points* form a priority search tree $\mathcal{T}$ [10]. In a standard priority search tree of a set of points P, apart from the pointers to its left and right children, each node stores two keys (priority_key, search_key) [16], where

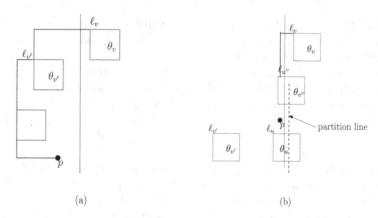

Fig. 4. Finding a path: p is below ℓ_v in the left side of the partition line

priority_key corresponds to the point having maximum y-coordinate among the points in P inside the partition corresponding to that node, and search_key is the x-coordinate of a vertical line that equi-partitions the points inside that partition. In the in-place version of the priority search tree, in each node we store a member of $\mathcal{R}$ corresponding to the priority key of that node. We do not explicitly store the search_key; it is computed on demand in $O(\log n)$ time. Moreover, we do not store the pointer indicating its left and right children using index computation in $O(1)$ time, and computed on demand. Thus, each node stores only a unit square, and the entire tree can be stored in the input array itself. While organizing the members of $\mathcal{R}$ in $\mathcal{T}$, we use $O(1)$ work-space. Reader is referred to [10] for the detail. In [10], it is also shown that all the standard queries on a priority search tree can be performed in the same time as that in [16] using $O(1)$ extra space. During the presentation of the algorithm, any reference of search_key implies computation of the search key in the data structure $\mathcal{T}$. The unit-square corresponding to a node $v \in \mathcal{T}$ will be denoted by R_v. Its center-point, left-point and right-point are denoted by θ_v, ℓ_v and ρ_v respectively.

We first present an algorithm for finding a y-monotone path from the *left-point* of the square R_r to the point p, where r denotes the root node of $\mathcal{T}$.

If p is above ℓ_r, then we join the points p and ℓ_r by an *L-path*[1], and exit. Otherwise, we traverse $\mathcal{T}$ (from node r) to report the path. At any node v on the traversal path, p may lie in the left- or right-partition defined by its search_key.

Without loss of generality assume that p is in the left partition of node v. The other case is similar, and omitted. Let v' be the left child of node v. If p is below $\ell_{v'}$, we join ℓ_v and $\ell_{v'}$ by an L-path (see Figure 4(a)). There exists no member of $\mathcal{R}$ in either the left or the right partition that can intersect this L-path from ℓ_v to $\ell_{v'}$. The reason is that, (i) all the members of $\mathcal{R}$ in the left-partition are below $\ell_{v'}$, and (ii) the members in the right-partition can not reach the vertical segment of the L-path since their center-points are to the right of $\theta_{v'}$. The process continues to connect $\ell_{v'}$ and p (by setting $v = v'$). Note that here

---

[1] A path is said to be an *L-path* if it consists of at most two line segments, a horizontal segment followed by a vertical segment or vice-versa.

the search is restricted to the subtree rooted at v', and the right sub-tree of v need not be considered since no vertical segment on the path from ℓ_v to p will be intersected by any square from the right subtree of v.

If p is above $\ell_{v'}$, no other vertex of the left-subtree of v needs to be considered. However, the L-path obtained by joining ℓ_v and p may be intersected by some other members of $\mathcal{R}$ in the right subtree of v as shown Figure 4(b). We use a temporary variable v^*, and initialize it with v to handle this case. Let v'' be the right child of v.

- If p is above $\ell_{v''}$, then we can join ℓ_v and p by an L-path, which will not be intersected by any other member of $\mathcal{R}$. The process stops.
- If p is below $\ell_{v''}$, then the vertical segment of the L-path from ℓ_v to p *may* or *may not* be intersected by the square corresponding to v''. In it intersects, then we connect ℓ_v and $\ell_{v''}$ by an L-path, and set $v = v^* = v''$. Otherwise, we set $v = v''$ and continue the search.

Proceeding this way, when after processing a square at a vertex $v \in \mathcal{T}$, p is observed to lie above the square corresponding to both of its children, then ℓ_{v^*} and p are connected by using an L path[2]. This leads to the following result.

Lemma 3. *For any query point $p \in I\!R^2 \setminus \mathcal{R}$, there exists a y-monotone obstacle avoiding orthogonal path with $O(\log n)$ links from the top-left corner of the rectangle attached to the root of $\mathcal{T}$ and the point p, and the path can be computed in $O(\log^2 n)$ time.*

The time complexity follows from the fact that in each level of the tree, we need to compute the search-key to determine the partition line. This can be reduced to $O(\log n)$ by using the technique of traversing the tree as in [10] for answering the standard queries of priority search tree in its in-place version.

Given a pair of query points (p, q), the search in $\mathcal{T}$ proceeds from the root until a node v is observed whose different partitions contain p and q respectively, or at least one among p and q having higher y-coordinate than the current node v in $\mathcal{T}$ is reached. In the first case, we report the path by concatenating two paths from ℓ_v to p and from ℓ_v to q. In the second case, let p be above v. We concatenate an L-path from ℓ_v to p, and a y-monotone path from ℓ_v to q. Thus, we have the following result.

Theorem 5. *Given a set of n disjoint unit square obstacles $\mathcal{R}$ in an array of size n, they can be preprocessed in an in-place manner in $O(n \log n)$ time and using $O(1)$ extra space, such that for any pair of query points $p, q \in I\!R^2 \setminus \mathcal{R}$, one can report a path from p to q avoiding the obstacles in $\mathcal{R}$ in $O(\log n)$ time.*

6 Conclusion

Several variations of reporting rectilinear paths between a pair of points among a set of axis-parallel rectangular obstacles are studied. An interesting question

---

[2] Where the edge of the L-path incident to p is a horizontal line segment.

still remains unanswered is the problem of computing the shortest rectilinear path among a pair of points in memory constrained environment.

While considering these path problems, we found that the query version of the problem is even interesting in an in-place setup. So it raises the following open question: *is it possible to report the shortest path between a pair of query points in an in-place setup in $O(k + polylog\ n)$ time using $O(1)$ extra space?*.

We could show the bound on the path length for axis-parallel unit squares only (see Theorem 5). It will also be interesting to investigate the other configurations where a similar bound can be proved.

References

1. Asano, T., Buchin, K., Buchin, M., Korman, M., Mulzer, W., Rote, G., Schulz, A.: Memory-constrained algorithms for simple polygons. Comput. Geom. 46(8), 959–969 (2013)
2. Asano, T., Doerr, B.: Memory-constrained algorithms for shortest path problem. In: Proc. CCCG (2011)
3. Asano, T., Mulzer, W., Rote, G., Wang, Y.: Constant-work-space algorithms for geometric problems. JoCG 2(1), 46–68 (2011)
4. Barba, L., Korman, M., Langerman, S., Sadakane, K., Silveira, R.I.: Space-time trade-offs for stack-based algorithms. In: STACS (2013)
5. Bint, G., Maheshwari, A., Smid, M.H.M.: xy-Monotone path existence queries in a rectilinear environment. In: Proc. CCCG, pp. 35-40 (2012)
6. Bose, P., Maheshwari, A., Morin, P., Morrison, J., Smid, M.H.M., Vahrenhold, J.: Space efficient geometric divide-and-conquer algorithms. Comput. Geom. 37(3), 209–227 (2007)
7. Brönnimann, H., Chan, T.M., Chen, E.Y.: Towards in-place geometric algorithms and data structures. In: Proc. Symp. on Comput. Geom., pp. 239-246 (2004)
8. Chan, T.M.: Comparison-based time-space lower bounds for selection. ACM Transactions on Algorithms 6(2) (2010)
9. Chan, T.M., Ian Munro, J., Raman, V.: Faster, space-efficient selection algorithms in read-only memory for integers. In: Cai, L., Cheng, S.-W., Lam, T.-W. (eds.) ISAAC 2013. LNCS, vol. 8283, pp. 405–412. Springer, Heidelberg (2013)
10. De, M., Maheshwari, A., Nandy, S.C., Smid, M.H.M.: An in-place min-max priority search tree. Comput. Geom. 46(3), 310–327 (2013)
11. Elmasry, A., Juhl, D.D., Katajainen, J., Satti, S.R.: Selection from read-only memory with limited workspace. In: Du, D.-Z., Zhang, G. (eds.) COCOON 2013. LNCS, vol. 7936, pp. 147–157. Springer, Heidelberg (2013)
12. Frederickson, G.: Upper bounds for time-space tradeoffs in sorting and selection. Journal of Computer and System Sciences 34(1), 19–26 (1987)
13. Goldreich, O.: Computational Complexity: A Computational Perspective, p. 182. Cambridge University Press (2008)
14. Munro, J.I., Paterson, M.: Selection and sorting with limited storage. Theoretical Computer Science 12, 315–323 (1980)
15. Munro, J.I., Raman, V.: Selection from read-only memory and sorting with minimum data movement. Theor. Comput. Sci. 165(2), 311–323 (1996)
16. McCreight, E.M.: Priority search trees. SIAM J. Comput. 14, 257–276 (1985)
17. de Rezende, P.J., Lee, D.T., Wu, Y.F.: Rectilinear shortest paths in the presence of rectangular barriers. Discrete Computational Geometry 4, 41–53 (1989)
18. Raman, V., Ramnath, S.: Improved upper bounds for time-space tradeoffs for selection with limited storage. In: Arnborg, S. (ed.) SWAT 1998. LNCS, vol. 1432, pp. 131–142. Springer, Heidelberg (1998)

New Polynomial Case for Efficient Domination in P_6-free Graphs

T. Karthick

Computer Science Unit,
Indian Statistical Institute, Chennai Centre,
Chennai-600113, India
karthick@isichennai.res.in

Abstract. In a graph G, an *efficient dominating set* is a subset D of vertices such that D is an independent set and each vertex outside D has exactly one neighbor in D. The EFFICIENT DOMINATING SET problem (EDS) asks for the existence of an efficient dominating set in a given graph G, and the WEIGHTED EFFICIENT DOMINATING SET problem (WEDS) asks for an efficient dominating set of minimum total weight in the given graph G with vertex weight function w on $V(G)$. The (W)EDS is known to be NP-complete for P_7-free graphs, and is known to be polynomial time solvable for P_5-free graphs. However, the computational complexity of the (W)EDS problem is unknown for P_6-free graphs. In this paper, we show that the WEDS problem can be solved in polynomial time for a subclass of P_6-free graphs, namely $(P_6,$ banner)-free graphs, where a *banner* is the graph obtained from a chordless cycle on four vertices by adding a vertex that has exactly one neighbor on the cycle.

Keywords: Graph algorithms, Domination in graphs, Efficient domination, Perfect code, P_6-free graphs, Square graph.

1   Introduction

Throughout this paper, let $G = (V, E)$ be a finite, undirected and simple graph. We follow West [22] for standard notations and terminology. If $\mathcal{F}$ is a family of graphs, a graph G is said to be $\mathcal{F}$-*free* if it contains no induced subgraph isomorphic to any graph in $\mathcal{F}$. Let P_t denotes the path on t vertices.

In a graph G, a subset $D \subseteq V$ is a *dominating set* if each vertex outside D has some neighbor in D. An *efficient dominating set* is a dominating set D such that D is an independent set and each vertex outside D has exactly one neighbor in D. Note that not all graphs have efficient dominating sets. For example, the cycle C_n has an efficient dominating set if and only if $n \equiv 0 \pmod 3$. Also, a set $D = \{v_1, v_2, \ldots, v_k\}$ is an efficient dominating set of G if and only if $\{N[v_1], N[v_2], \ldots, N[v_k]\}$ is a partition of $V(G)$ [11], where $N[v_i]$ $(1 \le i \le k)$ denotes the closed neighborhood of v_i. Bange et al. [1] showed that if G has an efficient dominating set, then the cardinality of any efficient dominating set equals the cardinality of a minimum dominating set in G, and thus all efficient

S. Ganguly and R. Krishnamurti (Eds.): CALDAM 2015, LNCS 8959, pp. 81–88, 2015.

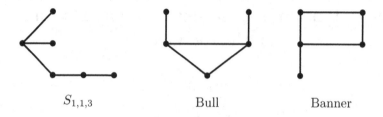

$S_{1,1,3}$ Bull   Banner

Fig. 1. Some special graphs

dominating sets have the same cardinality. The EFFICIENT DOMINATING SET problem (EDS) asks for the existence of an efficient dominating set in a given graph G, and the WEIGHTED EFFICIENT DOMINATING SET problem (WEDS) asks for an efficient dominating set of minimum total weight in the given graph G with vertex weight function w on $V(G)$.

Efficient dominating sets were independently introduced by Bange et al. [1] and Biggs [2], and are also called perfect codes, perfect dominating sets and independent perfect dominating sets in the literature. The study of efficient dominating sets are motivated by various interesting applications such as coding theory [2, 10], graph embedding [20, 21], facility location on geographical area [23, 24], and resource allocation in parallel computer networks [15]. We refer to [11] for more information, and [17] for a brief history and survey on efficient domination in graphs.

The (W)EDS is known to be NP-complete in general, and is known to be NP-complete for several restricted classes of graphs such as: chordal bipartite graphs [17], planar bipartite graphs [17], and planar graphs with maximum degree three [9]. However, the (W)EDS is solvable in polynomial time for split graphs [6], cocomparability graphs [8], interval graphs [7], circular-arc graphs [7], and for many more classes of graphs (see [5] and the references therein, and Figure 1 of [17]). In particular, EDS is NP-complete for $2P_3$-free chordal graphs [19], and hence EDS remains NP-complete for P_7-free graphs. Milanič [18] showed that the WEDS is solvable in polynomial time for P_5-free graphs. Brandstädt and Le [4] showed that the WEDS is solvable in polynomial time for (E, xNet)-free graphs, thereby extending the result on P_5-free graphs. However, the computational complexity of EDS is unknown for P_6-free graphs. In [5], Brandstädt et al. showed that the WEDS is solvable in polynomial time for $(P_6, S_{1,2,2})$-free graphs. Recently, the author showed that WEDS is solvable in polynomial time for $(P_6, S_{1,1,3})$-free graphs, and (P_6, bull)-free graphs [13]. We refer to Figure 1 of [5] for the recent complexity status of (W)EDS on several graph classes. See Figure 1 for some of the special graphs used in this paper.

In this paper, we show that the WEDS problem can be solved in polynomial time for (P_6, banner)-free graphs, where a *banner* is the graph obtained from a chordless cycle on four vertices by adding a vertex that has exactly one neighbor on the cycle (see Figure 1). A banner is also called as P, 4-apple and A_4 in the literature.

The class of banner-free graphs includes several well studied classes of graphs in the literature such as: P_4-free graphs (or co-graphs), $K_{1,3}$-free graphs (or claw-free graphs), and C_4-free graphs. Note that from the NP-completeness result for $K_{1,3}$-free graphs [16], it follows that for banner-free graphs, the EDS remains NP-complete.

If G is a graph, and if S is a subset of $V(G)$, then $G[S]$ denote the subgraph induced by S in G. For any two vertices u and v in G, $\text{dist}_G(u, v)$ denote the *distance* between u and v in G. The *square* of a graph $G = (V, E)$ is the graph $G^2 = (V, E^2)$ such that $uv \in E^2$ if and only if $\text{dist}_G(u, v) \in \{1, 2\}$.

The following lemma given in [18] (see also [4]) relates the EDS problem on G and the MAXIMUM WEIGHT INDEPENDENT SET (MWIS) problem on G^2.

Lemma 1. *[18] Let G be a graph with vertex weight $w(v)$ equal to the number of neighbors of v plus one. Then the following statements are equivalent for any subset $D \subseteq V$:*

(i) D is an efficient dominating set in G.
(ii) D is a minimum weight dominating set in G with $\sum_{v \in D} w(v) = |V|$.
(iii) D is a maximum weight independent set in G^2 with $\sum_{v \in D} w(v) = |V|$. $\square$

Thus, the WEDS problem on a graph class $\mathcal{G}$ can be reduced to the MWIS problem on the squares of graphs in $\mathcal{G}$.

In Section 2, we show that if G is a $(P_6$, banner)-free graph that has an efficient dominating set, then G^2 is also $(P_6$, banner)-free (Theorems 1 and 2). Since MWIS can be solved in polynomial time for $(P_6$, banner)-free graphs [3, 12], we deduce that the WEDS problem can be solved in polynomial time for $(P_6$, banner)-free graphs, by Lemma 1 (Theorem 3).

2   WEDS in $(P_6$, Banner)-Free Graphs

In this section, we show that the WEDS can be solved efficiently in $(P_6$, banner)-free graphs. We present below the proofs of the theorems briefly. For more detailed proofs, we refer the reader to the archiv version [14].

First, we prove the following:

Theorem 1. *Let $G = (V, E)$ be a $(P_6$, banner)-free graph. If G has an efficient dominating set, then G^2 is P_6-free.*

Proof of Theorem 1: Let G be a $(P_6$, banner)-free graph having an efficient dominating set D, and assume to the contrary that G^2 contains an induced P_6, say with vertices v_1, v_2, v_3, v_4, v_5 and v_6, and edges $v_1v_2, v_2v_3, v_3v_4, v_4v_5$, and v_5v_6. Then $\text{dist}_G(v_1, v_2) \leq 2, \text{dist}_G(v_2, v_3) \leq 2, \text{dist}_G(v_3, v_4) \leq 2, \text{dist}_G(v_4, v_5) \leq 2$, and $\text{dist}_G(v_5, v_6) \leq 2$ while $\text{dist}_G(v_1, v_3) \geq 3, \text{dist}_G(v_1, v_4) \geq 3, \text{dist}_G(v_1, v_5) \geq 3, \text{dist}_G(v_1, v_6) \geq 3, \text{dist}_G(v_2, v_4) \geq 3, \quad \text{dist}_G(v_2, v_5) \geq 3, \quad \text{dist}_G(v_2, v_6) \geq 3, \text{dist}_G(v_3, v_5) \geq 3, \text{dist}_G(v_3, v_6) \geq 3$, and $\text{dist}_G(v_4, v_6) \geq 3$. We often use these distance properties implicitly in the remaining proof.

Case 1: Suppose that $\text{dist}_G(v_5, v_6) = 1$.

Since $\text{dist}_G(v_4, v_6) \geq 3$, we have $\text{dist}_G(v_4, v_5) = 2$, and so there exists $d \in V$ such that $dv_4, dv_5 \in E$. If $\text{dist}_G(v_3, v_4) = 1$, then since $\text{dist}_G(v_2, v_4) \geq 3$, we have $\text{dist}_G(v_2, v_3) = 2$, and hence there exists $x \in V$ such that $xv_2, xv_3 \in E$. Now, (i) if $xd \in E$, then $\{v_4, v_3, x, d, v_5\}$ will induce a banner in G, and (ii) if $xd \notin E$ then $\{v_6, v_5, d, v_4, v_3, x\}$ will induce a P_6 in G, a contradiction.

So, assume that $\text{dist}_G(v_3, v_4) = 2$, and hence there exists $c \in V$ such that $cv_3, cv_4 \in E$. Then $cd \in E$ (otherwise, $G[\{v_6, v_5, d, v_4, c, v_3\}]$ is a P_6 in G). So, $\text{dist}_G(v_2, v_3) = 2$ (otherwise, $G[\{v_6, v_5, d, c, v_3, v_2\}]$ is a P_6 in G), and hence there exists $b \in V$ such that $bv_2, bv_3 \in E$. Then $bc \in E$ (otherwise, since $G[\{v_6, v_5, d, c, v_3, b\}]$ is not an induced P_6 in G, we have $bd \in E$. But, then $G[\{b, v_3, c, d, v_5\}]$ is a banner in G, a contradiction), and hence $bd \in E$ (otherwise, $G[\{v_6, v_5, d, c, b, v_2\}]$ is a P_6 in G). Then $\text{dist}_G(v_1, v_2) = 2$ (otherwise, $G[\{v_6, v_5, d, b, v_2, v_1\}]$ is a P_6 in G), and hence there exists $a \in V$ such that $av_1, av_2 \in E$. Then $ab \in E$ (otherwise, $\{a, v_2, b, d, v_5, v_6\}$ will induce a banner or P_6 in G according as $ad \in E$ or $ad \notin E$ respectively, a contradiction), and hence $ad \in E$ (otherwise, $G[\{v_1, a, b, d, v_5, v_6\}]$ is a P_6 in G).

Then we note that $a, v_2, b \notin D$, and hence there exists $v_2' \in D$ such that $v_2 v_2' \in E$. Then $v_2' a, v_2' b \in E$ (otherwise, $\{v_2', v_2, a, b, d, v_5, v_6\}$ will induce a banner or a P_6 in G).

Thus, $v_1 \notin D$, and hence there exists $v_1' \in D$ such that $v_1 v_1' \in E$, and $v_1' \neq v_2'$ (that is, $v_1 v_2' \notin E$).

Now, if $v_1' d \in E$, then $\{v_1', v_1, a, d, v_5\}$ will induce a banner in G, and if $v_1' d \notin E$, then $\{v_1', v_1, a, d, v_5, v_6\}$ will induce a P_6 in G, a contradiction.

Case 2: Suppose that $\text{dist}_G(v_1, v_2) = 2 = \text{dist}_G(v_5, v_6)$.

Then there exist $a, e \in V$ such that $av_1, av_2, ev_5, ev_6 \in E$. If $\text{dist}_G(v_2, v_3) = 1$, then since $\text{dist}_G(v_2, v_4) \geq 3$, we have $\text{dist}_G(v_3, v_4) = 2$, and hence exists $x \in V$ such that $xv_3, xv_4 \in E$. But, then $\{v_1, a, v_2, v_3, x, v_4\}$ will induce either a banner or a P_6 in G, according as $ax \in E$ or $ax \notin E$ respectively, a contradiction. So, $\text{dist}_G(v_2, v_3) = 2$. Similarly, $\text{dist}_G(v_4, v_5) = 2$. Hence, there exist $b, d \in V$ such that $bv_2, bv_3, dv_4, dv_5 \in E$. Then $\text{dist}_G(v_3, v_4) = 2$ (otherwise, $\{v_2, b, v_3, v_4, c, v_5\}$ will induce either a banner or a P_6 in G, according as $bd \in E$ or $bd \notin E$ respectively, a contradiction.) So, there exists $c \in V$ such that $cv_3, cv_4 \in E$. Then we see that $ab, bc, cd, de \in E$, and $v_2, b, v_5, d \notin D$.

Case 2.1: Suppose that $bd \notin E$.

We first note that $ac, ce \in E$ (otherwise, P_6 or banner is an induced subgraph of G). Then we see that $a, e \notin D$. Since $v_2, b, v_5, d \notin D$, using the distance reasons, we see that there exist $v_2', v_5' \in D, v_2' \neq v_5'$ such that $v_2 v_2', v_5 v_5' \in E$. Then we observe that $v_2' a, v_2' b, v_5' d, v_5' e \in E$. Also, since $a, v_1 \notin D$, there exists $v_1' \in D$ such that $v_1 v_1' \in E$.

Now, if $v_4 v_5' \in E$, then $v_2' c \notin E$ (otherwise, $G[\{v_2, v_2', c, v_4, v_5', v_5\}]$ is a P_6 in G), and hence $v_5' c \in E$ (otherwise, $G[\{v_2', b, c, v_4, v_5', v_5\}]$ is a P_6 in G). Then $v_2' v_3 \notin E$ (otherwise, $G[\{v_2, v_2', v_3, c, v_5', v_5\}]$ is a P_6 in G). Thus, since $b, v_3, c \notin D$, there exists $v_3' \neq v_2', v_5'$ such that $v_3 v_3' \in E$. Note that by the distance reason,

$v_1' \neq v_3'$, and $v_1' \neq v_2'$ (otherwise, $G[\{v_1, v_2', b, c, v_5', v_5\}]$ is a P_6 in G). Now, $G[\{v_1', v_1, a, b, v_3, v_3'\}]$ is a P_6 in G, a contradiction.

So, assume that $v_4 v_5' \notin E$. Since $c, v_4, d \notin D$ and by the distance reasons, there exists $v_4'(\neq v_1', v_2', v_5') \in D$ such that $v_4 v_4' \in E$. Next, we prove that $v_1 v_2' \notin E$. Suppose not. Then $v_2' c \in E$ (otherwise, $G[\{v_1, v_2', b, c, d, v_5\}]$ is a P_6 in G). Then $v_5' v_6 \notin E$ (otherwise, $G[\{v_1, v_2', c, d, v_5', v_6\}]$ is a P_6 in G). Since $e, v_6 \notin D$, there exists $v_6'(\neq v_5', v_4', v_2') \in D$ such that $v_6 v_6' \in E$. Then, $G[\{v_1, v_2', c, e, v_6, v_6'\}]$ is a P_6 in G, a contradiction. So, $v_1 v_2' \notin E$, and hence there exists $v_1' \neq v_2'$. Then $v_1' c \notin E$ (otherwise, $G[\{v_1', v_1, a, c, v_4\}]$ is a banner in G), and hence $v_4' c \in E$ (otherwise, $G[\{v_1', v_1, a, c, v_4, v_4'\}]$ is a P_6 in G). But, then $G[\{v_1', v_1, a, d, v_4, v_4'\}]$ or $G[\{v_1', v_1, a, c, d, v_5\}]$ is a P_6 in G according as $ad \in E$ or $ad \notin E$ respectively, a contradiction.

Case 2.2: Suppose that $bd \in E$.

Then either $ad \in E$ or $be \in E$ (otherwise, $G[\{a, b, d, e, v_6\}]$ is a banner in G, if $ae \in E$, and $G[\{v_1, a, b, d, e, v_6\}]$ is a P_6 in G, if $ae \notin E$, a contradiction). We may assume that $ad \in E$. Then in this case also we see that $a, e \notin D$.

Since $v_2, v_5, a, b, d, e \notin D$, there exist $v_2', v_5' \in D(v_2' \neq v_5')$ such that $v_2 v_2', v_5 v_5' \in E$. Then we see that $v_2' b \in E$. Since $b, c, v_3 \notin D$, there exists $v_3'(\neq v_5') \in D$ such that $v_3 v_3' \in E$. Also, we see that $v_5' d \in E$. Since $c, d, v_4 \notin D$, there exists $v_4'(\neq v_2') \in D$ such that $v_4 v_4' \in E$. Now, note that $v_2' a \in E$, and so $v_1 \notin D$. Hence, there exists $v_1' \in D$ such that $v_1 v_1' \in E$, and $v_1' \neq v_2'$.

Then $v_3' = v_2'$ and $v_4' = v_5'$ (otherwise, either $G[\{v_1', v_1, a, b, v_3, v_3'\}]$ or $G[\{v_1', v_1, a, d, v_4, v_4'\}]$ is a P_6 in G). That is, $v_3 v_2', v_4 v_5' \in E$. Then: (i) If $ae \in E$, then since $v_1' e \notin E$ (else, $G[\{v_1', v_1, a, e, v_6\}]$ is a banner in G), we have $v_5' e \in E$ (otherwise, $G[\{v_1', v_1, a, e, v_5, v_5'\}]$ is a P_6 in G). But, then $G[\{v_1', v_1, a, e, v_5', v_4\}]$ is a P_6, a contradiction. (ii) If $ae \notin E$, then $v_2' e \in E$ (else, $G[\{v_3, v_2', a, d, e, v_6\}]$ is a P_6 in G). But, then $G[\{v_4, v_5', v_5, e, v_2', v_2\}]$ is a P_6 in G, a contradiction.

Since the other cases are symmetric, we have proved the theorem. $\qquad\square$

Next, we prove the following:

Theorem 2. Let $G = (V, E)$ be a $(P_6,$ banner$)$-free graph. If G has an efficient dominating set, then G^2 is banner-free.

Proof of Theorem 2: Let G be a $(P_6,$ banner$)$-free graph having an efficient dominating set D, and assume to the contrary that G^2 contains an induced banner, say with vertices v_1, v_2, v_3, v_4, and v_5, and edges $v_1 v_2, v_2 v_3, v_3 v_4, v_4 v_1$, and $v_3 v_5$. Then $\text{dist}_G(v_1, v_2) \leq 2, \text{dist}_G(v_2, v_3) \leq 2, \text{dist}_G(v_3, v_4) \leq 2, \text{dist}_G(v_4, v_1) \leq 2$, and $\text{dist}_G(v_3, v_5) \leq 2$, while $\text{dist}_G(v_1, v_3) \geq 3, \text{dist}_G(v_1, v_5) \geq 3, \text{dist}_G(v_2, v_4) \geq 3, \text{dist}_G(v_2, v_5) \geq 3$, and $\text{dist}_G(v_4, v_5) \geq 3$. We often use these distance properties implicitly in the remaining proof.

Case 1: Suppose that $\text{dist}_G(v_3, v_5) = 1$.

Then since $\text{dist}_G(v_2, v_5) \geq 3$, $\text{dist}_G(v_2, v_3) = 2$. Again, since $\text{dist}_G(v_4, v_5) \geq 3$, we have $\text{dist}_G(v_3, v_4) = 2$. So, there exist vertices a and b in V such that $av_2, av_3, bv_3, bv_4 \in E$. Since $\text{dist}_G(v_2, v_4) \geq 3$, at least one of $\text{dist}_G(v_1, v_4)$, $\text{dist}_G(v_1, v_2)$ is equal to two. We may assume (wlog.) that $\text{dist}_G(v_1, v_2) = 2$.

Hence, there exists $c \in V$ such that $cv_1, cv_2 \in E$. Then $ac \in E$ (otherwise, $\{v_5, v_3, a, v_2, c, v_1\}$ will induce a P_6 in G). Then $\mathrm{dist}_G(v_1, v_4) = 2$ (otherwise, $\{v_5, v_3, a, c, v_1, v_4\}$ will induce a P_6 in G). Hence, there exists $d \in V$ such that $dv_1, dv_4 \in E$. As earlier, $bd \in E$, and hence $cd \in E$ (otherwise, either $G[\{v_5, v_3, b, d, v_1, c\}]$ is a P_6 in G or $G[\{v_3, b, d, v_1, c\}]$ is a banner in G according as $bc \notin E$ or $bc \in E$ respectively, a contradiction). Then $bc, ad \in E$ (otherwise, either $G[\{v_5, v_3, b, d, c, v_2\}]$ is a P_6 in G or $G[\{v_5, v_3, a, c, d, v_4\}]$ is a P_6 in G), and hence $ab \in E$ (otherwise, $G[\{v_5, v_3, b, d, a\}]$ is a banner in G). Then see that $v_2, v_4, a, b, c, d \notin D$.

So, there exist $v_2', v_4' \in D$ such that $v_2 v_2', v_4 v_4' \in E$. Note that by the distance reason, we have $v_2' \neq v_4'$. We then observe that $v_2' c, v_4' d \in E$.

Thus, $v_1 \notin D$, and hence there exists $v_1' \in D$ such that $v_1 v_1' \in E$. Also, we see that $v_1' \neq v_2', v_4'$.

Now, since $v_1' c \notin E$ (by the definition of D) and since $v_1' a \notin E$ (else, $G[\{v_1', v_1, c, a, v_3\}]$ is a banner in G), using the distance properties, we see that $G[\{v_1', v_1, c, a, v_3, v_5\}]$ is a P_6 in G, a contradiction.

Case 2: Suppose that $\mathrm{dist}_G(v_3, v_5) = 2$.

Then there exists $a \in V$ such that $av_3, av_5 \in E$. Then $\mathrm{dist}_G(v_2, v_3) = 2 = \mathrm{dist}_G(v_3, v_4)$ (otherwise, if $\mathrm{dist}_G(v_2, v_3) = 1$, then $\mathrm{dist}_G(v_1, v_2) = 1$ (else, there exists $x \in V$ such that $xv_1, xv_2 \in E$, and hence $\{v_5, a, v_3, v_2, x, v_1\}$ will induce either a P_6 or a banner in G, a contradiction). Hence, a contradiction to the fact that $\mathrm{dist}_G(v_1, v_3) \geq 3$. A similar contradiction can be arrived if $\mathrm{dist}_G(v_3, v_4) = 1$). So, there exist $b, c \in V$ such that $bv_2, bv_3, cv_3, cv_4 \in E$.

Since $\mathrm{dist}_G(v_2, v_4) \geq 3$, we may assume that $\mathrm{dist}_G(v_1, v_4) = 2$, and there exists $d \in V$ such that $dv_1, dv_4 \in E$. Then $\mathrm{dist}_G(v_1, v_2) = 2$ (otherwise, $\{v_4, d, v_1, v_2, b, v_3\}$ will induce either a P_6 or a banner in G), and hence there exists $e \in V$ such that $ev_1, ev_2 \in E$. Then it is not difficult to see that $ac, ab, bc, cd, de, be \in E$.

Also, we have $v_2, v_4, b, c, d, e \notin D$. So, there exist $v_2', v_4' \in D$ such that $v_2 v_2', v_4 v_4' \in E$. Note that by the distance reason, $v_2' \neq v_4'$, and then $v_1 \notin D$ (otherwise, P_6 is an induced subgraph of G).

Since $d, e, v_1 \notin D$, there exists $v_1' \in D$ such that $v_1 v_1' \in E$. Also, $v_1' \neq v_2'$ and $v_1' \neq v_4'$. So, either $v_4' d \in E$ or $v_2' e \in E$ (otherwise, since $G[\{v_4', v_4, d, e, v_2, v_2'\}]$ is not a P_6 in G, either $v_4' e \in E$ or $v_2' d \in E$. Then either $G[\{v_4, v_4', d, e, v_2\}]$ or $G[\{v_2, v_2', d, e, v_4\}]$ is a banner in G, a contradiction). We may assume that $v_4' d \in E$. Then $v_1' c \notin E$ (otherwise, $G[\{v_1', v_1, d, c, v_3\}]$ is a banner in G).

Now, if $ad \notin E$, then $v_1' a \in E$ (otherwise, $G[\{v_1', v_1, d, c, a, v_5\}]$ is a P_6 in G). But, then $G[\{v_5, a, v_1', v_1, d, v_4'\}]$ is a P_6 in G, a contradiction.

So, assume that $ad \in E$. Then $v_1' a \notin E$ (else, $G[\{v_1', v_1, d, a, v_5\}]$ is a banner in G). Then it is easy to see that $bd \in E$. Since $v_1' b \notin E$ (otherwise, $G[\{v_1', v_1, d, b, v_3\}]$ is a banner in G), we have $v_2' b \in E$ (otherwise, $G[\{v_1', v_1, d, b, v_2, v_2'\}]$ is a P_6 in G). Since $a, b, c, v_3 \notin D$, there exists $v_3'(\neq v_1') \in D$ such that $v_3 v_3' \in E$. Also, $v_3' \neq v_2', v_4'$ (that is, $v_3 v_2', v_3 v_4' \notin E$) (otherwise, P_6 or banner is an induced subgraph of G). Hence, $G[\{v_1', v_1, d, b, v_3, v_3'\}]$ is a P_6 in G, a contradiction. $\square$

Theorem 3. *The WEDS problem can be solved in polynomial time for (P_6, banner)-free graphs.*

Proof of Theorem 3 : Since the MWIS problem in (P_6, banner)-free graphs can be solved in polynomial time [3, 12], the theorem follows by Theorems 1 and 2, and Lemma 1.   □

Time bound: Let $T(n, m)$ be the best time bound for constructing G^2 from given graph G. Since MWIS can be solved in time $O(n^5 m)$ for in (P_6, banner)-free graphs [12], and since (P_6, banner)-free graphs can be recognized in time $O(n^6)$, we have, by Lemma 1, for a given (P_6, banner)-free graph G, the WEDS problem can be solved in time $T + O(n^5 \cdot |E(G^2)|)$.

Acknowledgement. The author thanks Mathew. C. Fransis for the fruitful discussions.

References

[1] Bange, D.W., Barkauskas, A., Slater, P.J.: Efficient dominating sets in graphs. In: Ringeisen, R.D., Roberts, F.S. (eds.) Application of Discrete Mathematics, pp. 189–199. SIAM, Philadelphia (1988)

[2] Biggs, N.: Perfect codes in graphs. Journal of Combinatorial Theory, Series B 159, 1–11 (1996)

[3] Brandstädt, A., Klembt, T., Lozin, V.V., Mosca, R.: On independent vertex sets in subclasses of apple-free graphs. Algorithmica 56, 383–393 (2010)

[4] Brandstädt, A., Le, V.B.: A note on efficient domination in a superclass of P_5-free graphs. Information Processing Letters 114, 357–359 (2014)

[5] Brandstädt, A., Milanič, M., Nevries, R.: New polynomial cases of the weighted efficient domination problem. In: Chatterjee, K., Sgall, J. (eds.) MFCS 2013. LNCS, vol. 8087, pp. 195–206. Springer, Heidelberg (2013)

[6] Chang, M.S., Liu, Y.C.: Polynomial algorithms for the weighted perfect domination problems on chordal graphs and split graphs. Information Processing Letters 48, 205–210 (1993)

[7] Chang, M.S., Liu, Y.C.: Polynomial algorithms for the weighted perfect domination problems on interval and circular-arc graphs. Journal of Information Sciences and Engineering 11, 215–222 (1994)

[8] Chang, G.J., Pandurangan, C., Coorg, S.R.: Weighted independent perfect domination on co-comparability graphs. Discrete Applied Mathematics 63, 215–222 (1995)

[9] Fellows, M.R., Hoover, M.N.: Perfect domination. Australasian Journal of Combinatorics 3, 141–150 (1991)

[10] Hammond, P., Smith, D.: Perfect codes in the graphs Q_k. Journal of Combinatorial Theory Series B 19, 239–255 (1975)

[11] Haynes, T.W., Hedetniemi, S.T., Slater, P.J.: Fundamentals of Dominiation in Graphs. Marcel Dekker, New York (1998)

[12] Karthick, T.: Maximum weight independent sets in (P_6, banner)-free graphs (2013) (submitted for publication)

[13] Karthick, T.: Efficient domination in certain classes of P_6-free graphs (2014) (submitted for publication)

[14] Karthick, T.: New polynomial case for efficient domination in P_6-free graphs. arXiv:1409.1676 [cs.DM] (2014)

[15] Livingston, M., Stout, Q.: Distributing resources in hypercube computers. In: Proceedings of Third Conference on Hypercube Concurrent Computers and Applications, pp. 222–231 (1988)

[16] Lu, C.L., Tang, C.Y.: Solving the weighted efficient edge domination problem on bipartite permutation graphs. Discrete Applied Mathematics 87, 203–211 (1998)

[17] Lu, C.L., Tang, C.Y.: Weighted efficient domination problem on some perfect graphs. Discrete Applied Mathematics 117, 163–182 (2002)

[18] Milanič, M.: Hereditary efficiently dominatable graphs. Journal of Graph Theory 73, 400–424 (2013)

[19] Smart, C.B., Slater, P.J.: Complexity results for closed neighborhood parameters. Congressus Numerantium 112, 83–96 (1995)

[20] Weichsel, P.M.: Distance regular subgraphs of a cube. Discrete Mathematics 109, 297–306 (1992)

[21] Weichsel, P.M.: Dominating sets in n-cubes. Journal of Graph Theory 18, 479–488 (1994)

[22] West, D.B.: Introduction to Graph Theory, 2nd edn. Prentice-Hall, Englewood Cliffs (2000)

[23] Yen, C.C., Lee, R.C.T.: The weighted perfect domination problem. Information Processing Letters 35, 295–299 (1990)

[24] Yen, C.C., Lee, R.C.T.: The weighted perfect domination problem and its variants. Discrete Applied Mathematics 66, 147–160 (1996)

Higher-Order Triangular-Distance Delaunay Graphs: Graph-Theoretical Properties*

Ahmad Biniaz, Anil Maheshwari, and Michiel Smid

Carleton University, Ottawa, Canada

Abstract. We consider an extension of the triangular-distance Delaunay graphs (TD-Delaunay) on a set P of points in general position in the plane. In TD-Delaunay, the convex distance is defined by a fixed-oriented equilateral triangle $\triangledown$, and there is an edge between two points in P if and only if there is an empty homothet of $\triangledown$ having the two points on its boundary. We consider higher-order triangular-distance Delaunay graphs, namely k-TD, which contains an edge between two points if the interior of the smallest homothet of $\triangledown$ having the two points on its boundary contains at most k points of P. We consider the connectivity, Hamiltonicity and perfect-matching admissibility of k-TD. Finally we consider the problem of blocking the edges of k-TD.

1 Introduction

The *triangular-distance Delaunay graph* of a point set P in the plane, TD-Delaunay for short, was introduced by Chew [12]. A TD-Delaunay is a graph whose convex distance function is defined by a fixed-oriented equilateral triangle. Let $\triangledown$ be a downward equilateral triangle whose barycenter is the origin and one of its vertices is on the negative y-axis. A *homothet* of $\triangledown$ is obtained by scaling $\triangledown$ with respect to the origin by some factor $\mu \geq 0$, followed by a translation to a point b in the plane: $b + \mu\triangledown = \{b + \mu a : a \in \triangledown\}$. In the TD-Delaunay graph of P, there is a straight-line edge between two points p and q if and only if there exists a homothet of $\triangledown$ having p and q on its boundary and whose interior does not contain any point of P. In other words, (p, q) is an edge of TD-Delaunay graph if and only if there exists an empty downward equilateral triangle having p and q on its boundary. In this case, we say that the edge (p, q) has the *empty triangle property*.

We say that P is in general position if the line passing through any two points from P does not make angles $0°$, $60°$, or $120°$ with horizontal. In this paper we consider point sets in general position and our results assume this pre-condition. If P is in general position, the TD-Delaunay graph of P is planar, see [7]. We define $t(p, q)$ as the smallest homothet of $\triangledown$ having p and q on its boundary. See Figure 1(a). Note that $t(p, q)$ has one of p and q at a vertex, and the other one on the opposite side. Thus,

---

* Research supported by NSERC.

Observation 1. *Each side of $t(p, q)$ contains either p or q.*

Since every homothet of $\triangledown$ with p and q on its boundary contains $t(p, q)$, the TD-Delaunay graph has an edge (p, q) iff the interior of $t(p, q)$ does not contain any point of P.

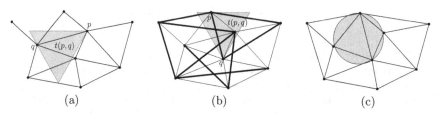

(a)   (b)   (c)

Fig. 1. (a) Triangular-distance Delaunay graph (0-TD), (b) 1-TD graph, the light edges belong to 0-TD as well, and (c) Delaunay triangulation

In [4], the authors proved a tight lower bound of $\lceil \frac{n-1}{3} \rceil$ on the size of a maximum matching in a TD-Delaunay graph. In this paper we study higher-order TD-Delaunay graphs. An *order-k TD-Delaunay graph* of a point set P, denoted by k-TD, is a geometric graph which has an edge (p, q) iff the interior of $t(p, q)$ contains at most k points of P; see Figure 1(b). The standard TD-Delaunay graph corresponds to 0-TD. We consider graph-theoretic properties of higher-order TD-Delaunay graphs (connectivity, Hamiltonicity, and perfect-matching admissibility). We also consider the problem of blocking TD-Delaunay graphs.

1.1 Previous Work

A *Delaunay triangulation* (DT) of P (which does not have any four co-circular points) is a graph whose distance function is defined by a fixed circle $\bigcirc$ centered at the origin. DT has an edge between two points p and q iff there exists a homothet of $\bigcirc$ having p and q on its boundary and whose interior does not contain any point of P; see Figure 1(c). In this case the edge (p, q) is said to have the *empty circle property*. An *order-k Delaunay Graph* on P, denoted by k-DG, is defined to have an edge (p, q) iff there exists a homothet of $\bigcirc$ having p and q on its boundary and whose interior contains at most k points of P. The standard Delaunay triangulation corresponds to 0-DG.

For each pair of points $p, q \in P$ let $D[p, q]$ be the closed disk having pq as diameter, and let $L(p, q)$ be the intersection of the two open disks with radius $|pq|$ centered at p and q, where $|pq|$ is the Euclidean distance between p and q. A *Gabriel Graph* on P is a geometric graph which has an edge between two points p and q iff $D[p, q]$ does not contain any point of $P \setminus \{p, q\}$. An *order-k Gabriel Graph* on P, denoted by k-GG, is defined to have an edge (p, q) iff $D[p, q]$ contains at most k points of $P \setminus \{p, q\}$. A *Relative Neighborhood Graph* on P is a geometric graph which has an edge between two points p and q iff $L(p, q)$

does not contain any point of P. An *order-k Relative Neighborhood Graph* on P, denoted by k-RNG, is defined to have an edge (p, q) iff $L(p, q)$ contains at most k points of P. It is obvious that for a fixed point set, k-RNG is a subgraph of k-GG, and k-GG is a subgraph of k-DG.

Let $K_n(P)$ be a complete edge-weighted geometric graph on a point set P which contains a straight-line edge between any pair of points in P. For an edge (p, q) in $K_n(P)$ let $w(p, q)$ denote the weight of (p, q). A *bottleneck matching* (resp. *bottleneck Hamiltonian cycle*) in P is defined to be a perfect matching (resp. Hamiltonian cycle) in $K_n(P)$ in which the weight of the maximum-weight edge is minimized. A graph is *biconnected* if there is a simple cycle between any pair of its vertices. A *bottleneck biconnected spanning graph* of P is a spanning subgraph, $G(P)$, of $K_n(P)$ which is biconnected and in which the weight of the longest edge is minimized. For $H \subseteq G$ we denote the bottleneck of H, i.e., the length of the maximum-weight edge in H, by $\lambda(H)$.

The problem of determining whether an order-k geometric graph always has a (bottleneck) perfect matching or a (bottleneck) Hamiltonian cycle is quite of interest. If for each edge (p, q) in $K_n(P)$, $w(p, q)$ is equal the Euclidean distance between p and q, then Chang et al. [10,11,9] proved that a bottleneck biconnected spanning graph, a bottleneck perfect matching, and a bottleneck Hamiltonian cycle of P are contained in 1-RNG, 16-RNG, 19-RNG, respectively. This implies that 16-RNG has a perfect matching and 19-RNG is Hamiltonian. Since k-RNG is a subgraph of k-GG, the same results hold for 16-GG and 19-GG. It is known that k-GG is $(k+1)$-connected [8] and 15-GG (and hence 15-DG) is Hamiltonian [1]. Recently, Kaiser et al. [15] proved that 10-GG is Hamiltonian. Dillencourt showed that any Delaunay triangulation (0-DG) admits a perfect matching [14] but it can fail to be Hamiltonian [13].

Given a geometric graph $G(P)$ on a set P of n points, we say that a set K of points *blocks* $G(P)$ if in $G(P \cup K)$ there is no edge connecting two points in P. Actually P is an independent set in $G(P \cup K)$. Aichholzer et al. [2] considered the problem of blocking the Delaunay triangulation (i.e. 0-DG) for P in general position. They show that $\frac{3n}{2}$ points are sufficient to block 0-DG and at least $n - 1$ points are necessary. To block 0-GG, $n - 1$ points are sufficient [3].

1.2 Our Results

In this paper we consider the bottleneck problems in P with respect to the triangular-distance. We assume that the weight of each edge (p, q) in $K_n(P)$ is equal to the area of $t(p, q)$. We define some geometric notions in Section 2. In Section 3 we prove that every k-TD graph is $(k + 1)$-connected. In addition we show that a bottleneck biconnected spanning graph of P is contained in 1-TD. Using a similar approach as in [1,9], in Section 4 we show that a bottleneck Hamiltonian cycle of P is contained in 8-TD. In Section 5 we prove that a bottleneck perfect matching of P is contained in 6-TD. In addition we prove that 2-TD has a matching of size $\lceil \frac{(n-1)}{2} \rceil$ and 1-TD has a matching of size at least $\lceil \frac{2(n-1)}{5} \rceil$. For some configurations of P, 5-TD fails to have any bottleneck

Hamiltonian cycle or bottleneck perfect matching. In Section 6 we consider the problem of blocking k-TD. We show that at least $\lceil \frac{n-1}{2} \rceil$ points are necessary and $n-1$ points are sufficient to block a 0-TD. Due to the space limitations, details of some proofs are omitted from this version of the paper.

2   Preliminaries

Bonichon et al. [6] showed that the half-Θ_6 graph of a point set P in the plane is equal to the TD-Delaunay graph of P. A half-Θ_6 graph on a point set P can be constructed in the following way. For each point p in P, let l_p be the horizontal line through p. Define l_p^γ as the line obtained by rotating l_p by γ-degrees in counter-clockwise direction around p. Actually $l_p^0 = l_p$. Consider three lines l_p^0, l_p^{60}, and l_p^{120} which partition the plane into six disjoint cones with apex p. Let $C_p^1, \ldots, C_p^6$ be the cones in counter-clockwise order around p as shown in Figure 2. C_p^1, C_p^3, C_p^5 will be referred to as *odd cones*, and C_p^2, C_p^4, C_p^6 will be

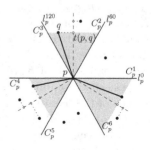

Fig. 2. Construction of the TD-Delaunay graph

referred to as *even cones*. For each even cone C_p^i, connect p to the "nearest" point q in C_p^i. The *distance* between p and q, $d(p,q)$, is defined as the Euclidean distance between p and the orthogonal projection of q onto the bisector of C_p^i. See Figure 2. The resulting graph is the half-Θ_6 graph which is defined by even cones [6]. Moreover, the resulting graph is the TD-Delaunay graph defined with respect to homothets of $\bigtriangledown$. By considering the odd cones, another half-Θ_6 graph is obtained. The well-known Θ_6 graph is the union of half-Θ_6 graphs defined by odd and even cones. To construct k-TD, for each point $p \in P$ we connect p to its $(k+1)$ nearest neighbors in each even cone around p.

Recall that $t(p,q)$ is the smallest homothet of $\bigtriangledown$ having p and q on its boundary, i.e., $t(p,q)$ is the smallest downward equilateral triangle through p and q. Similarly we define $t'(p,q)$ as the smallest upward equilateral triangle through p and q. Clearly, the even cones correspond to downward triangles and odd cones correspond to upward triangles. We define an order on the equilateral triangles: for each two equilateral triangles t_1 and t_2 we say that $t_1 \prec t_2$ if the area of t_1 is less than the area of t_2. Since the area of $t(p,q)$ is directly related to $d(p,q)$,

$$d(p,q) < d(r,s) \quad \text{if and only if} \quad t(p,q) \prec t(r,s).$$

Observation 2. *If $t(p,q)$ contains a point r, then $t(p,r)$ and $t(q,r)$ are contained in $t(p,q)$ (see Figure 3).*

As a direct consequence of Observation 2, if a point r is contained in $t(p,q)$, then $\max\{t(p,r), t(q,r)\} \prec t(p,q)$. It is obvious that,

Observation 3. *For each two points $p, q \in P$, the area of $t(p,q)$ is equal to the area of $t'(p,q)$.*

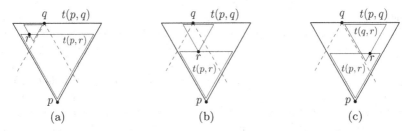

Fig. 3. Illustration of Observation 2: the triangles $t(p, r)$ and $t(q, r)$ are inside $t(p, q)$

Thus, we define $X(p, q)$ as a regular hexagon centred at p which has q on its boundary, and its sides are parallel to l_p^0, l_p^{60}, and l_p^{120}.

Observation 4. *If $X(p, q)$ contains a point r, then $t(p, r) \prec t(p, q)$.*

For a graph $G = (V, E)$ and $K \subseteq V$, let $G - K$ be the subgraph obtained from G by removing the vertices in K, and let $o(G - K)$ be the number of odd components in $G - K$. Tutte [16] gave a characterization of the graphs which have a perfect matching. Berge [5] extended Tutte's result to a formula (known as Tutte-Berge formula) for the maximum size of a matching in a graph. In a graph G, the *deficiency*, $\mathrm{def}_G(K)$, is $o(G - K) - |K|$. Let $\mathrm{def}(G) = \max_{K \subseteq V} \mathrm{def}_G(K)$.

Theorem 1 (Tutte-Berge formula; Berge [5]). *The size of a maximum matching in G is $(n - \mathrm{def}(G))/2$.*

For an edge-weighted graph G we define the *weight sequence* of G, $\mathrm{WS}(G)$, as the sequence containing the weights of the edges of G in non-increasing order. For two graphs G_1 and G_2 we say that $\mathrm{WS}(G_1) \prec \mathrm{WS}(G_2)$ if $\mathrm{WS}(G_1)$ is lexicographically smaller than $\mathrm{WS}(G_2)$. A graph G_1 is said to be less than a graph G_2 if $\mathrm{WS}(G_1) \prec \mathrm{WS}(G_2)$.

3   Connectivity

For a set P of points in general position in the plane, the TD-Delaunay graph, i.e., 0-TD, is not necessarily a triangulation [12], but it is connected and internally triangulated [4]. As shown in Figure 1(a), 0-TD may not be biconnected.

Theorem 2. *For every point set P in general position, k-TD is $(k + 1)$-connected. In addition, for every k, there exists a point set P such that k-TD is not $(k + 2)$-connected.*

By Theorem 2, 0-TD may not be biconnected, but 1-TD is biconnected. We show that a bottleneck biconnected spanning graph of P is contained in 1-TD.

Theorem 3. *For every point set P in general position, 1-TD contains a bottleneck biconnected spanning graph of P.*

Proof. Let $\mathcal{G}$ be the set of all biconnected spanning graphs with vertex set P. We define a total order on the elements of $\mathcal{G}$ by their weight sequence. If two elements have the same weight sequence, we break the ties arbitrarily to get a total order. Let $G^* = (P, E)$ be a graph in $\mathcal{G}$ with minimal weight sequence. Clearly, G^* is a bottleneck biconnected spanning graph of P. We will show that all edges of G^* are in 1-TD. By contradiction suppose that some edges in E do not belong to 1-TD, and let $e = (a, b)$ be the longest one (by the area of the triangle $t(a, b)$). If the graph $G^* - \{e\}$ is biconnected, then by removing e, we obtain a biconnected spanning graph G with $WS(G) \prec WS(G^*)$; this contradicts the minimality of G^*. Thus, there is a pair $\{p, q\}$ of points such that any cycle between p and q in G^* goes through e. Since $(a, b) \notin$ 1-TD, $t(a, b)$ contains at least two points of P, say x and y. Let G be the graph obtained from G^* by removing the edge (a, b) and adding the edges $(a, x), (b, x), (a, y), (b, y)$. We show that in G there is a cycle C between p and q which does not go through e. Consider a cycle C^* in G^* between two points p and q (which goes through e). If none of x and y belong to C^*, then $C = (C^* - \{(a, b)\}) \cup \{(a, x), (b, x)\}$ is a cycle in G between p and q. If one of x or y, say x, belongs to C^*, then $C = (C^* - \{(a, b)\}) \cup \{(a, y), (b, y)\}$ is a cycle in G between p and q. If both x and y belong to C^*, w.l.o.g. assume that x is between b and y in the path $C^* - \{(a, b)\}$. Consider the partition of C^* into four parts: (a) edge (a, b), (b) path δ_{bx} between b and x, (c) path δ_{xy} between x and y, and (d) path δ_{ya} between y and a. There are four cases:

1. None of p and q are on δ_{xy}. Let $C = \delta_{bx} \cup \delta_{ya} \cup \{(a, x), (b, y)\}$.
2. Both p and q are on δ_{xy}. Let $C = \delta_{xy} \cup \{(a, x), (a, y)\}$.
3. One of p, q is on δ_{xy} while the other is on δ_{bx}. Let $C = \delta_{bx} \cup \delta_{xy} \cup \{(b, y)\}$.
4. One of p, q is on δ_{xy} while the other is on δ_{ya}. Let $C = \delta_{xy} \cup \delta_{ya} \cup \{(a, x)\}$.

In all cases, C is a cycle in G between p and q. Thus, between any pair of points in G there exists a cycle, and hence G is biconnected. Since x and y are inside $t(a, b)$, by Observation 2, $\max\{t(a, x), t(a, y), t(b, x), t(b, y)\} \prec t(a, b)$. Therefore, $WS(G) \prec WS(G^*)$; contradicting the minimality of G^*.   □

4   Hamiltonicity

In this section we show that 8-TD contains a bottleneck Hamiltonian cycle. For some point sets, 5-TD does not contain any bottleneck Hamiltonian cycle.

Theorem 4. *For every point set P in general position, 8-TD has a bottleneck Hamiltonian cycle.*

Proof. Let $\mathcal{H}$ be the set of all Hamiltonian cycles through the points of P. Define a total order on the elements of $\mathcal{H}$ by their weight sequence. If two elements have exactly the same weight sequence, break ties arbitrarily to get a total order. Let $H^* = a_0, a_1, \ldots, a_{n-1}, a_0$ be a cycle in $\mathcal{H}$ with minimal weight sequence. It is obvious that H^* is a bottleneck Hamiltonian cycle of P. We will show that all the edges of H^* are in 8-TD. Consider any edge $e = (a_i, a_{i+1})$ in H^* and let $t(a_i, a_{i+1})$ be the triangle corresponding to e (all the index manipulations are modulo n).

Claim 1: None of the edges of H^* can be completely in the interior $t(a_i, a_{i+1})$. Suppose there is an edge $f = (a_j, a_{j+1})$ inside $t(a_i, a_{i+1})$. Let H be a cycle obtained from H^* by deleting e and f, and adding (a_i, a_j) and (a_{i+1}, a_{j+1}). By Observation 2, $t(a_i, a_{i+1}) \succ \max\{t(a_i, a_j), t(a_{i+1}, a_{j+1})\}$, and hence $\mathrm{WS}(H) \prec \mathrm{WS}(H^*)$. This contradicts the minimality of H^*.

Therefore, we may assume that no edge of H^* lies completely inside $t(a_i, a_{i+1})$. Suppose there are w points of P inside $t(a_i, a_{i+1})$. Let $U = u_1, u_2, \ldots, u_w$ represent these points indexed in the order we would encounter them on H^* starting from a_i. Let $R' = r_1, r_2, \ldots, r_w$ represent the vertices where r_i is the vertex succeeding u_i in the cycle. All the vertices in R', probably except r_w, are different from a_i and a_{i+1}. Let $R = R' - \{r_w\}$. Without loss of generality assume that $a_i \in C_{a_{i+1}}^4$, and $t(a_i, a_{i+1})$ is anchored at a_{i+1}, as shown in Figure 4.

Claim 2: For each $r_j \in R$, $t(r_j, a_{i+1}) \succeq \max\{t(a_i, a_{i+1}), t(u_j, r_j)\}$. Suppose there is a point $r_j \in R$ such that $t(r_j, a_{i+1}) \prec \max\{t(a_i, a_{i+1}), t(u_j, r_j)\}$. Construct a new cycle H by removing the edges (u_j, r_j), (a_i, a_{i+1}) and adding the edges (a_{i+1}, r_j) and (a_i, u_j). Since the two new edges have length strictly less than $\max\{t(a_i, a_{i+1}), t(u_j, r_j)\}$, $\mathrm{WS}(H) \prec \mathrm{WS}(H^*)$; which is a contradiction.

Claim 3: For each $r_j, r_k \in R$, $t(r_j, r_k) \succeq \max\{t(a_i, a_{i+1}), t(u_j, r_j), t(u_k, r_k)\}$. Suppose there is a pair r_j and r_k such that $t(r_j, r_k) \prec \max\{t(a_i, a_{i+1}), t(u_j, r_j), d(u_k, r_k)\}$. Construct a cycle H from H^* by first deleting (u_j, r_j), (u_k, r_k), (a_i, a_{i+1}). This results in three paths. One of the paths must contain both a_i and either r_j or r_k. W.l.o.g. suppose that a_i and r_k are on the same path. Add the edges (a_i, u_j), (a_{i+1}, u_k), (r_j, r_k). Since $\max\{t(u_j, r_j), t(u_k, r_k), d(a_i, a_{i+1})\} \succ \max\{t(a_i, u_j), t(a_{i+1}, u_k), t(r_j, r_k)\}$, $\mathrm{WS}(H) \prec \mathrm{WS}(H^*)$; we get a contradiction.

We use Claim 2 and Claim 3 to show that the size of R is at most seven, and consequently $w \leq 8$. Consider the lines $l_{a_{i+1}}^0$, $l_{a_{i+1}}^{60}$, $l_{a_{i+1}}^{120}$, and $l_{a_i}^{120}$ as shown in Figure 4. Let l_1 and l_2 be the rays starting at the corners of $t(a_i, a_{i+1})$ opposite to a_{i+1} and parallel to $l_{a_{i+1}}^0$ and $l_{a_{i+1}}^{60}$ respectively. These lines and rays partition the plane into 12 regions, as shown in Figure 4. We will show that each of the regions $D_1, D_2, D_3, D_4, C_1, C_2$, and $B = B_1 \cup B_2$ contains at most one point of R, and the other regions do not contain any point of R. Consider the hexagon $X(a_{i+1}, a_i)$. By Claim 2 and Observation 4, no point of R can be inside $X(a_{i+1}, a_i)$. Moreover, no point of R can be inside the cones A_1, A_2, or A_3, because if $r_j \in \{A_1 \cup A_2 \cup A_3\}$, the (upward) triangle $t'(u_j, r_j)$ contains a_{i+1}. Then by Observation 4, $t(r_j, a_{i+1}) \prec t(u_j, r_j)$; which contradicts Claim 2.

We show that each of the regions D_1, D_2, D_3, D_4 contains at most one point of R. Consider the region D_1; by similar reasoning we can prove the claim for D_2, D_3, D_4. Using contradiction, let r_j and r_k be two points in D_1, and w.l.o.g. assume that r_j is the farthest to $l_{a_{i+1}}^{60}$. Then r_k can lie inside any of the cones $C_{r_j}^1, C_{r_j}^5$, and $C_{r_j}^6$ (but not in X). If $r_k \in C_{r_j}^1$, then $t'(r_j, r_k)$ is smaller than $t'(a_i, a_{i+1})$ which means that $t(r_j, r_k) \prec t(a_i, a_{i+1})$. If $r_k \in C_{r_j}^5$, then $t'(u_j, r_j)$ contains r_k, that is $t(r_j, r_k) \prec t(u_j, r_j)$. If $r_k \in C_{r_j}^6$, then $t(u_j, r_j)$ contains r_k, that is $t(r_j, r_k) \prec t(u_j, r_j)$. All cases contradict Claim 3.

Now consider the region C_1 (or C_2). By contradiction assume that it contains two points r_j and r_k. Let r_j be the farthest from $l^0_{a_{i+1}}$. It is obvious that $t'(u_j, r_j)$ contains r_k, that is $t(r_j, r_k) \prec t(u_j, r_j)$; which contradicts Claim 3.

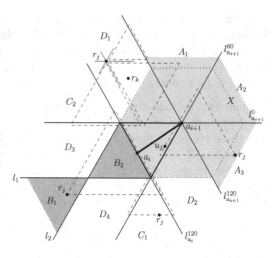

Fig. 4. Illustration of Theorem 4

Consider the region $B = B_1 \cup B_2$. If both r_j and r_k belong to B_2, then $t'(r_j, r_k)$ is smaller that $t(a_i, a_{i+1})$. If $r_j \in B_1$ and $r_k \in B_2$, then $t'(u_j, r_j)$ contains r_k, and hence $t(r_j, r_k) \prec t(u_j, r_j)$. If both r_j and r_k belong to B_1, let r_j be the farthest from $l^{120}_{a_i}$. Clearly, $t(u_j, r_j)$ contains r_k and hence $t(r_j, r_k) \prec t(u_j, r_j)$. All cases contradict Claim 3.

Therefore, any of the regions D_1, D_2, D_3, D_4, C_1, C_2, and $B = B_1 \cup B_2$ contains at most one point of R. Thus, $|R| \leq 7$ and $w \leq 8$, and $t(a_i, a_{i+1})$ contains at most 8 points of P. Therefore, $e = (a_i, a_{i+1})$ is an edge of 8-TD. $\square$

5 Perfect Matching Admissibility

In this section we consider the matching problem in k-TD graphs.

Theorem 5. *For a set P of an even number of points in general position in the plane, 6-TD contains a bottleneck perfect matching.*

For some point sets, 5-TD does not contain any bottleneck perfect matching. As for the maximum matching, in [4] the authors proved a tight lower bound of $\lceil \frac{n-1}{3} \rceil$ on the size of a maximum matching in 0-TD. We prove that 1-TD has a matching of size at least $\lceil \frac{2(n-1)}{5} \rceil$ and 2-TD has a matching of size $\lceil \frac{n-1}{2} \rceil$.

Let $\mathcal{P} = \{P_1, P_2, \dots\}$ be a partition of the points in P. Let $G(\mathcal{P})$ be the complete graph with vertex set $\mathcal{P}$. For each edge $e = (P_i, P_j)$ in $G(\mathcal{P})$, let $w(e)$ be equal to the area of the smallest triangle between a point in P_i and a point in P_j, i.e. $w(e) = \min\{t(a, b) : a \in P_i, b \in P_j\}$. That is, the weight of an edge $e \in G(\mathcal{P})$ corresponds to the size of the smallest triangle $t(e)$ defined by the endpoints of e. Let $\mathcal{T}$ be a minimum spanning tree of $G(\mathcal{P})$. Let T be the set of triangles corresponding to the edges of $\mathcal{T}$, i.e. $T = \{t(e) : e \in \mathcal{T}\}$.

Lemma 1. *The interior of any triangle in T does not contain any point of P.*

Lemma 2. *Each point in the plane can be in the interior of at most three triangles in T.*

The following two theorems are based on Lemma 1, Lemma 2, and Theorem 1.

Theorem 6. *For every set P of n points in general position in the plane, 2-TD has a matching of size $\lceil \frac{n-1}{2} \rceil$.*

Proof. First we show that by removing a set K of k points from 2-TD, at most $k + 1$ components are generated. Let K be a set of k vertices removed from 2-TD, and let $\mathcal{C} = \{C_1, \ldots, C_{m(k)}\}$ be the resulting $m(k)$ components, where m is a function depending on k. Actually, $\mathcal{C} = $ 2-TD $- K$ and $\mathcal{P} = \{V(C_1), \ldots, V(C_{m(k)})\}$ is a partition of the vertices in $P \setminus K$.

Claim 1. $m(k) \leq k + 1$. Let $G(\mathcal{P})$ be a complete graph with vertex set $\mathcal{P}$ which is constructed as described above. Let $\mathcal{T}$ be a minimum spanning tree of $G(\mathcal{P})$ and let T be the set of triangles corresponding to the edges of $\mathcal{T}$. It is obvious that $\mathcal{T}$ contains $m(k) - 1$ edges and hence $|T| = m(k) - 1$. Let $F = \{(p, t) : p \in K, t \in T, p \in t\}$ be the set of all (point, triangle) pairs where $p \in K$, $t \in T$, and p is inside t. By Lemma 2 each point in K can be inside at most three triangles in T. Thus, $|F| \leq 3 \cdot |K|$. Now we show that each triangle in T contains at least three points of K. Consider any triangle $\tau \in T$. Let $e = (V(C_i), V(C_j))$ be the edge of $\mathcal{T}$ which is corresponding to τ, and let $a \in V(C_i)$ and $b \in V(C_j)$ be the points defining τ. By Lemma 1, τ does not contain any point of $P \setminus K$ in its interior. Therefore, τ contains at least three points of K, because otherwise (a, b) is an edge in 2-TD which contradicts the fact that a and b belong to different components in $\mathcal{C}$. Thus, each triangle in T contains at least three points of K in its interior. That is, $3 \cdot |T| \leq |F|$. Therefore, $3(m(k) - 1) \leq |F| \leq 3k$, and hence $m(k) \leq k + 1$.

Note that $o(\mathcal{C}) \leq |\mathcal{C}| = m(k)$. By Claim 1, $m(k) \leq k + 1$. Thus, $o(\mathcal{C}) \leq k + 1$. This implies that $\text{def}(2\text{-}TD) \leq 1$. Therefore, by Theorem 1, the size of a maximum matching, M^*, is $\frac{n-1}{2}$. Since $|M^*|$ is an integer number, $|M^*| = \lceil \frac{n-1}{2} \rceil$. $\square$

Theorem 7. *For every set P of n points in general position in the plane, 1-TD has a matching of size at least $\lceil \frac{2(n-1)}{5} \rceil$.*

Proof. Let K be a set of k vertices removed from 1-TD, and let $\mathcal{C} = \{C_1, \ldots, C_{m(k)}\}$ be the resulting $m(k)$ components. Actually, $\mathcal{C} = $ 1-TD $- K$ and $\mathcal{P} = \{V(C_1), \ldots, V(C_{m(k)})\}$ is a partition of the vertices in $P \setminus K$. Note that $o(\mathcal{C}) \leq m(k)$. Let M^* be a maximum matching in 1-TD. By Theorem 1,

$$|M^*| = \frac{1}{2}(n - \text{def}(1\text{-}TD)), \tag{1}$$

where

$$\text{def}(1\text{-}TD) = \max_{K \subseteq P}(o(\mathcal{C}) - |K|) \leq \max_{K \subseteq P}(|\mathcal{C}| - |K|) = \max_{0 \leq k \leq n}(m(k) - k). \tag{2}$$

Define $G(\mathcal{P})$, $\mathcal{T}$, T, and F as in the proof of Theorem 6. By Lemma 2, $|F| \leq 3 \cdot |K|$. By the same reasoning as in the proof of Theorem 6, each triangle in

T has at least two points of K in its interior. Thus, $2 \cdot |T| \leq |F|$. Therefore, $2(m(k) - 1) \leq |F| \leq 3k$, and hence

$$m(k) \leq \frac{3k}{2} + 1. \tag{3}$$

In addition, $k + m(k) = |K| + |\mathcal{C}| \leq |P| = n$, and hence

$$m(k) \leq n - k. \tag{4}$$

By Inequalities (3) and (4),

$$m(k) \leq \min\{\frac{3k}{2} + 1, n - k\}. \tag{5}$$

Thus, by (2) and (5)

$$
\begin{aligned}
\operatorname{def}(1\text{-TD}) &\leq \max_{0 \leq k \leq n} (m(k) - k) \\
&\leq \max_{0 \leq k \leq n} \{\min\{\frac{3k}{2} + 1, n - k\} - k\} \\
&= \max_{0 \leq k \leq n} \{\min\{\frac{k}{2} + 1, n - 2k\}\} = \frac{n+4}{5}, \tag{6}
\end{aligned}
$$

where the last equation is obtained by setting $\frac{k}{2} + 1$ equal to $n - 2k$. Finally by substituting (6) in Equation (1) we have $|M^*| \geq \frac{2(n-1)}{5}$. Sine $|M^*|$ is an integer number, $|M^*| \geq \lceil \frac{2(n-1)}{5} \rceil$. $\square$

6   Blocking TD-Delaunay graphs

In this section we consider the problem of blocking TD-Delaunay graphs. Let P be a set of n points in general position in the plane. Recall that a point set K blocks k-TD(P) if in k-TD$(P \cup K)$ there is no edge connecting two points in P. That is, P is an independent set in k-TD$(P \cup K)$.

Theorem 8. *At least $\lceil \frac{(k+1)(n-1)}{3} \rceil$ points are necessary to block k-TD(P).*

Proof. Let K be a set of m points which blocks k-TD(P). Let $G(\mathcal{P})$ be the complete graph with vertex set $\mathcal{P} = P$. Let $\mathcal{T}$ be a minimum spanning tree of $G(\mathcal{P})$ and let T be the set of triangles corresponding to the edges of $\mathcal{T}$. It is obvious that $|T| = n - 1$. By Lemma 1 the triangles in T are empty, thus, the edges of $\mathcal{T}$ belong to any k-TD(P) where $k \geq 0$. To block each edge, corresponding to a triangle in T, at least $k + 1$ points are necessary. By Lemma 2 each point in K can lie in at most three triangles of T. Therefore, $m \geq \lceil \frac{(k+1)(n-1)}{3} \rceil$, which implies that at least $\lceil \frac{(k+1)(n-1)}{3} \rceil$ points are necessary to block all the edges of $\mathcal{T}$ and hence k-TD(P). $\square$

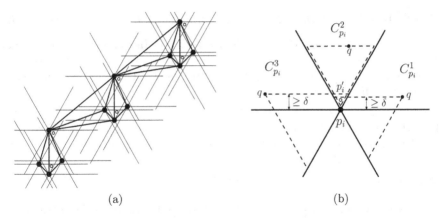

Fig. 5. (a) A 0-TD graph which is shown in bold edges is blocked by $\lceil \frac{n-1}{2} \rceil$ white points, (b) p'_i blocks all the edges connecting p_i to the vertices above $l^0_{p_i}$

By Theorem 8, at least $\lceil \frac{n-1}{3} \rceil$, $\lceil \frac{2(n-1)}{3} \rceil$, $n-1$ points are necessary to block 0-, 1-, 2-TD(P) respectively. Now we introduce another formula which gives a better lower bound for 0-TD. For a point set P, let $\nu_k(P)$ and $\alpha_k(P)$ respectively denote the size of a maximum matching and a maximum independent set in k-TD(P). For every edge in the maximum matching, at most one of its endpoints can be in the maximum independent set. Thus,

$$\alpha_k(P) \leq |P| - \nu_k(P). \tag{7}$$

Let K be a set of m points which blocks k-TD(P). By definition there is no edge between points of P in k-TD($P \cup K$). That is, P is an independent set in k-TD($P \cup K$). Thus,

$$n \leq \alpha_k(P \cup K). \tag{8}$$

By (7) and (8) we have

$$n \leq \alpha_k(P \cup K) \leq (n+m) - \nu_k(P \cup K). \tag{9}$$

Theorem 9. *At least $\lceil \frac{n-1}{2} \rceil$ points are necessary to block 0-TD(P).*

Proof. Let K be a set of m points which blocks 0-TD(P). Consider 0-TD($P \cup K$). It is known that $\nu_0(P \cup K) \geq \lceil \frac{n+m-1}{3} \rceil$; see [4]. By Inequality (9),

$$n \leq (n+m) - \lceil \frac{n+m-1}{3} \rceil \leq \frac{2(n+m)+1}{3},$$

and consequently $m \geq \lceil \frac{n-1}{2} \rceil$ (note that m is an integer number). □

Figure 5(a) shows a 0-TD graph on a set of 12 points which is blocked by 6 points. By removing the topmost point we obtain a set with odd number of points which can be blocked by 5 points.

Theorem 10. *There exists a set K of $(k+1)(n-1)$ points that blocks k-$TD(P)$.*

This bound is tight. Consider the case where $k = 0$. In this case 0-$TD(P)$ can be a path representing $n - 1$ disjoint triangles and for each triangle we need at least one point to block its corresponding edge. In k-$TD(P)$ we need at least $k + 1$ points to block each of these edges.

Acknowledgments. We thank the referees for helpful suggestions improving the quality of the paper.

References

1. Abellanas, M., Bose, P., García-López, J., Hurtado, F., Nicolás, C.M., Ramos, P.: On structural and graph theoretic properties of higher order Delaunay graphs. Int. J. Comput. Geometry Appl. 19(6), 595–615 (2009)
2. Aichholzer, O., Monroy, R.F., Hackl, T., van Kreveld, M.J., Pilz, A., Ramos, P., Vogtenhuber, B.: Blocking Delaunay triangulations. Comput. Geom. 46(2), 154–159 (2013)
3. Aronov, B., Dulieu, M., Hurtado, F.: Witness Gabriel graphs. Comput. Geom. 46(7), 894–908 (2013)
4. Babu, J., Biniaz, A., Maheshwari, A., Smid, M.: Fixed-orientation equilateral triangle matching of point sets. To appear in Theoretical Computer Science
5. Berge, C.: Sur le couplage maximum d'un graphe. C. R. Acad. Sci. Paris 247, 258–259 (1958)
6. Bonichon, N., Gavoille, C., Hanusse, N., Ilcinkas, D.: Connections between Theta-Graphs, Delaunay Triangulations, and Orthogonal Surfaces. In: Thilikos, D.M. (ed.) WG 2010. LNCS, vol. 6410, pp. 266–278. Springer, Heidelberg (2010)
7. Bose, P., Carmi, P., Collette, S., Smid, M.H.M.: On the stretch factor of convex Delaunay graphs. Journal of Computational Geometry 1(1), 41–56 (2010)
8. Bose, P., Collette, S., Hurtado, F., Korman, M., Langerman, S., Sacristan, V., Saumell, M.: Some properties of k-Delaunay and k-Gabriel graphs. Comput. Geom. 46(2), 131–139 (2013)
9. Chang, M.-S., Tang, C.Y., Lee, R.C.T.: 20-relative neighborhood graphs are Hamiltonian. Journal of Graph Theory 15(5), 543–557 (1991)
10. Chang, M.-S., Tang, C.Y., Lee, R.C.T.: Solving the Euclidean bottleneck biconnected edge subgraph problem by 2-relative neighborhood graphs. Discrete Applied Mathematics 39(1), 1–12 (1992)
11. Chang, M.-S., Tang, C.Y., Lee, R.C.T.: Solving the Euclidean bottleneck matching problem by k-relative neighborhood graphs. Algorithmica 8(3), 177–194 (1992)
12. Chew, P.: There are planar graphs almost as good as the complete graph. J. Comput. Syst. Sci. 39(2), 205–219 (1989)
13. Dillencourt, M.B.: A non-hamiltonian, nondegenerate Delaunay triangulation. Inf. Process. Lett. 25(3), 149–151 (1987)
14. Dillencourt, M.B.: Toughness and Delaunay triangulations. Discrete & Computational Geometry 5, 575–601 (1990)
15. Kaiser, T., Saumell, M., Cleemput, N.V.: 10-Gabriel graphs are Hamiltonian. arXiv: 1410.0309 (2014)
16. Tutte, W.T.: The factorization of linear graphs. Journal of the London Mathematical Society 22(2), 107–111 (1947)

Separator Theorems for Interval Graphs and Proper Interval Graphs

B.S. Panda

Computer Science and Application Group
Department of Mathematics
Indian Institute of Technology Delhi, Hauz Khas
New Delhi 110016, India
bspanda@maths.iitd.ac.in

Abstract. C.L.Monma and V.K.Wei [1986, J. Comb. Theory, Ser-B, 41, 141-181] proposed a unified approach to characterize several subclasses of chordal graphs using clique separator. The characterizations so obtained are called separator theorems. Separator theorems play an important role in designing algorithms in subclasses of chordal graphs. In this paper, we obtain separator theorems for interval graphs and proper interval graphs, which are subclasses of chordal graphs, following the framework of Monma and Wei.

Keywords: Chordal Graphs, Interval Graphs, Proper Interval Graphs.

1 Introduction

A graph $G = (V, E)$ is said to be an **intersection graph** if there exists a finite family F of non-empty sets and there is a one-to-one correspondence between the vertices of G and sets in F such that two vertices of G are adjacent if and only if the corresponding two sets have non empty intersection. If F is a family of intervals in a linearly ordered set (like the real line), then G is called an **interval graph**. If no interval of F properly contains the other set-theoretically, then G is called a **proper interval graph**.

A graph $G = (V, E)$ is said to be a chordal graph if every cycle in G of length at least four has a chord, that is, an edge joining two non-consecutive vertices of the cycle. Interval graphs and proper interval graphs are important subclasses of chordal graphs. Interval graphs arise in many application areas and have been extensively studied by researchers (see [2, 3, 7–9]).

A subset $C \subseteq V(G)$ of a graph $G = (V, E)$ is called a clique if the induced subgraph $G[C]$, that is, $G[C] = (C, E')$, where $E' = \{xy | x, y \in C \text{ and } xy \in E(G)\}$, is a complete subgraph of G. If C is a clique of G and no proper superset of C is a clique, then C is called a maximal clique of G. If $G \setminus C$ is disconnected by a maximal clique C into components $H_i = (V_i, E_i)$, $1 \leq i \leq r$, $r \geq 2$, then C is said to be a separating clique and $G_i = G[V_i \cup C]$, $1 \leq i \leq r$ is said to be a separated graph of G with respect to C. Monma and Wei [4] have characterized

S. Ganguly and R. Krishnamurti (Eds.): CALDAM 2015, LNCS 8959, pp. 101–110, 2015.

various subclasses of chordal graphs in terms of their separated subgraphs. The characterizations so obtained are called separator theorems. Separator theorems play an important role in designing algorithms in subclasses of chordal graphs.

In this paper, we obtain separator theorems for interval graphs and proper interval graphs, which are subclasses of chordal graphs.

The paper is organized as follows. Section 2 presents some pertinent definitions and preliminary results. Section 3 presents the separator theorem for interval graphs and Section 4 presents the separator theorem for proper interval graphs.

2   Definition and Preliminaries

Throughout the discussion our graph $G = (V, E)$ is assumed to be connected and a clique C of G is assumed to be a maximal clique unless otherwise specified.

In this section, we give some pertinent definitions and state some know results which will be used subsequently in the paper.

Definition 1. *Let G be a chordal graph, and G_i, $1 \leq i \leq r$, $r \geq 2$ be the separated graphs of G with respect to some separating clique C of G. Cliques which intersect C but not equal to C are called relevant cliques with respect to C. For a separated graph G_i, let $W(G_i) = \{v \in C|$ there is a vertex $w \in (V(G_i) \setminus C)$ such that $vw \in E(G_i)\}$. Relevant cliques of G_i which contain $W(G_i)$ are called principal cliques of G_i.*

The existence of a principal clique of any separated graph of a chordal graph is assured by the following result due to Panda et al.[5].

Proposition 1. *[5] Every separated graph G_i of a chordal graph G has a principal clique.*

Let $C(G)$ denote the set of all cliques of G, and for each $v \in V(G)$, $C_v(G)$ denote the set of all cliques of G containing v.

In the following definitions, only relevant cliques are considered.

Definition 2. *Let C_1 and C_2 be two cliques of G. We say*

(1) C_1 and C_2 are unattached, denoted $C_1|C_2$, if $C_1 \cap C \cap C_2 = \emptyset$; otherwise, they are attached,
(2) C_1 dominates C_2, denoted $C_1 \geq C_2$, if $C_1 \cap C \supseteq C_2 \cap C$,
(3) C_1 properly dominates C_2, denoted $C_1 > C_2$, if $C_1 \cap C \supset C_2 \cap C$,
(4) C_1 and C_2 are congruent, denoted $C_1 \sim C_2$, if they are attached and $C_1 \cap C = C_2 \cap C$, and
(5) C_1 and C_2 are antipodal, denoted $C_1 \Longleftrightarrow C_2$, if they are attached and neither dominates the other.

Definition 3. *Let G_1 and G_2 be two separated graphs of G with respect to C. We say*

1. G_1 and G_2 are *unattached*, denoted $G_1 \mid G_2$, if $C_1 \mid C_2$ for every clique C_1 in G_1 and for every clique C_2 in G_2; otherwise, they are *attached*,
2. G_1 *dominates* G_2, denoted $G_1 \geq G_2$, if they are attached and for every clique C_1 in G_1, $C_1 \geq C_2$ for all cliques C_2 in G_2 or $C_1 \mid C_2$ for all cliques C_2 in G_2,
3. G_1 *properly dominates* G_2, denoted $G_1 > G_2$, if $G_1 \geq G_2$ but not $G_2 \geq G_1$,
4. G_1 and G_2 are *congruent*, denoted $G_1 \sim G_2$, if G_1 dominates G_2 and G_2 dominates G_1; in this case, $C_1 \sim C_2$ for all C_1 in G_1 and for all C_2 in G_2, and
5. G_1 and G_2 are *antipodal*, denoted $G_1 \Longleftrightarrow G_2$, if they are attached and neither dominates the other.

The above concepts were introduced in [4].

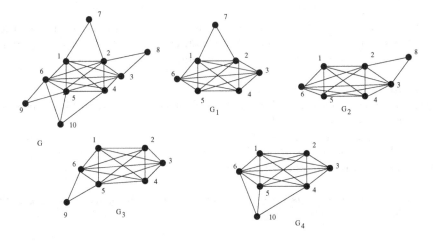

Fig. 1. Example of the separation of G into G_1, G_2, G_3, and G_4 by $C = \{1, 2, 3, 4, 5, 6\}$

Consider the graph $G = (V, E)$ separated by the clique $C = \{1, 2, 3, 4, 5, 6\}$ into separated subgraphs G_1, G_2, G_3, and G_4 as shown in Figure 1. We have, $W(G_1) = \{1, 2\}$, $W(G_2) = \{2, 3\}$, $W(G_3) = \{5, 6\}$, and $W(G_4) = \{4, 5, 6\}$. So, $G_1 \Longleftrightarrow G_2$, $G_4 > G_3$, and $G_i \mid G_j$ for $1 \leq i \leq 2$ and $3 \leq j \leq 4$.

For graph theoretic concepts not defined here and for a rich collection of results about subclasses of chordal graphs, we refer to Golumbic [3].

The following result characterizes the antipodality of two separated graphs of a chordal graph.

Lemma 1. *[6] Two separated graphs G_1 and G_2 of a chordal graph G are antipodal if and only if*

(1) $C_1 \Longleftrightarrow C_2$, *or*

(2) $C_1 > C_2'$, $C_2 > C_1'$, *for some cliques* C_1, C_1' *in* G_1 *and* C_2, C_2' *in* G_2.
(The cliques C_1 *and* C_2 *in condition (2) are principal cliques of* G_1 *and* G_2,
respectively.)

The following lemma of Monma and Wei [4] will be used in our characterization
theorem.

Proposition 2. *[4] Any collection of pair-wise non-antipodal separated graphs
of a (general) graph can be arranged in such a way that* $G_i \geq G_j$ *implies* $i < j$.

3   Separator Theorem for Interval Graphs

In this section, we present a characterization of interval graphs in terms of the
separated subgraphs with respect to a separating clique. The following theorem
characterizes interval graphs in terms of certain ordering of cliques.

Theorem 1. *[1] An undirected graph* G *is an interval graph if and only if the
cliques of* G *can be linearly ordered such that for each vertex* x *of* G, *the cliques
containing* x *occur consecutively.*

The following characterization of interval graphs as the intersection graphs of
subpaths in a path, follows from Theorem 1.

Theorem 2. *A graph* G *is an interval graph if and only if there exists a path
T such that* $V(T) = C(G)$ *and for every vertex* v *of* G, $T[C_v(G)]$ *is a subpath
of* T.

The path T satisfying Theorem 2 is called an **interval clique tree** of the
interval graph G.

Let C separate G into separated graphs $G_1, G_2, \ldots, G_r$, $r \geq 2$.

Proposition 3. *If* G *is an interval graph, then each* G_i *is an interval graph
with a clique tree* T_i *having* C *as an end vertex.*

Proof. Since every induced subgraph of an interval graph is an interval graph,
and each G_i is an induced subgraph $G[V_i \cup C]$, each G_i is an interval graph. Let
T be an interval clique tree for G. Let $\pi(V_i)$ be the subgraph of T consisting
of vertices traversed by paths corresponding to the vertices in V_i, where $V_i =$
$V(G_i) \backslash C$. Since $G[V_i]$ is connected, so is $\pi(V_i)$. Since, T is a path, $\pi(V_i)$ is a path.
So, there is a unique path $\pi^* = C, C_{i_1}, C_{i_2}, \ldots, C_{i_r}$ such that $C_{i_1}, \ldots, C_{i_{r-1}} \notin$
$\pi(V_i)$ and $C_{i_r} \in \pi(V_i)$. Construct T_i by augmenting $\pi(V_i)$ by a new vertex C
and a new edge CC_{i_r}. Then T_i is an interval clique tree for G_i having C as an
end vertex.   □

A separated graph G_i with respect to C is said to be **strictly relevant** if it has no nonrelevant cliques. A separated graph is called a non relevant separated graph if it is not strictly relevant. Let $X(G_i)$ be the set of cliques of G_i excluding C. Then, $\pi(V_i) = T[X(G_i)]$. Let $\pi(v)$ denote the path (consisting of vertices) in T corresponding to $v \in V$.

We now present the separator theorem for interval graphs.

Theorem 3 (Separator Theorem). *A chordal graph G is an interval graph if and only if*

(i) *each G_i is an interval graph,*
(ii) *the set of separated graphs can be 2-colored such that no two antipodal graphs receive the same color,*
(iii) *in each color class no two relevant cliques are unattached, and*
(iv) *in each color class all separated graphs, except possibly one, are strictly relevant. The exceptional graph in (iv), should it exist, must be dominated by every separated graphs of like color.*

Proof. **Necessity**

The condition (1) follows from Proposition 3. Let T be a clique tree of G. Since C is a separating clique of G, C corresponds to an internal vertex of T. Consider T as a rooted tree with C as the root. Then there are two branches, say B_1 and B_2, of T emanating from C. As we have seen in Proposition 3, for each separated graph G_i, $\pi(V_i)$, where $V_i = V(G_i) \setminus C$, is a subpath of T. So, $\pi(V_i)$ will be a subpath of the either B_1 or B_2 for each i. Color the separated graphs according to which branch of the tree it is on. Since T has two branches, $G_i's$ are 2-colored. Assuming subgraphs $G_1 \Longleftrightarrow G_2$, and that G_1 and G_2 have the same color, we will obtain a contradiction. By Lemma 1, there are two cases to consider.

Case 1: $C_1 \Longleftrightarrow C_2$ for some C_i in G_i, $i = 1, 2$.

Let $x \in (C_1 \cap C) \setminus C_2$ and $y \in (C_2 \cap C) \setminus C_1$. Now $\pi(x)$ and $\pi(y)$ are paths of T each containing the vertex C of T. since, G_1 and G_2 have the same color, either $\pi(x)$ is a subpath of $\pi(y)$ or conversely. Without loss of generality, $\pi(x)$ is a subpath of $\pi(y)$. since $y \notin C_1$ and $\pi(y)$ contains C_1, T is not an interval clique tree for G, a contradiction.

Case 2: $C_1 > C_2$, $C_1' < C_2'$ for some C_1, C_1' in G_1 and C_2, C_2' in G_2.

Let $x \in (C_1 \cap C) \setminus C_2$ and $y \in (C_2' \cap C) \setminus C_1'$. Without loss of generality, $\pi(x)$ is a subpath of $\pi(y)$. So, $\pi(C_1)$ is a subpath of $\pi(y)$. This yields a contradiction as $y \notin C_1'$.

So, no two antipodal separated graphs receive the same color. Hence, condition(2) is true.

Next, we claim that no two relevant cliques of the separated graphs with the same color are unattached. If possible, let $C_1|C_2$, where $C_i \in G_i$, $i = 1, 2$, and G_1 and G_2 have the same color (G_1 and G_2 may be the same separated graph). Let $x \in (C_1 \cap C)$ and $y \in (C_2 \cap C)$. Then, $\pi(x)$ is a subpath of $\pi(y)$ or conversely. Since, $x \notin C_2$ and $y \notin C_1$, and T is an interval clique tree for G, neither $\pi(x)$ can be a subpath of $\pi(y)$ nor $\pi(y)$ can be a subpath of $\pi(x)$. So, we have got a contradiction. So, condition (3) is true.

Now we assume that G_1 and G_2 are two non relevant separated graphs with the same color and exhibit a contradiction.

Now G_1 and G_2 are attached by condition (3). Since, $\pi(V(G_1))$ and $\pi(V(G_2))$ are paths in T, without loss of generality, $\pi(V(G_1))$ is a subpath of $\pi(V(G_2))$. Let $y \in W(G_2)$. So, $\pi(y)$ contains a vertex corresponding to a nonrelevant clique, say C_1', of G_1. But $y \notin C_1'$. So, we get a contradiction. Hence, at most one separated graph in each color class has a nonrelevant clique.

Let G_1 be a non relevant separated graph and G_2 be a strictly relevant separated graphs and G_1 and G_2 have the same colors. If possible, let G_2 do not dominate G_1. Since, G_1 is neither antipodal nor unattached with G_2, $G_1 > G_2$. So, there exist a clique C_1 in G_1 and a clique C_2 in G_2 such that $C_1 > C_2$. Let $x \in (C_1 \cap C) \setminus C_2$ and $y \in (C_2 \cap C)$. Since, x is not in C_2, $\pi(x)$ is a subpath of $\pi(y)$. So, $\pi(V(G_1))$ is a subpath of $\pi(V(G_2))$. But, $C(G_1)$ contains a nonrelevant clique, say C_1''. Since, $y \notin C_1''$, $\pi(x)$ can not be a subpath of $\pi(y)$. This contradicts the fact that T is an interval clique tree of G. So if there is a non relevant separated graph, then it must be dominated by all relevant separated graphs of like color. So, Condition (4) is true.

Hence, necessity is proved.

Sufficiency

Assume that each separated subgraph is an interval graph and that they can be 2-colored satisfying the conditions of the theorem. Let $X = \{G_i \mid$ color of G_i is 1$\}$. Let $X' = X - \{G_i'\}$ if $G_i' \in X$ and G_i' is a non relevant separated graph, otherwise $X' = X$. Since no two $G_i's$ of X' are antipodal and no two are unattached, by Proposition 2, the separated graphs of X' can be ordered $G_1, G_2, \ldots, G_k$ such that $G_i \geq G_j$ if and only if $i < j$. If $X = X' \cup \{G_i'\}$, call $G_i' = G_{k+1}$ and consider the ordering $G_1, G_2, \ldots, G_k, G_r$. So, $r = k$ or $k + 1$. By Proposition 3, each G_i has an interval clique tree T_i having C as an end vertex. Assume that $G[C \cup V_1 \cup V_2 \cup \cdots \cup V_i]$ has an interval clique tree T^{i-1} with end vertices C and C^{i-1}. Let C_i be the vertex adjacent to C in T_i. Delete the vertex C from T_i and add an edge between C^{i-1} of T^{i-1} and C_i of T_i to obtain T_i. Clearly, T^i is an interval clique tree for $G[C \cup V_1 \cup V_2 \cup \cdots \cup V_i]$. Applying the above procedure finally construct T^r. Let $T(1) = T^r$. Similarly construct a tree $T(2)$ for the separated graphs having color 2. Now merge the vertex C of $T(1)$ and the vertex C of $T(2)$ to obtain the tree T. It is easy to see that T is an interval clique tree for G. Hence G is an interval graph. $\square$

4 Separator Theorem for Proper Interval Graphs

In this section, we present the separator theorem for proper interval graphs. A family F of intervals is said to be a **proper interval representation, PIR,** of a proper interval graph G if G is the intersection graph of F, and no interval in F properly contains another interval in F. For any interval $I = [a, b]$, let $L(I) = a$ and $R(I) = b$. For a finite collection of intervals F, let $min(L(F)) = min\{L(I)|I \in F\}$, and $max(R(F)) = max\{R(I)|I \in F\}$.

The following lemma proves the existence of proper interval representation of a proper interval graph having certain property.

Lemma 2. *Let C be a nonseparating clique of a proper interval graph G having at least three cliques. Let $X_1 = \{v_1, v_2, \ldots, v_r\}$ be the set of all vertices in C such that v_i lies in exactly one clique of G, $1 \le i \le r$. Let $X_2 = \{v_{r+1}, v_{r+2}, \ldots, v_t\} \subseteq C$ such that v_j lies in exactly two cliques of G, $r+1 \le j \le t$, and $X_3 = C \setminus (X_1 \cup X_2) = \{v_{t+1}, v_{t+2}, \ldots, v_s\}$. Then in any proper interval representation of G, the intervals corresponding to the vertices of C occur consecutively. Furthermore,*

(i) there exits a proper interval representation of G satisfying the following:
$L(I_{v_1}) < L(I_{v_2}) < \ldots < L(I_{v_s}) < L(I_y)$ *for every $y \in V(G) \setminus C$, and*
(ii) there exists a proper interval representation satisfying the following:
$R(I_y) < R(I_{v_s}) < R(I_{v_{s-1}}) < \ldots < R(I_{v_1})$, *for every $y \in V(G) \setminus C$.*

Proof. Let G be a proper interval graph and $F = \{I_v, v \in V(G)\}$ be a proper interval representation of G. Let $p \in (I_{v_1} \cap I_{v_2} \cap \ldots \cap I_{v_s})$. If possible, suppose there exists $y \in V(G)$ such that $L(I_{v_i}) < L(I_{v_y}) < L(I_{v_j})$ for some i, j, $1 \le i, j \le s$. Then, $L(I_y) < L(I_{v_j}) < p$, and $R(I_y) > R(I_{v_i}) > p$. So, $C \cup \{y\}$ is a clique of G, which contradicts the maximality of C. So the intervals corresponding to the vertices of C occur consecutively.

Since C is not a separating clique of G, for every proper interval representation F of G, either $min(L(F)) = min(L(\{I_v, v \in C\}))$ or $max(R(F)) = max(R(\{I_v, v \in C\}))$. If $max(R(F)) = max(R(\{I_v, v \in C\}))$, then without loss of generality, assume that $min(L(F)) = 0$. Now $F' = \{I' = [-d, -c]$, where $I = [c, d] \in F\}$. Then F' is a proper interval representation of G such that $min(L(F')) = min(L(\{I_v, v \in C\}))$. So, without loss of generality, let $min(L(F)) = min(L(\{I_v, v \in C\}))$. Since each vertex of X_1 lies in exactly one clique of G, we can construct a proper interval representation F_1 of G from F by interchanging the intervals if needed, satisfying the following: $L(I_{v_1}) < L(I_{v_2}) < \ldots < L(I_{v_r})$. Since each vertex of X_2 lies in exactly two cliques of G, and the intervals corresponding to the vertices of C occur consecutively, we construct a proper interval representation F_2 from F_1 satisfying Lemma 2(i).

We construct a proper interval representation F_3 from F_2 as follows:

Let $a = min(L(F_2))$. Without loss of generality, we take $a = 0$. Let $F_3 = \{I' = [-d, -c]$ such that $I = [c, d] \in F_2\}$. Clearly F_3 is a proper interval representation of G satisfying Lemma 2(ii).   □

Below, we present the separator theorem for proper interval graphs.

Theorem 4. *[Separator Theorem] Let $G_1, G_2 \ldots, G_r$, $r > 1$ be the separated graphs with respect to a separating clique C. Then G is a proper interval graph if and only if*

(i) *Each G_i is a proper interval graph,*
(ii) *If $W(G_1) \cap W(G_2) \neq \emptyset$, then $W(G_1) \cup W(G_2) = C$, and there is exactly one clique C_i in G_i intersecting $W(G_1) \cap W(G_2)$, $i = 1, 2$, and*
(iii) *$r = 2$.*

Proof. **Necessity:**
(i) This follows from the fact that every induced subgraph of a proper interval graph is a proper interval graph.
(ii) If possible, let there be G_1 and G_2 such that $W(G_1) \cap W(G_2) \neq \emptyset$, and $W(G_1) \cup W(G_2) \neq C$. Let $x \in W(G_1) \cap W(G_2)$, and $x_1 \in (C \backslash (W(G_1) \cup W(G_2)))$, $y_1 \in (C_1 \setminus W(G_1))$, and $z_1 \in (C_2 \setminus W(G_2))$, where C_i is a principal clique of G_i, $i = 1, 2$. Now, $\{x_1, y_1, z_1\}$ forms an independent set in G. Let I_x, I_{x_1}, I_{y_1}, I_{z_1} be the intervals corresponding to x, x_1, y_1, and z_1, respectively in some proper interval representation F of G. Since, I_{x_1}, I_{y_1} and I_{z_1} are mutually disjoint, without loss of generality assume that $R(I_{x_1}) < L(I_{y_1})$ and $R(I_{y_1}) < L(I_{z_1})$. Then as $I_x \cap I_{x_1} \neq \emptyset$,and $I_x \cap I_{z_1} \neq \emptyset$, $I_{y_1} \subset I_x$, which contradicts the fact that F is a proper interval representation of G.

Suppose $W(G_1) \cap W(G_2) \neq \emptyset$, and $W(G_1) \cup W(G_2) = C$. But there are two relevant cliques C_1 and C_1' of G_1 intersecting $W(G_1) \cap W(G_2)$. Let $x_1 \in (C_1 \backslash C_1')$, $y_1 \in (C_1' \setminus C_1)$, $x \in W(G_1) \cap W(G_2)$, and $z_1 \in (C_2 \setminus W(G_2))$, where C_2 is a relevant clique of G_2. Now x is adjacent to all of x_1, y_1, and z_1, and $\{x_1, y_1, z_1\}$ is an independent set in G. So, as above, G is not a proper interval graph. So, (ii) is true.

(iii) Suppose $r > 2$. If $W(G_1)$, $W(G_2)$, and $W(G_3)$ are pair wise disjoint, then by Theorem 2, G is not an interval graph, and hence is not a proper interval graph. If $W(G_i) \subseteq W(G_j)$ for some $1 \leq i \leq j \leq 3$, then $W(G_i) \cap W(G_j) \neq C$, which contradicts Theorem 2(ii). So, $W(G_1)$, $W(G_2)$, and $W(G_3)$ are pair wise not comparable, and $W(G_i) \cap W(G_j) \neq \emptyset$ for $1 \leq i \leq j \leq 3$. So, G_1, G_2, and G_3 are pair wise antipodal. So, by Theorem 2, G is not an interval graph, and hence not a proper interval graph. So, (iii) is true.

Sufficiency
Assume that the separated graphs satisfy the conditions of Theorem 4. We claim that G is a proper interval graph.

Case 1: $G_1|G_2$.

Let $W(G_1) = \{v_1, v_2, \ldots, v_r\}$, $X = C \setminus (W(G_1) \cup W(G_2)) = \{v_{r+1}, \ldots, v_{t-1}\}$, and $W(G_2) = \{v_t, v_{t+1}, \ldots, v_s\}$. Now C is not a separating clique of G_i, $i = 1, 2$. Consider the graph G_1. Now, $\{v_{r+1}, , v_s\}$ is the set of vertices of C such that v_i lies in exactly one clique of G_1, namely C, $r + 1 \leq i \leq s$. So, by Lemma 2, there is a proper interval representation F_1 of G_1 satisfying the following:

$$R(I_y) < R(I_{v_1}) < R(I_{v_2}) < \cdots < L(I_{v_s}) \text{ for every } y \in V(G_1) - C.$$

Similarly, by Lemma 2, there is a proper interval representation F_2 of G satisfying the following:

$$L(I_{v_1}) < L(I_{v_2}) < \cdots < L(I_{v_s}) < L(I_y) \text{ for every } y \in V(G_2) \setminus C$$

Without loss of generality, no intervals in F_1 intersect any interval in F_2. Let $l_i = L(I_{v_i})$, where $I_{v_i} \in F_1$, and $r_i = R(I_{v_i})$, where $I_{v_i} \in F_2$. Let $F_3 = ((F_1 \cup F_2) \setminus \{I_{v_i}, v_i \in C\}) \cup (\{I'_{v_i} = [l_i, r_i], v_i \in C\})$. Then F_3 is a proper interval representation of G, and hence G is a proper interval graph.

Case 2: $W(G_1) \cap W(G_2) \neq \emptyset$.

Let $X = \{v_1, v_2, \ldots, v_r\} = W(G_1) \setminus W(G_2)$, $Y = \{v_{r+1}, v_{r+2}, \ldots, v_t\} = W(G_1) \cap W(G_2)$, and $Z = C - (X \cup Y) = \{v_{t+1}, v_{t+2}, , v_s\}$. Then by Lemma 2, there is a proper interval representation F_1 of G_1 satisfying the following:

$$R(I_y) < R(I_{v_1}) < R(I_{v_2}) < \ldots < L(I_{v_s}) \text{ for every } y \in V(G_1) - C.$$

Again by Lemma 2, there is a proper interval representation F_2 of G satisfying the following:

$$L(I_{v_1}) < L(I_{v_2}) < \ldots < L(I_{v_s}) < L(I_y) \text{ for every } y \in V(G_2) - C$$

Without loss of generality, no intervals in F_1 intersect any interval in F_2. Let $l_i = L(I_{v_i})$, where $I_{v_i} \in F_1$, and $r_i = R(I_{v_i})$, where $I_{v_i} \in F_2$. Let $F_3 = ((F_1 \cup F_2) - \{I_{v_i}, v_i \in C\}) \cup (\{I'_{v_i} = [l_i, r_i], v_i \in C\})$. Then F_3 is a proper interval representation of G, and hence G is a proper interval graph. $\square$

References

1. Gilmore, P.C., Hoffman, A.J.: A Characterization of Comparability Graphs and Interval Graphs. Canadian J. Maths. 16, 539–548 (1964)
2. Golumbic, M.C. (ed.): Interval Graphs and Related Topics. Discrete Math. 55 (1985)
3. Golumbic, M.C.: Algorithmic Graph Theory and Perfect Graphs. Academic Press, New York (1980)
4. Monma, C.L., Wei, V.K.: Intersection Graphs of Paths in a Tree. J. Combin. Theory Ser-B 41, 141–181 (1986)

5. Panda, B.S., Mohanty, S.P.: Intersection Graphs of Vertex Disjoint Paths in a Tree. Discrete Math. 146, 179–209 (1995)
6. Panda, B.S.: The forbidden Subgraph Characterization of Directed Vertex Graphs. Discrete Math 196, 239–256 (1999)
7. Papadimitriou, C., Yanakakis, M.M.: Scheduling Interval Ordered Tasks. SIAM J. Comput. 8, 405–409 (1979)
8. Roberts, F.S.: Discrete Mathematical Model with Applications to Social, Biological and Environmental Problems. Prentice Hall, Englewood Cliff (1976)
9. Roberts, F.S.: Indifference Graphs. In: Harary, F. (ed.) Proof Techniques in Graph Theory, pp. 139–146. Academic Press, New York (1971)

Bounds for the *b*-chromatic Number
of Induced Subgraphs and $G - e$

P. Francis and S. Francis Raj

Department of Mathematics, Pondicherry University, Puducherry-605014, India
selvafrancis@gmail.com
francisraj_s@yahoo.com

Keywords: b-coloring, b-chromatic number.

1 Introduction

All graphs considered in this paper are simple, finite and undirected. Let G be a graph with vertex set V and edge set E. The order of G will be denoted by n and size (number of edges) by m. A b-coloring of a graph is a proper coloring of the vertices of G such that each color class contains a color dominating vertex (c.d.v.), that is, a vertex adjacent to at least one vertex of every other color class. The largest positive integer k for which G has a b-coloring using k colors is the b-chromatic number $b(G)$ of G. A b-chromatic coloring of G denotes a b-coloring using $b(G)$ colors. From the definition of $\chi(G)$, we observe that each color class of a χ-coloring contains a c.d.v. Thus $\omega(G) \leq \chi(G) \leq b(G)$, where $\omega(G)$ is the size of a maximum clique of G.

The concept of b-coloring was introduced by Irving and Manlove [4] in analogy to the achromatic number of a graph G. They have shown that determination of $b(G)$ is NP-hard for general graphs, but polynomial for trees. There has been an increasing interest in the study of b-coloring since the publication of [4]. Some of the references are [2,3], [5,6], and [7].

Let e be any edge of a graph G. We know that for the chromatic number of the edge-deleted subgraph $G - e$ of G, $\chi(G - e) = \chi(G)$ or $\chi(G) - 1$. Similarly for the achromatic number $\psi(G)$, $\psi(G - e) = \psi(G)$ or $\psi(G) - 1$. Surprisingly, a similar statement does not hold good for the b-chromatic number $b(G)$ of G. Indeed, the gap between $b(G-e)$ and $b(G)$ can be arbitrarily large. For example, consider the graphs in Fig. 1.

In this paper, for any connected graph with at least 5 vertices, we find a lower bound for the b-chromatic number of any induced subgraphs of G with at least 4 vertices and find an upper bound for the b-chromatic number of any induced subgraphs of G with at least 3 vertices. This turns out to be a generalization of the result due to R. Balakrishnan et.al. [1]. Also we show that for any connected graph G and any $e \in E(G)$, $b(G) - 1 \leq b(G - e) \leq b(G) + \left\lceil \frac{n}{2} \right\rceil - 2$. Further, we determine all graphs which attain the upper bound.

Note that in the figures, dotted lines indicate consecutive vertices and broken lines indicate possible edges.

S. Ganguly and R. Krishnamurti (Eds.): CALDAM 2015, LNCS 8959, pp. 111–116, 2015.

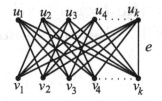

Fig. 1. $b(G) = 2$ and $b(G - e) = k$

2   Bounds for the b-chromatic Number of Induced Subgraphs

In this section, let us find bounds for the b-chromatic number of induced subgraphs of a connected graph G in terms of $b(G)$. Note that if H is an induced subgraph of G then there exist a subset S of $V(G)$, such that H is isomorphic to the subgraph induced by $V(G) - S$, which we denote by $G - S$. Let us start by finding an lower bound.

Theorem 21. *For any connected graph G with $n \geq 5$ vertices and for any $S \subset V(G)$ such that $1 \leq |S| \leq n - 3$,*

$$b(G - S) \leq b(G) + \left\lceil \frac{n - |S|}{2} \right\rceil - 2.$$

Proof. Let $S \subset V(G)$ such that $1 \leq |S| \leq n - 3$. Suppose $b(G - S) > b(G) + \left\lceil \frac{n-|S|}{2} \right\rceil - 2$, then

$$b(G - S) = b(G) + \left\lceil \frac{n - |S|}{2} \right\rceil - 2 + k, (k \geq 1). \tag{2.1}$$

Let c' be a b-chromatic coloring of $G - S$. Also let P' denote the singleton classes and Q' denote the remaining classes of c'. We can easily observe that P' induces a clique in $G - S$. Since $n \geq 5$ and $3 \leq n - |S| \leq n - 1$, from Equation (2.1) we get $b(G - S) - b(G) = \left\lceil \frac{n-|S|}{2} \right\rceil - 2 + k \geq 1$. Thus there exists at least one vertex $v \in S$ which has a neighbor in every color class of c'. Hence $|P'| \leq b(G) - 1$, otherwise $\omega(G) > b(G)$ a contradiction. Therefore from Equation (2.1) and $b(G - S) = |P'| + |Q'|$, we get $|Q'| \geq \left\lceil \frac{n-|S|}{2} \right\rceil - 1 + k$.

Case (i) Both n and $|S|$ are of Same Parity

Here $|Q'| \geq \frac{n-|S|}{2} - 1 + k \geq \frac{n-|S|}{2}$, and thus $|V(Q')| \geq n - |S|$. We also know that $|V(Q')| \leq |V(G - S)| = n - |S|$. Hence $|V(Q')| = n - |S|$, which in turn yields $|P'| = 0, |Q'| = \frac{n-|S|}{2}$ and $b(G - S) = \frac{n-|S|}{2}$. Now by again using Equation (2.1), we get $b(G - S) \geq 2 + \frac{n-|S|}{2} - 2 + 1 = \frac{n-|S|}{2} + 1$, a contradiction.

Case (ii) Both n and $|S|$ are of Different Parity

Here $|Q'| \geq \frac{n-|S|+1}{2} - 1 + k \geq \frac{n-|S|+1}{2}$ and thus $|V(Q')| \geq n - |S| + 1$, which is a contradiction to the fact that $|V(Q')| \leq n - |S|$.

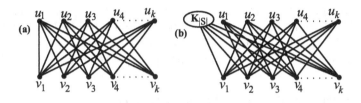

Fig. 2. Graphs which attain the bounds

The bound given in Theorem 21 is sharp. For instance, consider G to the graph given in Fig. 2(a). Let ℓ be an integer such that $1 \leq \ell \leq 2k - 3$. For $S = \{u_1, u_2, \ldots, u_{\lceil \frac{\ell}{2} \rceil}, v_1, v_2, \ldots, v_{\lfloor \frac{\ell}{2} \rfloor}\}$, we see that the upper bound is attained.

Next let us find a lower bound for the b-chromatic number of induced subgraphs of G in terms of $b(G)$.

Theorem 22. *For any connected graph G with $n \geq 5$ vertices and for any $S \subset V(G)$ such that $1 \leq |S| \leq n - 4$,*

$$b(G - S) \geq b(G) - \left\lfloor \frac{n + |S|}{2} \right\rfloor + 2.$$

Proof. Let us first consider the case when $b(G - S) = 1$. For $b(G)$ to be greater than or equal to $|S| + 2$, there should be at least $|S| + 2$ vertices of degree at least $|S| + 1$. But vertices of $G - S$ have degree at most $|S|$ in G and hence the number of vertices with degree at least $|S| + 1$ can be at most $|S|$, the vertices of S. Thus $b(G) \leq |S| + 1$. Also $n \geq |S| + 4$. Therefore $b(G) - \left\lfloor \frac{n+|S|}{2} \right\rfloor + 2 \leq 1$. Hence the bound is true for $b(G - S) = 1$. Let us next consider the case when $b(G - S) \geq 2$. Suppose $b(G - S) < b(G) - \left\lfloor \frac{n+|S|}{2} \right\rfloor + 2$, then

$$b(G - S) = b(G) - \left\lfloor \frac{n + |S|}{2} \right\rfloor + 2 - k, k \geq 1 \tag{2.2}$$

Let c be a b-chromatic coloring of G and P denote the set of singleton classes of c and Q denote the remaining classes of c, so that $|V(P)| = |P|$ and $|V(Q)| \geq 2|Q|$. Further $b(G) - b(G - S) = \left\lfloor \frac{n+|S|}{2} \right\rfloor - 2 + k \geq |S| + 1$. As c is a b-coloring, the vertices of P induces a clique in G and hence $|Q| \geq 1$.

Case (i) Both n and $|S|$ are of Same Parity
Suppose $|Q| > \frac{n-|S|}{2} - b(G - S) + 1$, say $|Q| = \frac{n-|S|}{2} - b(G - S) + 1 + l, \ l \geq 1$. Then by Equation (2.2), $|P| = 2b(G - S) + |S| - 3 + k - l$ and hence $|V(G)| = |V(P)| + |V(Q)| \geq n + 1$, a contradiction. Therefore

$$|Q| \leq \frac{n - |S|}{2} - b(G - S) + 1, \text{ and} \tag{2.3}$$

$$|P| \geq 2b(G - S) - 2 + |S|. \tag{2.4}$$

Rewrite Equation (2.4) as $|P| \geq b(G-S)+(b(G-S)-2+|S|)$. Since $b(G-S) \geq 2$, $|P| \geq b(G-S)+|S|$. If all the vertices of S belongs to P, then $b(G-S) \geq b(G)-|S|$ which implies $b(G)-b(G-S) \leq |S|$, a contradiction to $b(G)-b(G-S) \geq |S|+1$. If one of the vertex of S belongs to Q, then $|P \backslash S| \geq b(G-S)+1$ and P forms a clique in G. Therefore $\omega(G-S) \geq b(G-S)+1$, a contradiction.

Case (ii) Both n and $|S|$ are of Different Parity
By arguments similar to Case (i), we can prove that

$$|Q| \leq \frac{n-|S|-1}{2} - b(G-S) + 2, \text{ and} \qquad (2.4)$$

$$|P| \geq 2b(G-S) - 3 + |S|. \qquad (2.5)$$

If $b(G-S) \geq 3$ or $|P| > 2b(G-S)-3+|S|$, then $|P| \geq b(G-S)+|S|$. Again we get the same contradiction as mentioned in the Case (i). Therefore $b(G-S) = 2$ and $|P| = 1 + |S|$. Now by using Equation (2.2), we get $|Q| \geq \frac{n-|S|-1}{2}$. Also by Equation (2.4) we get $|Q| \leq \frac{n-|S|-1}{2}$. Thus $|Q| = \frac{n-|S|-1}{2}$. Here since n and $|S|$ are of different parity, $n - |S| \geq 5$ which in turns implies that $|Q| \geq 2$. If all the vertices of S belongs to P then $b(G) - b(G-S) \leq |S|$, a contradiction to $b(G)-b(G-S) \geq |S|+1$. If more than one vertex of S belongs to Q then in $G-S$, $|P \backslash S| \geq 3$, which induce a clique of size ≥ 3, a contradiction to $b(G-S) = 2$. Thus the only remaining possibility is $|S| - 1$ vertices of S belongs to P and one vertex belong to Q. Since $|Q| \geq 2$, in this case also we get a K_3 in $G - S$, a contradiction to $b(G-S) = 2$.

Here also we see that, the bound given in Theorem 22 is sharp. For instance, consider G to the graph given in Fig. 2(b). In Fig. 2(b), the circle denotes the clique with vertices $w_1, w_2, \ldots, w_{|S|}$ and every vertex in this clique is adjacent to every v_i, $i \in \{1, 2, \ldots, k\}$. For $S = \{u_1, w_2, \ldots, w_{|S|}\}$, we see that the lower bound is attained.

As a corollary to Theorem 21 and Theorem 22, we get the bounds for $b(G-v)$ in terms of $b(G)$ which was determined in [1].

Corollary 23 ([1]). *For any connected graph G with $n \geq 5$ vertices and for any $v \in V(G)$,*

$$b(G) - \left\lceil \frac{n}{2} \right\rceil + 2 \leq b(G-v) \leq b(G) + \left\lfloor \frac{n}{2} \right\rfloor - 2.$$

Proof. Let $S = \{v\}$. By applying Theorem 21, we get $b(G-v) \leq b(G) + \left\lceil \frac{n-1}{2} \right\rceil - 2 = b(G) + \left\lfloor \frac{n}{2} \right\rfloor - 2$. Now by applying Theorem 22 we get $b(G-v) \geq b(G) - \left\lfloor \frac{n+1}{2} \right\rfloor + 2 = b(G) - \left\lceil \frac{n}{2} \right\rceil + 2$.

3 Bounds for $b(G - e)$ in Terms of $b(G)$

Let us recall the result due to J. Kratochvil et.al. [7] on graphs for which $b(G) = 2$.

Lemma 31 ([7]). *Let G be a non-trivial connected graph. Then $b(G) = 2$ if and only if G is bipartite and has a full vertex in each part of the bipartition, where full vertex denotes a vertex adjacent to all the vertices in the other part.*

Next we shall see the bounds for the b-chromatic number of an edge-deleted subgraphs.

Theorem 32. *For any connected graph G with n vertices and for any $e \in E(G)$,*

$$b(G) - 1 \leq b(G - e) \leq b(G) + \left\lceil \frac{n}{2} \right\rceil - 2.$$

Proof. When $n = 2$, it is easy to observe that the result is true. Now let us consider $n \geq 3$ and $e \in E(G)$, where $e = uv$. Clearly since G is connected, $G - e$ contains at least one edge and hence $b(G - e) \geq 2$. Let c be a b-chromatic coloring of G with color classes $V_1, V_2, \ldots, V_{b(G)}$. Also let V_i and V_j be the color classes which contains u and v respectively. In $G - e$, if both V_i and V_j contains c.d.v., then $b(G - e) \geq b(G) > b(G) - 1$. If not, let V_i be a color class with no c.d.v then all the vertices in V_i can be moved to the other classes of c. Hence $b(G - e) \geq b(G) - 1$.

Next we establish the upper bound. Suppose $b(G - e) > b(G) + \lceil \frac{n}{2} \rceil - 2$. Then

$$b(G - e) = b(G) + \left\lceil \frac{n}{2} \right\rceil - 2 + k, \quad k \geq 1. \tag{3.1}$$

Let c' be a b-chromatic coloring of $G - e$. Let S' denote the singleton classes and T' denote the remaining classes of c'. Since $b(G - e) - b(G) \geq 1$, u and v must be in the same class of $G - e$ and hence $|T'| \geq 1$. Also $|S'| \leq b(G) - 1$. Thus from Equation (3.1), we get $|T'| \geq \lceil \frac{n}{2} \rceil$ and hence $|S'| = 0$. These yield us contradictions while considering n to be even or odd.

4 Extremal Graphs

For $n = 2, 3$ and 4, the extremal graphs are obtained immediately. We now consider the graphs G with $n \geq 5$, which attains the upper bound

$$b(G) = b(G - e) - \left\lceil \frac{n}{2} \right\rceil + 2 \tag{4.1}$$

Case (i) n is Odd
Here $|T'| \geq \frac{n+1}{2} - 1$, and thus $|V(T')| \geq n - 1$.
    If $|V(T')| = n - 1$, then $|S'| = 1$, $|T'| = \frac{n+1}{2} - 1$, and by using Equation (4.1), we get $b(G) = 2$. Let $S' = \{x\}$, $T' = \{u_i, v_i\}$, and let u_i be the c.d.v. of the color classes of T' for all $i \in \{1, 2, \ldots, b(G - e) - 1\}$. Here no two u_i's or no two

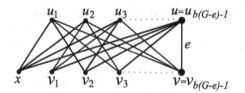

Fig. 3. n is odd and $|V(T')| = n - 1$

v_j's can be adjacent, $i, j \in \{1, 2, \ldots, b(G - e) - 1\}$. Thus G is isomorphic to the graph of Fig. 3 (where u and v are the full vertices). The case $|V(T')| = n$ is not possible, since $b(G) \geq 2$.

In a similar fashion (but with a little more involvement), in the case when n is even we have completely characterized the graphs G which attain the upper bound. It turned out that there are 14 different families of graphs which attain the bound.

Acknowledgments. For the first author, this research was supported by the Council of Scientific and Industrial Research, Government of India, File no: 09/559(0096)/2012-EMR-I.

References

1. Balakrisnan, R., Francis Raj, S.: Bounds for the b-chromatic number of $G - v$. Discrete Appl. Math. 161, 1173–1179 (2013)
2. Barth, D., Cohen, J., Faik, T.: On the b-continuity property of graphs. Discrete Appl. Math. 155, 1761–1768 (2007)
3. Corteel, S., Valencia-Pabon, M., Vera, J.C.: On approximating the b-chromatic number. Discrete Appl. Math. 146, 106–110 (2005), 997, 27-56
4. Irving, R.W., Manlove, D.F.: The b-Chromatic number of a graph. Discrete Appl. Math. 91, 127–141 (1999)
5. Kouider, M., Mahéo, M.: Some bounds for the b-Chromatic number of a graph. Discrete Math. 256, 267–277 (2002)
6. Kouider, M., Zaker, M.: Bounds for the b-Chromatic number of some families of graphs. Discrete Math. 306, 617–623 (2006)
7. Kratochvíl, J., Tuza, Z., Voigt, M.: On the b-chromatic number of graphs. In: Kučera, L. (ed.) WG 2002. LNCS, vol. 2573, pp. 310–320. Springer, Heidelberg (2002)

New Characterizations of Proper Interval Bigraphs and Proper Circular Arc Bigraphs

Ashok Kumar Das* and Ritapa Chakraborty

Department of Pure Mathematics, University of Calcutta
35, Ballygunge Circular Road, Kolkata-700019, India
{ashokdas.cu,ritapa.chakroborty}@gmail.com

Abstract. An interval bigraph B is a proper interval bigraph if there is an interval representation of B such that no interval of the same partite set is properly contained in the other. Similarly a circular arc bigraph B is a proper circular arc bigraph if there is a circular arc representation of B such that no arc of the same partite set is properly contained in the other. In this paper, we characterize proper interval bigraphs and proper circular arc bigraphs using two linear orderings of their vertex set.

Keywords: interval bigraphs, circular arc bigraphs, proper interval bigraphs, proper circular arc bigraphs, linear ordering.

1 Introduction

A bipartite graph (in short, bigraph) $B = (X, Y, E)$ is an intersection bigraph if there exists a family $\mathcal{F} = \{S_v : v \in X \cup Y\}$ of sets such that $uv \in E$, if and only if $S_u \bigcap S_v \neq \emptyset$ where u and v belong to opposite partite sets. An intersection bigraph is an *interval bigraph* (or *circular arc bigraph*) if $\mathcal{F}$ is a family of intervals (or arcs of a circle). These classes of bigraphs have been extensively studied by several researchers [5, 9, 8, 10, 12, 14, 13, 16]. The *biadjacency matrix* $A(B)$ of a bigraph B is the submatrix of the adjacency matrix consisting of the rows indexed by the vertices of one partite set and columns by the vertices of other.

A binary matrix, (i.e. a $0, 1$ - matrix) has the *partitionable zeros property*, if each 0 can be replaced by one of $\{R, C\}$ in such a way that every R has only $R's$ to its right and every C has only $C's$ below it. In [13], an interval bigraph B was characterized in terms of partitionable zeros property of its biadjacency matrix $A(B)$.

An interval bigraph $B = (X, Y, E)$ is a *proper interval bigraph* if its vertices have one to one correspondence with a family of intervals with the property that no interval corresponding to X - partite (or, Y - partite) set properly contains the other and two vertices in the opposite partite sets are adjacent when the corresponding intervals intersect. Recently Lundgren and Brown [2] generalized this definition as follows.

---

* Corresponding author.

S. Ganguly and R. Krishnamurti (Eds.): CALDAM 2015, LNCS 8959, pp. 117–125, 2015.

A bigraph $B = (X, Y, E)$ is a *proper interval bigraph* if its vertices can be represented by family of intervals $\{I_v, v \in X \cup Y\}$, with the property that no intervals properly contains another and, for $x \in X$ and $y \in Y$, x and y are adjacent if and only if I_x and I_y intersect.

A $(0, 1)$ - matrix is said to have a *monotone consecutive arrangement (MCA)* if there exists independent row and column permutations exhibiting the following structure : the $0's$ of the resulting matrix can be labeled R or C such that every position above and to the right of an R is an R, and every position below and left of a C is a C. The proper interval bigraphs were characterized by Sen and Sanyal [15] in terms of MCA of their biadjacency matrices. Other characterizations of proper interval bigraphs and their relation to the other families of graphs can be found in [2–4, 9, 15].

In [4], we characterized proper interval bigraphs via the absence of astral triple of edges. A set of three edges e_1, e_2, e_3 in a graph G is an *astral triple of edges* if for any two edges there is a path connecting them which **(1)** contains no vertex of the third and **(2)** does not contain adjacent vertices which are neighbors of the third. In a bigraph B we define a pair (u, v) of vertices as a dominating pair if u and v are non adjacent and all u, v - paths are dominating. In the same paper we showed that an interval bigraph may not contain dominating pair of vertices but a proper interval bigraph must contains a dominating pair of vertices.

A graph $G = (V, E)$ where $V = \{v_1, v_2, ..., v_n\}$ is a *permutation graph* if there is a labelling $l : V \to \{1, 2, ..., n\}$ and a permutation π of $1, 2, ..., n$ such that $v_i v_j \in E$ if and only if $l(v_i) < l(v_j)$ and $l(v_j)$ precedes $l(v_i)$ in π. Permutation graphs have been well studied. For a overview of these topics see [7]. A graph which is both bipartite and permutation graph is called *bipartite permutation graph*. Bipartite permutation graphs were introduced by Spinard, Brandstadt and Stewart [17]. In the same paper they characterized a bipartite permutation graph $B = (X, Y, E)$ in terms of ordering of the vertices of X and Y satisfying certain conditions. Steiners [18] showed that if the vertices of X and Y in the biadjacency matrix $A(B)$ of B can be arranged as stated in [17] then it shows a MCA. Actually he showed that B is a bipartite permutation graph only if and only if $A(B)$ has a MCA. Thus by the Theorem 4 of Sen and Sanyal [15] biaprtite permutation graphs are equivalent to proper interval bigraphs. But in [15] Sen and Sanyal did not provide any method how to construct the interval representation of a proper interval bigraph. From the MCA of the adjacency matrix of B we have two linear orderings of the vertex set of B. In theorem 1 of this paper we present another characterization of proper interval bigraphs using these linear orderings of the vertex set which satisfy certain conditions. This theorem also provides a method of constructing interval representation of a proper interval bigraph and also shows the equivalence between bipartite permutation graphs and proper interval bigraphs.

A circular arc bigraph $B = (X, Y, E)$ is a *proper circular arc bigraph* if there is a circular arc representation of B such that no two arcs of the same partite set is properly contained in another arc. The notion of proper circular arc bigraphs were introduced by Basu et al. in [1]. In that paper they characterized proper circular arc bigraphs in terms of a particular arrangement of their biadjacency matrices, called a monotone circular arrangement.

Motivated by the monotone circular arrangement of biadjacency matrices of circular arc bigraphs we define two linear orderings of the vertex satisfying the conditions of definition 2. Then we characterize proper circular arc bigraphs using two linear orderings of their vertex set.

2   Characterization of Proper Interval Bigraphs

In this section we characterize proper interval bigraphs using two linear orderings of their vertex set.

Definition 1. *Let $B = (X, Y, E)$ be a bipartite graph. Suppose the vertex $X \cup Y$ of B has two linear orderings $(<)$, L_1 and L_2 satisfying the following conditions:*
1) For any $x_i \in X$, if x_i is at the k_{i_1} th position of L_1 and at the k_{i_2} th position of L_2 then $k_{i_1} < k_{i_2}$;
2) For any $y_j \in Y$, if y_j is at the m_{j_1} th position of L_1 and at the m_{j_2} th position of L_2 then $m_{j_1} > m_{j_2}$;
3) $x_i y_j \in E$ if and only if $x_i < y_j$ in L_1 and $x_i > y_j$ in L_2,
then the vertex set $X \cup Y$ of B is said to have a weak bilinear ordering.

Theorem 1. *The bigraph $B = (X, Y, E)$ is a proper interval bigraph if and only if the vertex set $X \cup Y$ of B have a weak bilinear ordering.*

Proof. (Necessity) Suppose B has an interval representation $I_v, v \in X \cup Y$ such that no two members of $I_x, x \in X$ or $I_y, y \in Y$ are properly contained in the another and $xy \in E$ if and only if $I_x \cap I_y \neq \emptyset$. Let $l(x)$ and $r(x)$ are respectively denote the left end point and the right end point of I_x. Similarly $l(y)$ and $r(y)$ are respectively denote the left end point and the right end point of I_y. We index the vertices of X (Y) by natural numbers according as the increasing order of values of $l(x)'s$ $(l(y)'s)$. Let A_1 be the set $\{l(x_i) : x_i \in X\} \cup \{r(y_j) : y_j \in Y\}$. Also we may consider that all the elements of the set A_1 are distinct. If $l(x_i) = r(y_j)$ then we extend $r(y_j)$ rightward to a sufficiently small distance (which is less than the two consecutive end points of intervals in the interval representation of B). Next let $A_2 = \{l(y_j) : y_j \in Y\} \cup \{r(x_i) : x_i \in X\}$ and similarly we may consider that all the elements of the set A_2 are distinct. Now we construct L_1, a linear ordering $(<)$, of $X \cup Y$ according as the increasing order of values of the members of the set A_1. Next we construct L_2, a linear ordering $(<)$, of $X \cup Y$ according as the increasing order of values of the members of A_2.

Since the interval representation of B is proper so the vertices of the set X (or Y) occur in the same order in both the linear orderings L_1 and L_2. For any $x_i \in X$, $l(x_i) < r(x_i)$. And as x_i is not an isolated vertex so there must exists a $y_j \in Y$ such that, $l(x_i) < l(y_j) / r(y_j) < r(x_i)$ or $l(y_j) < l(x_i) < r(x_i)$. So, if x_i is at the k_{i_1} th position of L_1 and at the k_{i_2} th position of L_2, so we must have $k_{i_1} < k_{i_2}$. Similarly we can say for any $y_j \in Y$, if y_j is at the m_{j_1} th position of L_1 and at the m_{j_2} th position of L_2, so we must have $m_{j_1} > m_{j_2}$. To prove the third condition let $x_i y_j \in E$ and $x_i < y_j$ in L_1 thus $l(x_i) < r(y_j)$, and the intervals I_{x_i} and I_{y_j} intersect, so $r(x_i) > l(y_j)$ implying $y_j < x_i$ in L_2. If $x_i y_j \in E$ the other possibility can be similarly proved.

(*Sufficiency*) Let the vertex set $X \cup Y$ of B has two linear orderings $(<)$, L_1 and L_2 satisfying the conditions of definition 1. We shall construct a proper interval representation for B. For any $x_i \in X$, if x_i is at the k_{i_1} th position of L_1 and k_{i_2} th position of L_2, then we consider the interval $[k_{i_1}, k_{i_2}]$ for x_i. Similarly for $x_j \in X$ we have the interval $[k_{j_1}, k_{j_2}]$. Since in L_1 (or L_2) the vertices of X partite set occur in the increasing order of their suffix, thus if $i < j$ then $k_{i_1} < k_{j_1}$ and $k_{i_2} < k_{j_2}$. therefore the interval representation for the vertex set X is proper.

Next for $y_j \in Y$, if y_j is at the m_{j_1} th position of L_1 and m_{j_2} th position of L_2, then we consider the interval $[m_{j_2}, m_{j_1}]$ for y_j. Similarly as before we can prove that these intervals constitute a proper interval representation for the vertex set Y.

Finally, we show that when $x_i < y_j$ in L_1 and $x_i > y_j$ in L_2 then the intervals corresponding to x_i and y_j intersect. Now $x_i < y_j$ in L_1 implies $k_{i_1} < m_{j_1}$ and $x_i > y_j$ in L_2 implies $m_{j_2} < k_{i_2}$ thus the intervals $[k_{i_1}, k_{i_2}]$ and $[m_{j_2}, m_{j_1}]$ intersect as the left end point of each interval is less than the right end point of the other interval. So $x_i y_j \in E$. In other case if $x_i < y_j$ in L_1 but $x_i < y_j$ in L_2 then the second inequality implies that $k_{i_2} < m_{j_2}$. Again if $x_i > y_j$ in L_1 but $x_i < y_j$ in L_2 then the first inequality implies $k_{i_1} > m_{j_1}$. Which further implies that the intervals $[k_{i_1}, k_{i_2}]$ and $[m_{j_2}, m_{j_1}]$ do not intersect. So $x_i y_j$ does not belong to E. In the remaining case we can similarly prove the existence of non edge.   ■

3   Characterization of Proper Circular Arc Bigraphs

Motivated by the result of Golumbic [6] that a proper circular arc graph has a representation in which no two arcs share a common end point and no two arcs together cover the entire circle(i.e. they do not intersect at both ends), we prove an analogous result for proper circular arc bigraph. Before that we recall from [6] that it is useful to regard a collection of arcs as a sequence σ of its end points listed clockwise and we may assume that no two arcs share a common end point. In σ the symbol x denotes the anticlockwise end point of arc A_x and $\hat{x}$ denotes its clockwise end point. Any cyclic permutation of σ would be an equally valid representation. The manner in which two arcs A_x and A_y intersect is uniquely determined by the pattern of the subsequence of σ involving $\{x, \hat{x}, y, \hat{y}\}$. Some examples are shown in table 1. We utilize this model in proving Theorem 2.

Theorem 2. *If a bigraph $B = (X, Y, E)$ is a proper circular arc bigraph then B has proper circular arc representation in which no two arcs share a common end point and no two arcs corresponding to opposite partite set cover the whole circle (i.e. they do not intersect at both ends).*

Table 1. Coding a family of arcs as a sequence of letters. Any cyclic permutations of these patterns leave the interpretation unchanged.

Pattern of subsequence	Interpretation
$[x, \hat{x}, y, \hat{y}]$	$A_x \bigcap A_y = \emptyset$
$[\hat{y}, y, x, \hat{x}]$	$A_x \subset A_y$
$[x, y, \hat{x}, \hat{y}]$ or $[y, x, \hat{y}, \hat{x}]$	A_x, A_y intersect at one end
$[x, \hat{y}, y, \hat{x}]$	A_x, A_y intersect at both ends

Proof. Let $\mathcal{A} = \{A_v, v \in X \bigcup Y\}$ be a proper circular arc representation of B. We may assume that no two arcs share a common end point [Theorem 3]. We prove the theorem by induction on the number of pair of arcs of the opposite partite sets which cover the whole circle. Suppose A_{x_1} and A_{y_1} cover the entire circle, i.e. they intersect at both ends.

Let σ be the sequence of end points of the arcs going clockwise from the counterclockwise end points of A_{x_1}. Thus $[x_1, \hat{y_1}, y_1, \hat{x_1}]$ is a subsequence of σ involving $\{x_1, \hat{x_1}, y_1, \hat{y_1}\}$. Now σ may be expressed as the concatenation $\sigma = \tau\rho$, where $\tau = [x_1, ..., \hat{y_1}]$ and $\rho = [..., y_1, ..., \hat{x_1}, ...]$ i.e. $\sigma = [x_1, ..., \hat{y_1}, ..., y_1, ..., \hat{x_1}, ...]$.

For any $v \in V$, it is impossible for v and $\hat{v}$ to appear in τ in the order $[v, \hat{v}]$ since if $v = x$ then such an appearance would imply $A_x \subset A_{x_1}$ if $v = y$ we would have $A_y \subset A_{y_1}$, contradicting the supposition that $\mathcal{A}$ is proper.

Consider the sequence $\sigma' = \tau'\rho$, where τ' is obtained from τ by listing those entries of τ with hats followed by those without hats but preserving the relative order of each type. For any $x_i, y_j \in V$, this unshuffling operation will unchange the subsequence of σ involving $\{x_i, \hat{x_i}, y_j, \hat{y_j}\}$ unless either $[x_i, \hat{y_j}]$ or $[y_j, \hat{x_i}]$ is a subsequence of τ. We assume $[x_i, \hat{y_j}]$ is in τ. The other case may be similarly studied.

We allow the possibility that x_i may equals x_1 or that y_j may equals y_1. Now $\hat{x_i}$ may proceed x_i or succeed $\hat{x_1}$ in σ as otherwise $\mathcal{A}$ would not be proper. Similarly y_j must lies between $\hat{y_1}$ and y_1 otherwise $\mathcal{A}$ would not be proper. Thus either $[\hat{x_i}, x_i, \hat{y_j}, y_j]$ or $[x_i, \hat{y_j}, y_j, \hat{x_i}]$ is a subsequence of σ, implying that A_{x_i} and A_{y_j} intersect at both ends. After the transformation from σ to σ' these become respectively $[\hat{x_i}, \hat{y_j}, x_i, y_j]$ or $[\hat{y_j}, x_i, y_j, \hat{x_i}]$, which correspond to arcs which intersect at only one end.

Let $\mathcal{A}'$ be a set of arcs corresponding to σ'. We have just shown that some doubly intersecting arcs of opposite partite sets in $\mathcal{A}$ transformed into singly intersecting arcs of opposite partite sets in $\mathcal{A}'$. And other intersection relations of arcs of opposite partite sets are remaining unchanged. Thus $\mathcal{A}'$ is a proper circular arc representation of B with fewer circle covering pair of arcs corresponding to opposite partite sets and the theorem follows by induction. ∎

Now in the following definition we have extended Definition 1.

Definition 2. *Let $B = (X, Y, E)$ be a bipartite graph and $X = X_1 \cup X_2$, $Y = Y_1 \cup Y_2$ also $X_1 \cap X_2 = \emptyset$, $Y_1 \cap Y_2 = \emptyset$. Suppose L_1 and L_2 are two linear orderings $(<)$ of the vertex set $X \cup Y$ satisfying the following conditions:*

1) *For any $x_i, x_j \in X_1(X_2)$ if $x_i < x_j$ in L_1 then also $x_i < x_j$ in L_2. And for $x_i \in X_1$, $x_j \in X_2$ if $x_i < x_j$ in L_1 then $x_i > x_j$ in L_2. Similar condition hold for $y_i, y_j \in Y_1(Y_2)$. And if $y_i \in Y_1$, $y_j \in Y_2$ with $y_i > y_j$ in L_1 then $y_i < y_j$ in L_2;*

2) *For any $x_i \in X_1$, if x_i is at the k_{i_1} th position of L_1 and at the k_{i_2} th position of L_2 then $k_{i_1} < k_{i_2}$. Also if $x_i \in X_2$, if x_i is at the k_{i_1} th position of L_1 and at the k_{i_2} th position of L_2 then $k_{i_1} > k_{i_2}$;*

3) *For any $y_j \in Y_1$, if y_j is at the m_{j_1} th position of L_1 and at the m_{j_2} th position of L_2 then $m_{j_1} > m_{j_2}$. Also if $y_j \in Y_2$, if y_j is at the m_{j_1} th position of L_1 and at the m_{j_2} th position of L_2 then $m_{j_1} < m_{j_2}$;*

4a) *If $x_i \in X_1, y_j \in Y_1$, then $x_i y_j \in E$ if and only if $x_i < y_j$ in L_1 and $x_i > y_j$ in L_2 and $x_i y_j \in E$ for all $x_i \in X_2, y_j \in Y_2$;*

4b) *Otherwise, if $x_i \in X_1(X_2), y_j \in Y_2(Y_1)$ then $x_i y_j \in E$ if and only if $x_i < y_j$ in L_1 and $x_i < y_j$ in L_2 or $x_i > y_j$ in L_1 and $x_i > y_j$ in L_2,*
then the vertex set $X \cup Y$ of B is said to have a strong bilinear ordering.

Theorem 3. *$B = (X, Y, E)$ is a proper circular arc bigraph if and only if the vertex set $X \cup Y$ of B have a strong bilinear ordering.*

Proof. (*Necessity*) Let $B = (X, Y, E)$ is a proper circular arc bigraph. Then B has a circular arc representation, say, $\{A_v, v \in X \cup Y\}$ such that (i) no two arcs of the same partite set are properly contained in another and (ii) $xy \in E$ if and only if $A_x \cap A_y \neq \emptyset$. Also no two arcs corresponding to opposite partite set cover the whole circle. Let $a(x)$ and $c(x)$ respectively denotes the anticlockwise end point and the clockwise end point of the arc A_x. Similarly let $a(y)$ and $c(y)$ respectively denotes the anticlockwise end point and the clockwise end point of the arc A_y. We index the vertices of X (or Y) by natural number according as the increasing order of $a(x)$'s (or $a(y)$'s). Now let $X_1 = \{x : a(x) < c(x)\}$ and $X_2 = \{x : a(x) > c(x)\}$. Then $X = X_1 \cup X_2$ and $X_1 \cap X_2 = \phi$. Similarly $Y_1 = \{y : a(y) < c(y)\}$ and $Y_2 = \{y : a(y) > c(y)\}$. Then $Y = Y_1 \cup Y_2$ and $Y_1 \cap Y_2 = \phi$. Suppose $A_1 = \{a(x) : x \in X\} \cup \{c(y) : y \in Y\}$ and $A_2 = \{a(y) : y \in Y\} \cup \{c(x) : x \in X\}$. We may choose $a(x)$'s and $c(y)$'s such that all the elements of A_1 are distinct. If $a(x_i) = c(y_j)$ then we extend $c(y_j)$ to a (sufficiently small) arc distance in clockwise direction (which is less than the smallest of the distances between any two consecutive end points of arcs in the circular arc representation of B). Similarly for A_2 we may choose $a(y)$'s and $c(x)$'s such that all its elements are distinct.

Now we construct L_1, a linear ordering $(<)$ of $X \cup Y$ according as the increasing order of values of $a(x)$'s and $c(y)$'s (i.e. the elements of the set A_1). Similarly we construct L_2, a linear ordering $(<)$ of $X \cup Y$ according as the increasing order of values of $a(y)$'s and $c(x)$'s (i.e. the elements of the set A_2).

Now we verify the first condition for these two linear orderings. Let $x_i, x_j \in X_1(X_2)$ and $x_i < x_j$ in L_1 which implies $a(x_i) < a(x_j)$. Since the circular arc representation of B is proper so $c(x_i) < c(x_j)$ and which implies $x_i < x_j$ in L_2. Next let $x_i \in X_1, x_j \in X_2$ with $x_i < x_j$ in L_1 and this implies $a(x_i) < a(x_j)$. And as the circular arc representation of B is proper so $c(x_i) > c(x_j)$ which implies $x_i > x_j$ in L_2. Next let $y_i, y_j \in Y_1(Y_2)$ and $y_i < y_j$ in L_1 which implies $c(y_i) < c(y_j)$. So $a(y_i) < a(y_j)$ as the circular arc representation is proper and thus $y_i < y_j$ in L_2. Similarly we can verify the remaining condition.

Next to verify the second and third conditions for these two linear orderings. Let $x_i \in X_1$ then $a(x_i) < c(x_i)$ and as x_i is not an isolated vertex there must exist $a(y_j)/c(y_j)$ such that $a(x_i) < a(y_j) < c(x_i)$ or $a(x_i) < c(y_j) < c(x_i)$ which implies $k_{i_1} < k_{i_2}$ as $a(x_i)$ determines the position of x_i in L_1 and $c(x_i)$ determines the position of x_i in L_2. Again if $x_i \in X_2$ then $a(x_i) > c(x_i)$ and obviously $k_{i_1} > k_{i_2}$ holds. Similarly if $y_j \in Y_1$ then $a(y_j) < c(y_j)$ so there must exist $c(x_i)/a(x_i)$ such that $a(y_j) < c(x_i) < c(y_j)$ or $a(y_j) < a(x_i) < c(y_j)$ which implies $m_{j_1} > m_{j_2}$ as $c(y_j)$ determines the position of y_j in L_1 and $a(y_j)$ determines the position of y_j in L_2. The other condition holds as before.

To verify the condition $4(a)$, let $x_i \in X_1$ and $y_j \in Y_1$. Now $x_i < y_j$ in L_1 implies $a(x_i) < c(y_j)$. Also since $x_i y_j \in E$ then A_{x_i} and A_{y_j} intersect, so we have $a(y_j) \leq c(x_i)$. Again as $a(y_j)$ and $c(x_i)$ are distinct so $a(y_j) < c(x_i)$ and it implies $x_i > y_j$ in L_2. Again if $x_i \in X_2, y_j \in Y_2$ then $a(x_i) > c(x_i)$ and $a(y_j) > c(y_j)$. This implies $x_i y_j \in E$.

To verify the other condition let $x_i \in X_1$ and $y_j \in Y_2$. Now if $x_i y_j \in E$ and the arcs A_{x_i} and A_{y_j} do not cover the whole circle then either $a(x_i) < c(y_j)$ and $c(x_i) < a(y_j)$ or $a(x_i) > c(y_j)$ and $c(x_i) > a(y_j)$. This implies that either $x_i < y_j$ in L_1 and $x_i < y_j$ in L_2 or $x_i > y_j$ in L_1 and $x_i > y_j$ in L_2.

(*Sufficiency*) Let L_1 and L_2 be two strong linear orderings of the vertex set $X \cup Y$ of the bigraph $B = (X, Y, E)$. We shall show that B is a proper circular arc bigraph. Consider a $k = |X \cup Y|$ hour clock. Let $x_i \in X_1$ and x_i is at the k_{i_1} th position of L_1 and at the k_{i_2} th position of L_2, we consider the arc $[k_{i_1}, k_{i_2}]$ for x_i. Again if $x_i \in X_2$ and x_i is at the k_{i_1} th position of L_1 and at the k_{i_2} th position of L_2, then similarly we have the arc $[k_{i_1}, k_{i_2}]$ for x_i but here $k_{i_1} > k_{i_2}$. Now if $x_i, x_j \in X_1(X_2)$ and $i < j$ then $k_{i_1} < k_{j_1}$ and $k_{i_2} < k_{j_2}$. Next let $x_i \in X_1$ and $x_j \in X_2$ with $i < j$ then the arcs $[k_{i_1}, k_{i_2}]$ and $[k_{j_1}, k_{j_2}]$ are such that $k_{i_1} < k_{j_1}$ and $k_{i_2} > k_{j_2}$. So the arcs A_{x_i} and A_{x_j} are such that none of the arcs contains the other. Thus the circular arc representation for the vertex set X is proper.

Again for $y_i \in Y_1(Y_2)$, if y_i is at the m_{i_1} th position of L_1 and at the m_{i_2} th position of L_2, then we have the arc $[m_{i_2}, m_{i_1}]$ for y_i where $m_{i_2} < m_{i_1}$ if $y_i \in Y_1$ and $m_{i_2} > m_{i_1}$ if $y_i \in Y_2$. Now similarly we can check that these arcs form a proper circular arc representation for the vertex set Y.

Now we shall show the existence of edges only when the conditions of 4 are satisfied. Let $x_i \in X_1$, $y_j \in Y_1$ and $x_i < y_j$ in L_1 also $x_i > y_j$ in L_2.

Now $x_i < y_j$ in L_1 implies $k_{i_1} < m_{j_1}$ and $x_i > y_j$ in L_2 implies $k_{i_2} > m_{j_2}$ thus the two arcs $[k_{i_1}, k_{i_2}]$ and $[m_{j_2}, m_{j_1}]$ where $k_{i_1} < k_{i_2}$ and $m_{j_2} < m_{j_1}$ must intersect. Thus $x_i y_j \in E$. Next let $x_i \in X_2$ and $y_j \in Y_2$. Now the two arcs $[k_{i_1}, k_{i_2}]$ and $[m_{j_2}, m_{j_1}]$ must intersect as $k_{i_1} > k_{i_2}$ and $m_{j_2} > m_{j_1}$. So $x_i y_j \in E$.

Next let $x_i \in X_1$ and $y_j \in Y_2$. Now if $x_i < y_j$ in L_1 and $x_i > y_j$ in L_2 implies $k_{i_1} < m_{j_1}$ and $k_{i_2} > m_{j_2}$. Also $k_{i_1} < k_{i_2}$ and $m_{j_2} > m_{j_1}$, thus the two arcs $[k_{i_1}, k_{i_2}]$ and $[m_{j_2}, m_{j_1}]$ intersect at both ends. In other words the two arcs $A_{x_i} = [k_{i_1}, k_{i_2}]$ and $A_{y_j} = [m_{j_2}, m_{j_1}]$ cover the whole circle. So all $x_k \in X$ must adjacent to y_j and all $y_p \in Y$ must adjacent to x_i.

Again let $x_i \in X_1$, $y_j \in Y_2$ and $x_i < y_j$ in L_1 and $x_i < y_j$ in L_2. Suppose the two arcs $[k_{i_1}, k_{i_2}]$ and $[m_{j_2}, m_{j_1}]$ where $k_{i_1} < k_{i_2}$ and $m_{j_2} > m_{j_1}$ do not cover the whole circle. Now $x_i < y_j$ in L_1 implies $k_{i_1} < m_{j_1}$ and $x_i < y_j$ in L_2 implies $k_{i_2} < m_{j_2}$. Otherwise $x_i > y_j$ in L_1 implies $k_{i_1} > m_{j_1}$ and $x_i > y_j$ in L_2 implies $k_{i_2} > m_{j_2}$. And this implies that in any case the two arcs must intersect. So $x_i y_j \in E$.

Again if $x_i \in X_2$, $y_j \in Y_1$ and $x_i < y_j$ in L_1 and $x_i < y_j$ in L_2 or $x_i > y_j$ in L_1 and $x_i > y_j$ in L_2 and the arcs A_{x_i} and A_{y_j} do not cover the whole circle then we can similarly prove that these two arcs must intersect. So $x_i y_j \in E$.

Finally, let $x_i \in X_1$, $y_j \in Y_1$. Now if $x_i < y_j$ in L_1 and $x_i < y_j$ in L_2 then we have $k_{i_1} < m_{j_1}$ and $k_{i_2} < m_{j_2}$. Also if $x_i > y_j$ in L_1 and $x_i > y_j$ in L_2 then we have $k_{i_1} > m_{j_1}$ and $k_{i_2} > m_{j_2}$ and in any of these cases the arcs $A_{x_i} = [k_{i_1}, k_{i_2}]$ and $A_{y_j} = [m_{j_2}, m_{j_1}]$ do not intersect implying $x_i y_j$ does not belong to E.

Next let $x_i \in X_1$, $y_j \in Y_2$ satisfying $x_i > y_j$ in L_1 and $x_i < y_j$ in L_2 then we have $k_{i_1} > m_{j_1}$ and $k_{i_2} < m_{j_2}$. Since $k_{i_1} < k_{i_2}$ and $m_{j_2} > m_{j_1}$ then the two arcs $[k_{i_1}, k_{i_2}]$ and $[m_{j_2}, m_{j_1}]$ do not intersect implying $x_i y_j$ does not belong to E. Similarly in the remaining cases we can prove the existence of non edges. This completes the proof of the theorem.   ∎

4   Conclusion

Since bipartite permutation graphs can be recognized in linear time, so is the class of proper interval bigraphs [18]. Hell and Huang in [8] also presented a linear time algorithm to decide whether a bipartite graph B is a proper interval bigraph or not. In case B is not a proper interval bigraph then the same algorithm delivers, in linear time, the bichordal forbidden subgraphs of proper interval bigraphs or an induced six cycle. The problem of recognizing of circular arc graphs and proper circular arc graphs has been solved in linear time [6, 11]. We close this paper with the hope that the present paper may be a motivating factor to find out the recognition algorithm for the class of proper circular arc bigraphs.

Acknowledgement. Authors are very much grateful to Prof. Malay Sen for stimulating discussion and preparation of the final draft of the paper.

References

1. Basu, A., Das, S., Ghosh, S., Sen, M.: Circular arc bigraphs and its subclasses. J. of Graph Theory (73), 361–376 (2013)
2. Brown, D.E., Lundgren, J.R.: Characterization for unit interval bigraphs. In: Proceedings of the Forty-First Southeastern International Conference on Combinatorics, Graph Theory and Computing. Congr. Number. 206, 517 (2010)
3. Brown, D.E., Lundgren, J.R., Flink, S.C.: Characterization of interval bigraphs and unit interval bigraphs. Congress Number (2002)
4. Das, A.K., Chakraborty, R.: New characterization of proper interval bigraphs. Accepted for Publication in International Journal of Graphs and Combinatorics 12(1) (June 2015)
5. Das, A.K., Das, S., Sen, M.K.: Forbidden substructure of interval (bi/di) graphs. Submitted to Discrete Math. (2014)
6. Deng, X., Hell, P., Huang, J.: Linear time representation of proper circular arc graphs and proper interval graphs. SIAM J. Comput. 25, 390–403 (1996)
7. Golumbic, M.C.: Algorithmic graph theory and perfect graphs. Annals of Discrete Mathematics (2014)
8. Hell, P., Huang, J.: Certifying LexBFS recognition algorithm for proper interval graphs and proper interval bigraphs. SIAM J. Discrete Math. 18(3), 554–570 (2004)
9. Hell, P., Huang, J.: Interval bigraphs and circular arc graphs. J. of Graph Theory (46), 313–327 (2004)
10. Lin, I.J., Sen, M.K., West, D.B.: Class of interval digraphs and 0, 1- matrices. Congressus Num. 125, 201–209 (1997)
11. McConnell, R.M.: Linear time recognition of circular arc graphs. Algorithms 37(2), 93–147 (2003)
12. Sanyal, B.K., Sen, M.K.: New characterization of digraphs represented by intervals. J. of Graph Theory 22, 297–303 (1996)
13. Sen, M., Das, S., West, D.B.: Circular arc digraphs: A characterization. J.of Graph Theory 13, 581–592 (1989)
14. Sen, M., Das, S., Roy, A.B., West, D.B.: An analogue of interval graphs. J.of Graph Theory 13, 189–202 (1989)
15. Sen, M., Sanyal, B.K.: Indifference Digraphs: A generalization of indifference graphs and semiorders. SIAM J. Discrete Math. 7, 157–165 (1994)
16. Sen, M., Sanyal, B.K., West, D.B.: Representing digraphs using intervals or circular arcs. Discrete Math. 147, 235–245 (1995)
17. Spinard, J., Brandstad, A., Stewart, L.: Bipartite permutation graphs. Discrete Applied Math. 18, 279–292 (1987)
18. Steiner, G.: The recognition of indifference digraphs and generalized Semiorders. J. of Graph Theory 21(2), 235–241 (1996)

On Spectra of Corona Graphs

Rohan Sharma[1], Bibhas Adhikari[1], and Abhishek Mishra[2]

[1] Centre for System Science
[2] Centre for Information and Communication Technology
Indian Institute of Technology Jodhpur, Jodhpur-342011, India
{rohan.sharma,bibhas,amishra}@iitj.ac.in
http://www.iitj.ac.in/neww/ss/

Abstract. Product graphs have been gainfully used in literature to generate mathematical models of complex networks which inherit properties of real networks. Realizing the duplication phenomena imbibed in the definition of corona product of two graphs, we define corona graphs. Given a small simple connected graph which we call basic graph, corona graphs are defined by taking corona product of the basic graph iteratively. We calculate the possibility of having a node of degree k in any corona graph which lead to obtain degree distribution of corona graphs. We determine explicit formulae of eigenvalues, Laplacian eigenvalues and signless Laplacian eigenvalues of corona graphs when the basic graph is regular. Computable expressions of eigenvalues and signless Laplacian eigenvalues are also obtained when the basic graph is a star graph.

Keywords: Corona product of graphs, Laplacian spectra, signless Laplacian spectra, complex networks, graph products.

1 Introduction

In this paper, we propose a method of network generation for modelling complex networks which we call corona graphs. In the literature, corona product is defined for two graphs, introduced by Frucht and Harary in 1970 [9]. We investigate the spectra, Laplacian spectra and signless Laplacian spectra of corona graphs which can provide the topological properties of those graphs.

Corona graphs can be used as models for investigating the duplication mechanism of an individual gene as explained for the formation of new proteins in [11] and the references therein. A similar phenomenon is being observed in corona graphs but instead of a single node, a unit of basic graph is being duplicated and connected to every node of the existing network. Hence, corona graphs can reveal more insights on duplication phenomena and with the help of proposed graph spectra, the properties of gene duplication and other real world complex networks can be investigated more deeply [15].

We organize the paper as follows. In Sect.2, we define the corona graphs and the properties of corona graphs. In Sect.3, we derive the spectrum of corona graphs generated by a regular graph and a star graph. Further we determine formula of Laplacian and signless Laplacian eigenvalues of corona graphs generated by a simple connected graph.

S. Ganguly and R. Krishnamurti (Eds.): CALDAM 2015, LNCS 8959, pp. 126–137, 2015.

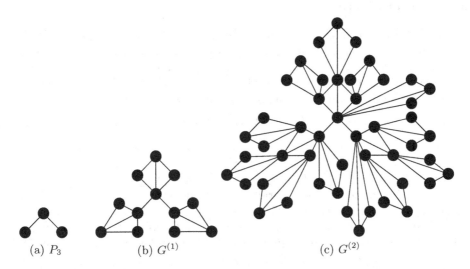

Fig. 1. Examples of the corona graphs: (a) A basic graph P_3 (b) Corona product of $P_3 = G$ with itself and hence generating $G^{(1)}$ (c) The graph of another successive corona product of P_3 with previously generated graph $G^{(1)}$ resulting in $G^{(2)}$

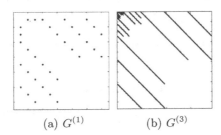

Fig. 2. Pattern of the adjacency matrices corresponding to (a) $G^{(1)}$ (b) $G^{(3)}$ for Figure 1(a). Dots represent $a_{v_i,v_j} = 1$ and white spaces represents $a_{v_i,v_j} = 0$.

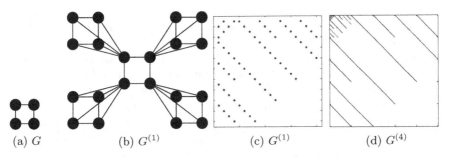

Fig. 3. Examples of the corona graphs: (a) 2-regular basic graph G (b) Corona graph for $G^{(1)}$. Pattern of the adjacency matrices in (c) $G^{(1)}$ (d) $G^{(4)}$.

2   Corona Graphs

Let $G = (V, E)$ be a graph having set of nodes $V = \{v_1, ..., v_n\}$ such that $|V| = n$, and E the set of edges. The adjacency matrix $A(G) = [a_{v_i v_j}]$ of G having dimension $|V| \times |V|$ is defined by $a_{v_i, v_j} = 1$ if $(v_i, v_j) \in E$ otherwise $a_{v_i, v_j} = 0$ when columns and rows are labelled by nodes of G. Laplacian matrix and signless Laplacian matrix of a graph G are given by

$$L(G) = D(G) - A(G) \tag{1}$$

$$Q(G) = D(G) + A(G) \tag{2}$$

respectively, where $D(G) = diag\{d_1, d_2, \dots, d_n\}$, $d_i = \sum_{i \neq j} a_{ij}$.

The corona product of the two graphs G_1, G_2, denoted by $G_1 \circ G_2$ is obtained by taking an instance of G_1 and $|V_{G_1}|$ instances of G_2 and hence connecting the ith node of G_1 to every node in the ith instance of G_2 for each i [2]. We extend this definition to define corona graphs. Let $G^{(0)} = G$. Given a basic graph G, the corona graphs generated by G are given by

$$G^{(m+1)} = G^{(m)} \circ G \tag{3}$$

where $m(\geq 0)$ is a large natural number. For instance, the corona graphs generated by P_3 are shown in Fig.1b and Fig.1c along with the pattern of their adjacency matrices in Fig.2a and Fig.2b respectively. Similarly, for a 2-regular graph with 4 nodes, the corona graph $G^{(1)}$ corresponding to G is shown in Fig.3b and the pattern of their adjacency matrices of $G^{(1)}$ and $G^{(4)}$ in Fig.3c and Fig.3d respectively. The following are some observations associated with corona graphs.

1. The number of nodes in $G^{(m)}$ is

$$|V^{(m)}| = n(n+1)^m. \tag{4}$$

2. If $|E|$ and n are the number of edges and nodes in the basic graph G respectively, then number of edges in $G^{(m)}$ is

$$|E^{(m)}| = (|E| + (|E| + n)((n+1)^m - 1)). \tag{5}$$

3. The number of nodes added in $i^{th}(i \leq m)$ step of the formation of $G^{(m)}$ is
$n(\sum_{j=1}^{i-1} |V^{(j)}|)$.

4. Connectivity of $G^{(m)}$: Since G is connected, corona graphs generated by G are connected graphs. Evidently, if basic graph G is disconnected, it generates disconnected corona graphs as shown in Fig.4. This feature is not observed in the case of Kronecker graphs given by Leskovec et al. in [12] where authors had taken multi-graph as a basic graph to ensure the connectivity of $G^{(m)}$.

5. Degree sequence of corona graphs: Assume that degree sequence of $G^{(0)} = G$ is given by $\{d_{i_1}^{(0)}, d_{i_2}^{(0)}, \dots, d_{i_n}^{(0)}\}$ where $d_{i_l}^{(j)}$ represents the degree of node i_l at j^{th} corona product, and x is the total distinct degrees. The degree sequence for $G^{(m)}$ is obtained as $\{D_{i_j}^{(1)}, (D_{i_j}^{(2)}, \dots, n \text{ times}), \dots, (D_{i_j}^{(x)}, \dots, n(n+1)^{m-1}$ times)$\}$, where $D_{i_j}^{(1)} = d_{i_j} + mn, D_{i_j}^{(2)} = d_{i_j} + (m-1)n, \dots, D_{i_j}^{(x)} = d_{i_j} + 1$.

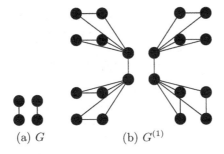

Fig. 4. A disconnected graph G of 4 nodes having two P_2 graphs as its components, along with its 1st corona product $G^{(1)}$ resulting in a disconnected graph

2.1   Degree Distribution

Now, we consider the degree distribution of corona graphs. In a network G, let $P(k)$ denote the fraction of vertices having degree k and hence the degree distribution is the probability distribution of the different degrees in the whole network [14],[1]. The total instances for a particular degree k in $G^{(m)}$ is given by

$$N_k = \sum_{j=1}^{n} \left(\delta_{k,D_{ij}^{(1)}} + n\delta_{k,D_{ij}^{(2)}} + n(n+1)\delta_{k,D_{ij}^{(3)}} + \ldots + n(n+1)^{(m-1)}\delta_{k,D_{ij}^{(x)}} \right) \quad (6)$$

where $\delta_{k,D_{ij}^{(b)}}$ is the Kronecker delta function. Therefore, the degree distribution $P(k)$ of corona graphs is given by

$$P(k)$$
$$= \frac{\sum_{j=1}^{n} \left(\delta_{k,D_{ij}^{(1)}} + n\delta_{k,D_{ij}^{(2)}} + n(n+1)\delta_{k,D_{ij}^{(3)}} + \ldots + n(n+1)^{(m-1)}\delta_{k,D_{ij}^{(x)}} \right)}{n(n+1)^m}$$
$$(7)$$

The degree distribution for K_3 and P_3 are shown in Fig.5. A crucial observation from the degree distribution is that there is huge multiplicities of degrees in the corona graphs and this can also be confirmed from the Fig.5a, Fig.5b. It is similar to the observation seen for Kronecker graphs as in [12]. The figure also shows that the degree distribution follows similar type of curve for different instances of corona graph i.e. for $G^{(i)}$ for each $i \in [1,m]$. The figure also shows the fat tailed degree distribution in both sub-figures of Fig.5. The fat tailed distributions are found abundantly in real world networks like data traffic on internet, return on financial markets etc. as mentioned in [13] and the references therein [4],[16].

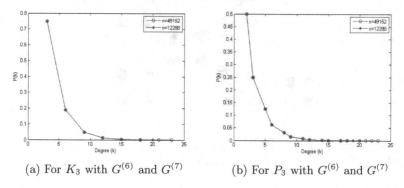

(a) For K_3 with $G^{(6)}$ and $G^{(7)}$ (b) For P_3 with $G^{(6)}$ and $G^{(7)}$

Fig. 5. Degree Distribution for basic graphs (a) K_3 with $G^{(7)}$ and $G^{(6)}$ having 49152 and 12288 nodes respectively. (b) P_3 with $G^{(7)}$ and $G^{(6)}$ having 49152 and 12288 nodes respectively.

3   Spectra of Corona Graphs

Let G be a simple connected graph. The adjacency matrix $A(G^{(m)})$ associated with $G^{(m)}$ is given by

$$\mathbf{A(G^{(m)})} = \begin{bmatrix} A(G^{(m-1)}) & \mathbf{1}^T_{n(n+1)^{m-1}} \otimes I_{n(n+1)^{m-1}} \\ \mathbf{1}_{n(n+1)^{m-1}} \otimes I_{n(n+1)^{m-1}} & A(G) \otimes I_{n(n+1)^{m-1}} \end{bmatrix}$$

where $A(G^{(m-1)})$ is the adjacency matrix of $G^{(m-1)}$, $I_{n(n+1)^{m-1}}$ is the identity matrix and $\mathbf{1}_{n(n+1)^{m-1}}$ is the column vector of $1s$ of length $n(n+1)^{m-1}$. We denote spectra of corona graphs i.e. spectrum of $A(G^{(m)})$ by

$$\sigma(G^{(m)}) = \{\lambda_1, \lambda_2, \ldots, \lambda_{n(n+1)^m}\} \tag{8}$$

where $\lambda_1 \leq \lambda_2 \leq \ldots \leq \lambda_{n(n+1)^m}$ and spectral radius of $A(G^{(m)})$ is denoted as $\rho(G^{(m)})$. The spectrum of corona product of two graphs $G = G_1 \circ G_2$ where G_1 is any graph and G_2 is a regular graph, and the Laplacian spectrum of corona product of any two graphs are provided by Barik et al. in [2]. Inspired by their work, we derive $\sigma(G^{(m)})$ when the basic graph G is regular.

In the next theorem, $\sigma(G^{(m)})$ is in terms of the eigenvalues of the basic graph, that is, $G^{(0)} = G$.

Theorem 1. *Let $G^{(0)} = G$ be a regular graph such that $\sigma(G) = (\mu_1, \mu_2, \ldots, \mu_n = r)$ (where, $\mu_1 \leq \mu_2 \leq \ldots \leq \mu_n = r$) and spectral radius of G be $\rho(G)$. Then, $\sigma(G^{(m)})$ is given by*

(a) $\lambda_i = \dfrac{\mu_i + r\sum_{a=0}^{m-1} 2^a \pm \left(\sum_{c=1}^{m-1} z_c + \sqrt{((r - \mu_i) \pm \sum_{c=1}^{m-1} z_c)^2 + 2^{2m}.n} \right)}{2^m} \in$

$\sigma(G^{(m)})$, *with multiplicity 1 for $i = 1, \ldots, n(n+1)^{m-1}$*

where,

$$z_1 = \sqrt{(r - \mu_i)^2 + 4n}, \dots, z_{m-1} = z_{m-2} + \sqrt{((r - \mu_i) \pm z_{m-2})^2 + n.2^{2(m-2)}}.$$

(b) $\mu_i \in \sigma(G^{(m)})$, *with multiplicity* $n(n + 1)^{m-1}$ *for* $i = 1, \dots, n - 1$.

The spectral radius of $G^{(m)}$ *is given by*

$$\rho(G^{(m)}) = \frac{\mu_n + r\sum_{a=0}^{m-1} 2^a + \left(\sum_{c=1}^{m-1} z_c + \sqrt{((r - \mu_i) - \sum_{c=1}^{m-1} z_c)^2 + 2^{2m}.n}\right)}{2^m}$$

where z_c *is defined above.*

Proof. We can prove the Part (a) of the theorem by induction and hence, the $\rho(G^{(m)})$ will be followed from the proof which is as follows

Base case: For $j = 1$, $G^{(1)} = G^{(0)} \circ G$, the $\sigma(G)$ can be defined according to Theorem 3.1 of [2] as

(i) $\frac{\mu_i + r \pm \sqrt{(r - \mu_i)^2 + 4n}}{2} \in \sigma(G)$ with multiplicity 1 for $i = 1, \dots, n$

(ii) $\mu_i \in \sigma(G)$ with multiplicity n for $i = 1, \dots, n - 1$

Inductive hypothesis: Let $j = m - 1$, $G^{(m-1)} = G^{(m-2)} \circ G$, then $\sigma(G^{(m-1)})$ can be defined as

(a) $\dfrac{\mu_i + r\sum_{a=0}^{m-2} 2^a \pm \left(\sum_{c=1}^{m-2} z_c + \sqrt{(b \pm \sum_{c=1}^{m-2} z_c)^2 + 2^{2(m-1)}.n}\right)}{2^{m-1}} \in \sigma(G^{(m-1)})$,

with multiplicity 1 for $i = 1, \dots, n(n + 1)^{m-2}$

where,

$$z_1 = \sqrt{(r - \mu_i)^2 + 4n}, \dots, z_{m-2} = z_{m-3} + \sqrt{((r - \mu_i) \pm z_{m-3})^2 + n.2^{2(m-3)}}.$$

(b) $\mu_i \in \sigma(G^{(m-1)})$, with multiplicity $n(n + 1)^{m-2}$ for $i = 1, \dots, n - 1$.

Inductive step: For $j = m$, $G^{(m)} = G^{(m-1)} \circ G$, the $\sigma(G^{(m)})$ can be obtained by substituting the eigenvalues of $\sigma(G^{(m-1)})$ in place of μ_i of the base step and we will get the eigenvalues as stated in theorem.

The remaining eigenvalues are $\mu_i \in G^{(m)}$ for $i = 1, \dots, n - 1$ with multiplicity $n(n + 1)^{m-1}$. □

We derive the spectra of star graphs S_k with $k \geq 3$ which is a special case of irregular graph.

Theorem 2. *Let* $k \geq 3$ *be an integer. The spectrum of the graph* $S_k \circ S_k$ *consists of the following eigenvalues*

(a) $\lambda_z = X_1^z + \frac{\mu_i}{3} \in \sigma(G^{(1)})$ *with multiplicity 1, and*

(b) $0 \in \sigma(G^{(1)})$ *with multiplicity* $k(k - 2)$.

where, λ_z *are the eigenvalues for* $G^{(1)}$ *with* $z = 1, 2, 3$ *for the 3 angles i.e.* $\frac{\theta}{3}, \frac{2\pi + \theta}{3}, \frac{4\pi + \theta}{3}$ *as shown in following sub-expressions*

$$X^z = \frac{2}{3} \cos \frac{y\pi + \theta}{3} \sqrt{\mu_i^2 + (6k - 3)}, \quad \theta = \cos^{-1}\left(\frac{2\mu_i^3 + \mu_i(18 - 9k) + (54k - 54)}{2(\mu_i^2 + (6k - 3))^{\frac{3}{2}}}\right) - y\pi$$

where $y = 0, 2, 4$. Here,
$$\lambda_z \in \left[-\mu_i + \frac{2\mu_i}{3} - \frac{2\sqrt{(6k-3)}}{3}, \mu_i + \frac{2\sqrt{(6k-3)}}{3}\right]$$

Proof. Let $Z_1, Z_2, \ldots, Z_n$ be the eigenvectors corresponding to basic input graph's eigenvalues $\mu_1, \ldots, \mu_n$. The eigenvalues corresponding to $G^{(1)}$ are as $\lambda_z = X_1^z + \frac{\mu_i}{3} \in \sigma(G^{(1)})$ for $z = 1, 2, 3$ and the eigenvectors corresponding to them are as

$$\begin{pmatrix} Z_i \\ (\frac{\lambda_i + k - 1}{\lambda_i^2 - k + 1})Z_i \\ (\frac{\lambda_i + 1}{\lambda_i^2 - k + 1})Z_i \\ \vdots \\ (\frac{\lambda_i + 1}{\lambda_i^2 - k + 1})Z_i \end{pmatrix}$$

The $|A(G^{(1)})| = 0$. Hence, $G^{(1)}$ has 0 as its one of the eigenvalue and its multiplicity is of $k(k-2)$.   □

Corollary 1. *Let G be the basic star graph S_k for each $k \geq 3$ such that $\sigma(G) = \{\mu_1, \mu_2, \ldots, \mu_n\}$. Let $m \geq 1$. Then $\sigma(G^{(m)})$ is given by*

(a) $\lambda_{z,1} = X_1^z + \frac{\mu_i}{3}, \ldots, \lambda_{z,m} = \sum_{j=0}^{m-1}(\frac{1}{3})^j X_{m-j}^z + (\frac{1}{3})^{m-1}(\frac{\mu_i}{3}) \in \sigma(G^{(m)})$ *with multiplicity 1 for each of them,*
(b) $0 \in \sigma(G^{(1)})$ *with multiplicity $k(k-2)(k+1)^{(m-1)}$.*

where, $\lambda_{z,j}$ are the eigenvalues for $G^{(m)}$ such that j represents j^{th} corona product with $z = 1, 2, 3$ for the 3 angles i.e. $\frac{\theta_j}{3}, \frac{2\pi + \theta_j}{3}, \frac{4\pi + \theta_j}{3}$ as shown in following subexpressions

$$X_l^z = \frac{2}{3}\cos\frac{y\pi + \theta_l}{3}\sqrt{\left(\sum_{j=1}^{l-1}(\frac{1}{3})^{j-1}X_{l-j}^z + (\frac{1}{3})^{l-1}\mu_i\right)^2 + (6k-3)},$$
$$X_1^z = \frac{2}{3}\cos\frac{y\pi + \theta_1}{3}\sqrt{\mu_i^2 + (6k-3)}$$
$$\theta_m = \cos^{-1}\left(\frac{2(\sum_{j=0}^{m-1}(\frac{1}{3})^j X_{m-j}^z + (\frac{1}{3})^{m-1}(\frac{\mu_i}{3}))^3 - (9k-18)(\sum_{j=0}^{m-1}(\frac{1}{3})^j X_{m-j}^z + (\frac{1}{3})^{m-1}(\frac{\mu_i}{3})) + (54k-54)}{2((\sum_{j=0}^{m-1}(\frac{1}{3})^j X_{m-j}^z + (\frac{1}{3})^{m-1}(\frac{\mu_i}{3}))^2 + (6k-3))^{\frac{3}{2}}}\right)$$
$$-y\pi,$$
where m is the mth corona product and $y = 0, 2, 4$ for the three angles.
Here, $\lambda_{z,j} \in \left[-\mu_i + \frac{2\mu_i}{3^j} - \frac{2j\sqrt{(6k-3)}}{3}, \mu_i + \frac{2j\sqrt{(6k-3)}}{3}\right]$, where $j = 1, \ldots, m$.

Proof. The proof follows by using similar arguments given in the proof of Theorem 1 and Theorem 2.   □

The Laplacian matrix $L(G^{(m)})$ associated with $G^{(m)}$ has the form

$$\mathbf{L(G^{(m)})} = \begin{bmatrix} \mathbf{L(G^{(m-1)})} + n\mathbf{I}_{n(n+1)^{m-1}} & -\mathbf{1}_{n(n+1)^{m-1}}^T \otimes \mathbf{I}_{n(n+1)^{m-1}} \\ -\mathbf{1}_{n(n+1)^{m-1}} \otimes \mathbf{I}_{n(n+1)^{m-1}} & (\mathbf{L(G)} + \mathbf{I}_n) \otimes \mathbf{I}_{n(n+1)^{m-1}} \end{bmatrix}$$

where $L(G^{(m-1)})$ is the Laplacian matrix of $G^{(m-1)}$, I_n and $I_{n(n+1)^{m-1}}$ are the identity matrices. We denote the Laplacian spectra $L(G^{(m)})$ of $G^{(m)}$ by

$$S(G^{(m)}) = \{\lambda_1, \lambda_2, \ldots, \lambda_{n(n+1)^{(m)}}\} \tag{9}$$

where $0 = \lambda_1 \le \lambda_2 \le \ldots \le \lambda_{n(n+1)^{(m)}}$. In the following theorem, we determine the elements of $S(G^{(m)})$ in terms of the Laplacian eigenvalues of the basic graph, $G^{(0)} = G$ where $S(G) = \{0 = \nu_1, \nu_2, \ldots, \nu_n\}$ (where, $\nu_1 \le \nu_2 \le \ldots \le \nu_n$) is the Laplacian spectra of G. The algebraic connectivity of a graph is defined as the second smallest eigenvalue of $L(G)$ [3].

Theorem 3. *Let G be a simple connected graph. We denote the algebraic connectivity of G and $G^{(m)}$ by $a(\nu_2)$ and $a(\lambda_2)$ respectively. The, Laplacian spectra $S(G^{(m)})$ of G is given by*

(a) $\dfrac{\nu_i + (n+1)\sum_{i=0}^{m-1} 2^i \pm \sum_{i=1}^{m} z_i}{2^m} \in S(G^{(m)})$ *with multiplicity 1 for $i = 1, \ldots,$*
$n(n+1)^{m-1}$. *where,*
$z_1 = \sqrt{(\nu_i + n + 1)^2 - 4\nu_i}$,
$\vdots$

$$z_m = \sqrt{\left(\nu_i + (n+1)\sum_{i=0}^{m-1} 2^i \pm \sum_{i=1}^{m-1} z_i\right)^2 - 2^{(m+1)}\left(\nu_i + (n+1)\sum_{i=0}^{m-2} 2^i \pm \sum_{i=1}^{m-1} z_i\right)}$$

(b) $\nu_i + 1 \in S(G^{(m)})$ *with multiplicity $n(n+1)^{m-1}$ for $i = 2, \ldots, n$.*

Hence, the algebraic connectivity of $S(G^{(m)})$ is

$$a(\lambda_2) = \frac{\nu_2 + (n+1)\sum_{i=0}^{m-1} 2^i - \sum_{i=1}^{m} z_i}{2^m} < 1$$

where z_i can be defined as above.

Proof. We can prove the Part (a) of the theorem by induction.
Base case: For $j = 1$, $G^{(1)} = G^{(0)} \circ G$, $S(G)$ can be defined according to Theorem 3.2 of [2] as

(i) $\frac{\nu_i + n + 1 \pm \sqrt{(\nu_i + n + 1)^2 - 4\nu_i}}{2} \in S(G)$ with multiplicity 1 for $i = 1, \ldots, n$
(ii) $\nu_i + 1 \in S(G)$ with multiplicity n for $i = 2, \ldots, n$

Inductive hypothesis: Let $j = m - 1$, $G^{(m-1)} = G^{(m-2)} \circ G$, the $S(G^{(m-1)})$ can be defined as

(a) $\dfrac{\nu_i + (n+1)\sum_{i=0}^{m-2} 2^i \pm \sum_{i=1}^{m-1} z_i}{2^{m-1}} \in S(G^{(m-1)})$ with multiplicity 1 for $i = 1, \ldots, n(n+1)^{m-2}$. where,

$$z_1 = \sqrt{(\nu_i + n + 1)^2 - 4\nu_i}\,,$$

$$\vdots$$

$$z_{m-1} = \sqrt{(\nu_i + (n+1)\sum_{i=0}^{m-2} 2^i \pm \sum_{i=1}^{m-2} z_i)^2 - 2^m(\nu_i + (n+1)\sum_{i=0}^{m-3} 2^i \pm \sum_{i=1}^{m-2} z_i)}$$

(b) $\nu_i + 1 \in S(G^{(m-1)})$ with multiplicity $n(n+1)^{m-2}$ for $i = 2, \ldots, n$.

Inductive step: For $j = m$, $G^{(m)} = G^{(m-1)} \circ G$, the $S(G^{(m)})$ can be obtained by substituting the eigenvalues of $S(G^{(m-1)})$ in place of ν_i of the base step and we will get the eigenvalues as stated in theorem. The other eigenvalues of $G^{(m)}$ are $\nu_i + 1$ for $i = 2, \ldots, n$ with multiplicity $n(n+1)^{m-1}$. $\square$

The signless Laplacian matrix $S_Q(G^{(m)})$ of $G^{(m)}$ is of the form

$$\mathbf{S_Q(G^{(m)})} = \begin{bmatrix} \mathbf{S_Q(G^{(m-1)})} + n\mathbf{I_{n(n+1)^{m-1}}} & \mathbf{1}^T_{n(n+1)^{m-1}} \otimes \mathbf{I_{n(n+1)^{m-1}}} \\ \mathbf{1_{n(n+1)^{m-1}}} \otimes \mathbf{I_{n(n+1)^{m-1}}} & \mathbf{(S_Q(G) + I_n)} \otimes \mathbf{I_{n(n+1)^{m-1}}} \end{bmatrix}$$

where $S_Q(G^{(m-1)})$ is the signless Laplacian matrix of $G^{(m-1)}$, I_n and $I_{n(n+1)^{m-1}}$ are the identity matrices. The spectrum of signless Laplacian of $Q(G^{(m)})$ for $G^{(m)}$ is denoted by

$$S_Q(G^{(m)}) = \{\lambda_1, \lambda_2, \ldots, \lambda_{n(n+1)^{(m)}}\} \tag{10}$$

where $\lambda_1 \le \lambda_2 \le \ldots \le \lambda_{n(n+1)^m}$. Recently, there is a lot of work on signless Laplacian matrices and their Q−spectra as in [5],[6],[10] and their authors think that it is more useful and hence a lot of work is going on to find their usefulness as [7],[8]. We will define the $S_Q(G^{(m)})$ inspired by the work of [5] on two graphs and here taking the basic initial graph G as the r-regular graph. In the following theorem, we derive the elements of $S_Q(G^{(i)})$ in terms of the signless Laplacian eigenvalues of a regular basic graph, $G^{(0)} = G$ such that $S_Q(G) = (q_1, q_2, \ldots, q_n = 2r)$ (where, $q_1 \le q_2 \le \ldots \le q_n = 2r$).

Theorem 4. *Let G be a simple connected graph. Then, $S_Q(G^{(m)})$ is given by*

(a) $\lambda_i = \dfrac{q_i + n\sum_{i=0}^{m-1} 2^i + r\sum_{i=1}^{m} 2^i + \sum_{i=0}^{m-1} 2^i \pm \sum_{j=1}^{m} z_j}{2^m}$ *with multiplicity of*
1, *for $i = 1, \ldots, n(n+1)^{m-1}$*
where,

$$z_j = \sqrt{(q_i + n\sum_{i=0}^{j-1} 2^i + r(\sum_{i=1}^{j-1} 2^i - 2^j) + (\sum_{i=0}^{j-2} 2^i - 2^{j-1}) \pm \sum_{i=1}^{j-1} z_i)^2 + 2^{2j}.n}$$

for $j = 2, \ldots, m$ and $z_1 = \sqrt{((q_i + n) - (2r + 1))^2 + 4n}$.
(b) $q_j + 1$ *with the multiplicity of $n(n+1)^{m-1}$ for $j = 1, \ldots, n - 1$*

Hence, spectral radius of $S_Q(G^{(m)})$ is

$$q(S_Q(G^{(m)})) = \frac{q_i + n\sum_{i=0}^{m-1} 2^i + r\sum_{i=1}^{m} 2^i + \sum_{i=0}^{m-1} 2^i + \sum_{j=1}^{m} z_j}{2m}$$

where z_j is defined as above.

Proof. We can prove the Part (a) of the theorem by induction and hence, the $\rho(G^{(m)})$ will be followed from the proof which is as follows

Base case: For $j = 1$, $G^{(1)} = G^{(0)} \circ G$, the $\sigma(G)$ can be defined according to Theorem 3.1 of [2] as

(i) $\frac{q_i+n+2r+1\pm\sqrt{((q_i+n)-(2r+1))^2+4n}}{2} \in S_Q(G)$ with multiplicity 1 for $i = 1, \ldots, n$

(ii) $q_i + 1 \in S_Q(G)$ with multiplicity n for $i = 1, \ldots, n-1$

Inductive hypothesis: Let $j = m - 1$, $G^{(m-1)} = G^{(m-2)} \circ G$, the $\sigma(G^{(m-1)})$ can be defined as

(a) $\lambda_i = \dfrac{q_i + n\sum_{i=0}^{m-2} 2^i + r\sum_{i=1}^{m-1} 2^i + \sum_{i=0}^{m-2} 2^i \pm \sum_{i=1}^{m-1} z_i}{2m-1}$ with multiplicity

   of 1, for $i = 1, \ldots, n(n + 1)^{m-2}$

where,

$$z_j = \sqrt{\left(q_i + n\sum_{i=0}^{j-1} 2^i + r\left(\sum_{i=1}^{j-1} 2^i - 2^j\right) + \left(\sum_{i=0}^{j-2} 2^i - 2^{j-1}\right) \pm \sum_{i=1}^{j-1} z_i\right)^2 + 2^{2j}.n}$$

   for $j = 2, \ldots, m - 1$ and $z_1 = \sqrt{((q_i + n) - (2r + 1))^2 + 4n}$.

(b) $q_i + 1$ with the multiplicity of $n(n + 1)^{m-2}$ for $i = 1, \ldots, n - 1$

Inductive step: For $j = m$, $G^{(m)} = G^{(m-1)} \circ G$, the $\sigma(G^{(m)})$ can be obtained by substituting the eigenvalues of $S_Q(G^{(m-1)})$ in place of q_i of the base step and we will get the eigenvalues as stated in theorem.

   The other eigenvalues are $q_i + 1$ with the multiplicity of $n(n + 1)^{m-1}$ for $j = 1, \ldots, n - 1$.   □

Consider star graph S_k which is an irregular graph. In the theorem below, we determine explicit formula of signless Laplacian elements of $G^{(m)} = S_k^{(m)}$.

Theorem 5. *Let $S_Q(G) = \{q_1, q_2, \ldots, q_n\}$ with $q_1 \leq q_2 \leq \cdots \leq q_n$. Then, $S_Q(G^{(1)})$ is given by*

(a) $\lambda_z = X_1^z + \frac{q_i+2k+2}{3} \in \sigma(G^{(1)})$ *with multiplicity 1, and*

(b) $q_j + 1 \in \sigma(G^{(1)})$ *where for $j = 2, \ldots, n - 1$ with multiplicity of each of them as k*

where, λ_z are the eigenvalues for $G^{(1)}$ with $z = 1, 2, 3$ for the 3 angles i.e. $\frac{\theta}{3}, \frac{2\pi+\theta}{3}, \frac{4\pi+\theta}{3}$ as shown in following sub-expressions

$X^z = \frac{2}{3}\cos\frac{y\pi+\theta}{3}\sqrt{q_i^2 + q_i(k - 2) + (k + 1)^2}$,

$$\theta = \cos^{-1}\left(\frac{2q_i^3 + (3k-6)q_i^2 - 3(k^2-k-2)q_i + (70k-94-12\sum_{a=1}^{k-2}(a+2)(k-a-1))}{2(q_i^2 + q_i(k-2) + (k+1)^2)^{\frac{3}{2}}} \right) - y\pi$$

where $y = 0, 2, 4$. Here, $\lambda_z \in \left[-q_i + \frac{2q_i}{3} + \frac{Y-(A+\sqrt{4B-A^2})}{3}, q_i + \frac{Y+A+\sqrt{4B-A^2}}{3} \right]$

where $A = (k-2), B = (k+1)^2, Y = (2k+2)$.

Proof. Let $Y_1, \ldots, Y_n$ be the eigenvectors corresponding to basic input graph's eigenvalues $q_1, \ldots, q_n$. The eigenvalues for $G^{(1)}$ are $\lambda_z = X_1^z + \frac{q_i+2k+2}{3} \in \sigma(G^{(1)})$ for $z = 1, 2, 3$ and the eigenvectors corresponding to them are as

$$\begin{pmatrix} Y_i \\ \frac{q_i+k-3}{q_i^2 - q_i(k+2)+(k+1)} Y_i \\ \frac{q_i-k+1}{q_i^2 - q_i(k+2)+(k+1)} Y_i \\ \vdots \\ \frac{q_i-k+1}{q_i^2 - q_i(k+2)+(k+1)} Y_i \end{pmatrix}$$

$q_j + 1 \in \sigma(G^{(1)})$ are the other eigenvalues for $j = 2, \ldots, n-1$ with multiplicity of each of them as k.   □

Corollary 2. *Let $S_Q(G)$ for star graph S_k for each $k \geq 3$ is $\{q_1, q_2, \ldots, q_n\}$ with $q_1 \leq q_2 \leq \ldots \leq q_n$. Then, $S_Q(G^{(m)})$ is given by*

(a) $\lambda_{z,1} = X_1^z + \frac{q_i+2k+2}{3}, \ldots, \lambda_{z,m} = \sum_{j=0}^{m-1}(\frac{1}{3})^j X_{m-j} + (\frac{1}{3})^m q_i + (2k+2)\sum_{j=1}^{m}(\frac{1}{3})^j$
  $\in \sigma(G^{(m)})$ *with multiplicity 1, and*

(b) $q_i + 1 \in \sigma(G^{(1)})$ *where for $i = 2, \ldots, n-1$ with multiplicity of each of them as $k(k+1)^{(m-1)}$.*

where, $\lambda_{z,j}$ are the eigenvalues for $G^{(1)}$ (for j^{th} corona product) with $z = 1, 2, 3$ for the 3 angles i.e. $\frac{\theta_j}{3}, \frac{2\pi+\theta_j}{3}, \frac{4\pi+\theta_j}{3}$ as shown in following sub-expressions
$X_1^z = \sqrt{(A)^2 + A + (k+1)^2}$
$X_1^z = \frac{2}{3}\cos\frac{y\pi+\theta}{3}\sqrt{q_i^2 + q_i(k-2) + (k+1)^2}$,
where $A = (\sum_{j=0}^{m-2}(\frac{1}{3})^j X_{m-j-1} + (\frac{1}{3})^{m-1}q_i + (2k+2)\sum_{j=1}^{m-1}(\frac{1}{3})^j)$
$$\theta = \cos^{-1}\left(\frac{2\lambda_{z,m-1}^3 + (3k-6)\lambda_{z,m-1}^2 - 3(k^2-k-2)\lambda_{z,m-1} + (70k-94-12\sum_{a=1}^{k-2}(a+2)(k-a-1))}{2(\lambda_{z,m-1}^2 + \lambda_{z,m-1}(k-2) + (k+1)^2)^{\frac{3}{2}}} \right)$$
$-y\pi$
where $y = 0, 2, 4$ and $\lambda_{z,m-1}$ is as defined in part(a) of corollary. Here,

$\lambda_{z,1} \in \left[-q_i + \frac{2q_i}{3} + \frac{Y-(A+\sqrt{4B-A^2})}{3}, q_i + \frac{Y+A+\sqrt{4B-A^2}}{3} \right], \ldots, \lambda_{z,m} \in \left[-q_i + \frac{2q_i}{3^m} - \right.$

$\left. \frac{m(A+\sqrt{4B-A^2})}{3} + Y(-2\sum_{i=1}^{m}\frac{m-i}{3^{i+1}} + \sum_{i=1}^{m}3^{-i}), q_i + m(\frac{Y+A+\sqrt{4B-A^2}}{3}) \right]$

where $A = (k-2), B = (k+1)^2, Y = (2k+2)$.

Proof. The proof follows by using similar arguments given in proof of Theorem 5.   □

4   Conclusion

We proposed a model for generation of complex networks inspired by the phenomena of duplication of genes. We defined corona graphs by taking corona product of a simple graph, which we call a basic graph, finite number of times. We determined spectra, Laplacian spectra and signless Laplacian spectra for corona graphs when the basic graph is regular. We also derived the spectra and signless Laplacian spectra of corona graphs when the basic graph is a star graph.

References

1. Albert, R., Barabási, A.-L.: Statistical mechanics of complex networks. Rev. Mod. Phys. 74, 47 (2002)
2. Barik, S., Pati, S., Sarma, B.: The spectrum of the corona of two graphs. SIAM J. Discrete Math. 21, 47–56 (2007)
3. Bapat, R.B.: Graphs and matrices. Springer (2010)
4. Crovella, M.E., Taqqu, M.S.: Estimating the heavy tail index from scaling properties. Methodol. Comput. Appl. 1, 55–79 (1999)
5. Cui, S.-Y., Tian, G.-X.: The spectrum and the signless Laplacian spectrum of coronae. Linear Algebra Appl. 437, 1692–1703 (2012)
6. Cvetković, D., Simić, S.K.: Towards a spectral theory of graphs based on the signless Laplacian, I. Publ. Inst. Math(Beograd)(NS) 85, 19–33 (2009)
7. Cvetković, D., Simić, S.K.: Towards a spectral theory of graphs based on the signless Laplacian, II. Linear Algebra Appl. 432, 2257–2272 (2010)
8. Cvetković, D., Simić, S.K.: Towards a spectral theory of graphs based on the signless Laplacian, III. Appl. Anal. Discrete Math. 4, 156–166 (2010)
9. Frucht, R., Harary, F.: On the corona of two graphs. Aequationes Math. 4, 322–325 (1970)
10. Haemers, W.H., Spence, E.: Enumeration of cospectral graphs. European J. Combin. 25, 199–211 (2004)
11. Ispolatov, I., Krapivsky, P.L., Yuryev, A.: Duplication-divergence model of protein interaction network. Phys. Rev. E 71, 061911 (2005)
12. Leskovec, J., Chakrabarti, D., Kleinberg, J., Faloutsos, C., Ghahramani, Z.: Kronecker graphs: An approach to modeling networks. J. Mach. Learn. Res. 11, 985–1042 (2010)
13. Misiewicz, J.: Fat-Tailed Distributions: Data, Diagnostics, and Dependence (2011)
14. Newman, M.E.J.: The structure and function of complex networks. SIAM Rev. 45, 167–256 (2003)
15. Parsonage, E., Nguyen, H.X., Bowden, R., Knight, S., Falkner, N., Roughan, M.: Generalized graph products for network design and analysis. In: 19th IEEE International Conference on Network Protocols (ICNP), pp. 79–88 (2011)
16. Rachev, S.T.: Handbook of Heavy Tailed Distributions in Finance: Handbooks in Finance 1. Elsevier (2003)

Axiomatic Characterization of the Median and Antimedian Functions on Cocktail-Party Graphs and Complete Graphs

Manoj Changat[1,*], Divya Sindhu Lekha[1,**], Henry Martyn Mulder[2,***], and Ajitha R. Subhamathi[3]

[1] Department of Futures Studies, University of Kerala, Trivandrum - 695 581, India
{mchangat,divi.lekha}@gmail.com
[2] Econometrisch Instituut, Erasmus Universiteit, P.O. Box 1738, 3000 DR Rotterdam, Netherlands
hmmulder@ese.eur.nl
[3] Department of Computer Applications, N.S.S College Rajakumari, Idukki, India
ar.subhamathi@gmail.com

Abstract. A median (antimedian) of a profile of vertices on a graph G is a vertex that minimizes (maximizes) the remoteness value, that is, the sum of the distances to the elements in the profile. The median (or antimedian) function has as output the set of medians (antimedians) of a profile. It is one of the basic models for the location of a desirable (or obnoxious) facility in a network. The median function is well studied. For instance it has been characterized axiomatically by three simple axioms on median graphs. The median function behaves nicely on many classes of graphs. In contrast the antimedian function does not have a nice behavior on most classes. So a nice axiomatic characterization may not be expected. In this paper an axiomatic characterization is obtained for the median and antimedian functions on cocktail-party graphs. In addition a characterization of the antimedian function on complete graphs is presented.

Keywords: median, antimedian, consensus function, consistency, cocktail-party graph, complete graph, consensus axiom.

1 Introduction

Facility location problems in discrete location theory deal with functions that find an appropriate location for a common facility or resource in a discrete network. The main objective is to minimize the cost of accessing a facility or

---

* Research work is supported by NBHM/DAE under grant No.2/48(2)/2010/ NBHM-R & D.
** Department of Information Technology, College of Engineering and Management Punnapra, Alappuzha - 688 003, India.
*** This research was initiated while the third author visited the University of Kerala in January 2011 under the Erudite Scheme of the Government of Kerala, India.

S. Ganguly and R. Krishnamurti (Eds.): CALDAM 2015, LNCS 8959, pp. 138–149, 2015.

sharing a resource in the network. Placing a common resource at a median position minimizes the cost of sharing the resource with other locations. Thus the algorithms for locating at medians in a graph are very often useful and form the basic models of discrete facility location problems. Typical problems of this kind that have been studied extensively are: (i) The *median problem*: finding a vertex that minimizes the distance sum to the clients. (ii) The *mean problem*: finding a vertex that minimizes the sum of the squares of the distances to the clients. (iii) the *center problem*: minimizing the maximum distance to the clients. The first two problems can be used to model finding the optimal location for a distribution center. The last problem can be used to model finding the optimal location for a fire station. The *antimedian problem* is a different type of location problem in which the facility is of obnoxious nature (i.e. the clients want to have it as far away as possible), for example a garbage dump. In this case the 'cost' is being maximized.

A consensus function is used to model consensus problems. These are problems in which one wants to reach consensus amongst agents or clients in a rational way. The input of the consensus function is information on the clients and the output concerns the issue on which consensus should be reached. To guarantee the rationality of the process, the consensus function satisfies certain "rational" rules called "consensus axioms". Such axioms should be appealing and simple. But this depends on the consensus function. A function with nice properties might be characterized by simple axioms. But a function that behaves badly might need more complicated or less appealing axioms. K. Arrow initiated the study of the axiomatics of consensus functions in his seminal paper [1] of 1951. For more references in this area see [2], [3], [17].

Location problems can also be viewed as consensus problems. Then one wants to characterize the location function by a set of axioms that are as nice and simple as possible. Holzman [10] was the first to study location functions from this perspective. His focus was on the mean function on a tree network (the continuous variant of a tree). Then Vohra [26] characterized the median function axiomatically on tree networks (continuous case). The discrete case was first dealt with by McMorris, Mulder & Roberts [16]: the median function on cube-free median graphs was characterized using three simple and appealing axioms, see below. The mean function on trees (discrete case) was first characterized by McMorris, Mulder & Ortega [14], [15]. The center function on trees has been characterized by McMorris, Roberts & Wang [18], see also [24]. The center function is also studied on some other graph classes, see [27]. Recently the median function has been characterized on hypercubes and median graphs by Mulder & Novick [22], [23] using the same three simple axioms as in [16]. In the case of the median function and the center function all axioms satisfy the criterion of being appealing and natural at first sight. The characterizations for the mean function are more complex than those for the median function or the center function. But except for one complex axiom they still satisfy the criterion of being simple and appealing. All above results for the center function and the mean function so far are on trees. The characterization for the median function is on a much wider class, viz. that of median graphs. The reason for this is the very nice behavior of the median function on these graphs. For more information on median graphs see e.g. [11], [20], [21].

We focus on the characterization of two location functions: the median function and the antimedian function. The antimedian function maximizes the sum of the distances to the clients, see e.g. [19], [4], [5], [6], [7], [25]. The differences between these two functions are quite striking. A first inspection of the antimedian function already shows that, even on trees, it does not behave nicely at all, let alone on arbitrary graphs. Only on special classes, such as paths, hypercubes and complete graphs, does it seem to have a nice behavior. The axiomatization of the antimedian function on hypercubes and paths is well studied in [8]. In this paper we focus on the cocktail-party graphs. A cocktail-party graph is a complete graph of even order minus a perfect matching. Besides we also study the antimedian function on complete graphs.

In Section 2 we set the stage. In Section 3 we characterize the median function on cocktail-party graphs by a set of four axioms. In Section 4 we characterize the antimedian function on the same graphs by another set of four axioms. In our view they are all simple and natural. In Section 5, we characterize the antimedian function on complete graphs, again by a set of four axioms. For axiomatic characterizations of the median function on complete graphs we refer to [13].

2   Preliminaries

Let $G = (V, E)$ be a finite, connected, simple graph with vertex set V and edge set E. The distance function of G is denoted by d, where $d(u, v)$ is the length of a shortest u, v-path. The interval $I(u, v)$ between two vertices u and v in G consists of all vertices on shortest u, v-paths, that is:

$$I(u, v) = \{x \mid d(u, x) + d(x, v) = d(u, v)\} \tag{1}$$

A profile π of length $k = |\pi|$ on G is a non-empty sequence $\pi = (x_1, x_2, \ldots, x_k)$ of vertices of V with repetitions allowed. We define V^* to be the set of all profiles of finite length. We call $x_1, x_2, \ldots, x_k$ the *elements* of the profile. A *vertex of* π is a vertex that occurs as an element in π. By $\{\pi\}$ we denote the set of all vertices of π. Note that a vertex may occur more than once as element in π. If we say that x is an *element* of π, then we mean an element in a certain position, say $x = x_j$ in the j-th position. A *subprofile* of π is just a non-empty subsequence of π. The concatenation of profiles π and ρ is denoted by $\pi\rho$. The profile consisting of the concatenation of m copies of π is denoted by π^m. Let π be a profile on G. A vertex in π with highest occurrence in π is called a *plurality vertex* of π. We denote the set of plurality vertices of π by $Pl(\pi)$.

A *consensus function* on G is a function $F : V^* \to 2^V - \emptyset$ that gives a non-empty subset of V as output for each profile on G. For convenience, we write $F(x_1, \ldots, x_k)$ instead of $F((x_1, \ldots, x_k))$, for any function F defined on profiles, but will keep the brackets where needed.

The *remoteness* of a vertex v to profile π is defined as

$$r(v, \pi) = \sum_{i=1}^{k} d(x_i, v). \tag{2}$$

A vertex minimizing $r(v, \pi)$ is called a *median* of the profile. The set of all medians of π is the *median set* of π and is denoted by $M(\pi)$. A vertex maximizing $r(v, \pi)$ is called an *antimedian* of the profile. The set of all antimedians of π is the *antimedian set* of π and is denoted by $AM(\pi)$. We can also think of M and AM as functions from V^* to $2^V - \emptyset$, and then call them the *Median Function* and *Antimedian Function*. Note that we have

$$M(x) = \{x\}, \tag{3}$$

and

$$M(x, y) = I(x, y). \tag{4}$$

Moreover, if $I(u, v) \cap I(v, w) \cap I(w, u) \neq \emptyset$, then

$$M(u, v, w) = I(u, v) \cap I(v, w) \cap I(w, u). \tag{5}$$

The median function has been studied extensively, especially on median graphs. A median graph is defined by the property that $|I(u, v) \cap I(v, w) \cap I(w, u)| = 1$, for any three vertices u, v, w. Equivalently, a median graph is a graph such that any profile of length 3 has a unique median. See e.g. [20], [11], [21] for a rich structure theory on median graphs. Also nice axiomatic characterizations are available for the median function on median graphs, see e.g. [16], [17], [23]. Three simple and natural axioms suffice for the characterization of the median function in this case. We present these here. The first two axioms are defined without any reference to metric.

(A) Anonymity: $F(\pi) = F(x_{\chi(1)}, x_{\chi(2)}, \ldots, x_{\chi(k)})$, for any profile $\pi = (x_1, x_2, \ldots, x_k)$ on V and for any permutation χ of $\{1, 2, \ldots, k\}$.

(C) Consistency: If $F(\pi) \cap F(\rho) \neq \emptyset$, for profiles π and ρ, then $F(\pi\rho) = F(\pi) \cap F(\rho)$.

(B) Betweenness: $F(u, v) = I(u, v)$, for all u, v in V.

Clearly, the median function satisfies axioms (A) and (B) on any graph. It is part of folklore that the median function also satisfies (C). Anyway, a proof of this can be found in [16].

The cocktail-party graph $K_{(n \times 2)}$ is obtained from the complete graph K_{2n} with vertex set $V = \{v_1, \ldots, v_n, v_{n+1}, \ldots, v_{2n}\}$ by deleting the perfect matching $v_1 v_{n+1}, \ldots, v_n v_{2n}$, see e.g. [9]. It arises in the handshake problem. It is distance-transitive, and hence also distance-regular. For each i with $1 \leq i \leq n$, we call $\{v_i, v_{n+i}\}$ a pair of *mates*. For any vertex v, we denote its mate by $\tilde{v}$. For any profile π, the profile $\tilde{\pi}$ is obtained from π by replacing each element by its mate.

For v in V, the profile $(v, \tilde{v})$ is called a *mating pair*. A *mating profile* is the concatenation of mating pairs. Note that, going from left to right through such a profile, each vertex in an odd position is followed by its mate in the next position. Finally, a *mate-free profile* π is such that if v is in π, then $\tilde{v}$ is not in π.

The following lemma is obvious but quite helpful in the sequel.

Lemma 1. *Let G be a cocktail-party graph with vertex set V, and let $\pi = (v, \tilde{v})$ be a mating pair. Then $r(u, \pi) = 2$, for all v in V.*

An immediate consequence of this lemma is that we can compute the median and antimedian function quite simply. Let π be a profile on the cocktail-party graph. Assume that π contains two elements that form a pair of mates, say $v, \tilde{v}$. Let π' be the profile obtained from π by removing the two elements v and $\tilde{v}$. Consider the remoteness of any vertex u with respect to π. Now u minimizes (maximizes) $r(u, \pi)$ if and only if it minimizes (maximizes) $r(u, \pi')$. So we have $M(\pi) = M(\pi')$ and $AM(\pi) = AM(\pi')$. Hence, in computing the median set or antimedian set of π, we can delete any pair of mates. Thus a subprofile ρ remains that is mate-free. Now the median vertices are precisely the vertices with highest occurrence in ρ, so $M(\pi) = Pl(\rho)$. For this fact we present an argument in the next section. The antimedian vertices are precisely the mates of the vertices with highest occurrence, so $AM(\pi) = Pl(\tilde{\rho})$.

3   Axiomatic Characterization of the Median Function on Cocktail-Party Graphs

In this section we characterize the median function on cocktail-party graphs. The next two lemmata are presented to put forward two basic properties of the median function. They are the motivation for the two additional axioms besides *Anonymity* and *Consistency* that we need for the median function. The first lemma is a trivial consequence of Lemma 1.

Lemma 2. *Let F be the median function defined on the vertex set V of a cocktail-party graph G. Then $F(v, \tilde{v}) = V$, for any $v \in V$.*

The next lemma is also simple.

Lemma 3. *Let F be the median function defined on the vertex set V of a cocktail-party graph G. Then $F(\pi) = Pl(\pi)$, for all mate-free profiles π.*

Proof. Let $\pi = (x_1, x_2, \ldots, x_k)$ be a mate-free profile. Let $\{\pi\} = \{y_1, y_2, \ldots, y_\ell\}$, and let f_j be the number of occurrences of y_j in π. Then, for any vertex w outside the profile π, we have $d(w, y_j) \geq 1$, for each vertex y_j in π. Write $f = \sum_{j=1}^{\ell} f_j$. So we have $r(w, \pi) \geq f$.

Let u be any vertex in π. Then we have $d(u, x_i) = 1$, for any $x_i \neq u$. Clearly $r(u, \pi) = f - f_j$, for $u = y_j$. So the vertices that minimize remoteness are all in π. Note that $r(u, \pi) = f - f_j$ is minimum when f_j is maximum. So the vertices that minimize remoteness are precisely those that occur most often in π. □

By Lemma 2 and Lemma 3, the median function on cocktail-party graphs satisfies the following two axioms.

(A_1): $F(v, \tilde{v}) = V$, for all $v \in V$.
(A_2): $F(\pi) = Pl(\pi)$, for all mate-free profiles π.

Let F be any consensus function satisfying the axioms (A_1) and (A_2). Note that for any vertex v in a cocktail-party graph, we have $I(v, \tilde{v}) = V$. Now consider any other profile $\pi = (u, v)$ such that $u \neq v, \tilde{v}$. Then, clearly, u and v are adjacent, whence axiom (A_2) implies $F(\pi) = \{u, v\} = I(u, v)$. We put these observations in the following remark.

Remark 1. Let F be a consensus function defined on the vertex set V of a cocktail-party graph G such that F satisfies A_1 and A_2. Then F satisfies the *Betweenness* axiom (B).

Theorem 1. *Let F be a consensus function on a cocktail-party graph G with vertex set V. Then F is the median function if and only if F satisfies axioms (A), (C), (A_1) and (A_2).*

Proof. It is straightforward to check that the median function satisfies all the four axioms.

Let F be a function that satisfies the four axioms. Take any profile π. If it contains a pair of mates $v, \tilde{v}$, then we can permute π such that v and $\tilde{v}$ are moved to the front two positions, thus getting the profile $(v, \tilde{v})\rho$, where ρ is the subprofile of π obtained by deleting the elements v and $\tilde{v}$ from their respective positions. By (A_1), we have $F(v, \tilde{v}) = V$. So $F(v, \tilde{v}) \cap F(\rho) \neq \emptyset$. Hence, by *Consistency*, we have $F((v, \tilde{v})\rho) = F(v, \tilde{v}) \cap F(\rho) = F(\rho)$. Finally, by *Anonymity*, we have $F(\pi) = F(v, \tilde{v}) \cap F(\rho) = F(\rho)$. We can repeat this process until we end up with a subprofile σ of π that is either a mating pair or mate-free. In the latter case, we have $F(\pi) = F(\sigma)$. From axiom (A_2), it follows that $F(\sigma) = Pl(\sigma) = M(\sigma) = M(\pi)$. If σ is a mating pair, then we have $F(\sigma) = V = F(\pi) = M(\pi)$. This completes the proof.   □

For any axiomatic characterization, we want to know whether the axioms involved are independent. We present some examples. In all cases G is a cocktail-party graph with vertex set V having at least 4 vertices.

Example 1. (A_1) excluded. Define the function F on G by $F(\pi) = Pl(\pi)$, for all profiles π. It is straightforward to check that F satisfies (A), (C) and (A_2). Since $F(v, \tilde{v}) = \{v, \tilde{v}\} \neq V$, for any vertex v, the function F does not satisfy (A_1).

Example 2. (A_2) excluded. Define the function F on G by $F(\pi) = V$, for all profiles π. Obviously, F satisfies axioms (A), (C) and (A_1). Take any two adjacent vertices u and v in G. Then

$$F(u, v) = V \neq \{u, v\} = Pl(u, v). \tag{6}$$

So F does not satisfy (A_2).

Example 3. (C) excluded. Define the function F on G by

(c1): $F(v, \tilde{v}) = V$, for all vertices v in V,

(c2): $F(\pi) = Pl(\pi)$, for all profiles π that are not a mating pair.

Clearly, F satisfies (A), (A_1) and (A_2). Take two vertices u and v that are not mates, and let $\pi = (u, \tilde{u}, v, \tilde{v})$. Then, by $(c2)$, we have

$$F(\pi) = Pl(\pi) = \{u, \tilde{u}, v, \tilde{v}\} \neq V = F(u, \tilde{u}) \cap F(v, \tilde{v}). \tag{7}$$

So F does not satisfy *Consistency*.

The case of *Anonymity* seems to be different. First we observe that the independence of *Anonymity* from other axioms is a non-trivial issue for other sets of axioms. In [12] two examples of sets are given where it is highly non-trivial that *Anonymity* is independent from the other axioms. One instance is the above mentioned case of the set (A), (B), (C) that characterizes the median function on median graphs. A rather intricate example was needed to show independence of (A). In our case we do not yet have an example that shows independence of *Anonymity*. On the other hand one would not expect that it follows from the other axioms. So we leave it as an open problem here.

4   Axiomatic Characterization of Antimedian Function on Cocktail-Party Graphs

First we present the analogue of axiom (A_2) that we need for the antimedian case. We skip the analogue of Lemma 3 and its proof. An obvious adaptation does the trick.

(A_3): $F(\pi) = Pl(\tilde{\pi})$, for all mate-free profiles π.

Theorem 2. *Let F be a consensus function on a cocktail-party graph G with vertex set V. Then F is the antimedian function if and only if F satisfies axioms (A), (C), (A_1) and (A_3).*

Proof. Let F be the antimedian function. Then F satisfies all the above four axioms.

Let F be a function that satisfies the four axioms. Take any profile π. If it contains a pair of mates $v, \tilde{v}$, then we can permute π such that v and $\tilde{v}$ are moved to the front two positions, thus getting the profile $(v, \tilde{v})\rho$, where ρ is the subprofile of π obtained by deleting the elements v and $\tilde{v}$ from their respective positions. By (A_1), we have $F(v, \tilde{v}) = V$. So $F(v, \tilde{v}) \cap F(\rho) \neq \emptyset$. Hence, by Consistency, we have $F((v, \tilde{v})\rho) = F(v, \tilde{v}) \cap F(\rho) = F(\rho)$. Finally, by Anonymity, we have $F(\pi) = F(v, \tilde{v}) \cap F(\rho) = F(\rho)$. We can repeat this process until we end up with a subprofile σ of π that is either a mating pair or mate-free. In the latter case, we have $F(\pi) = F(\sigma)$. From axiom (A_3) it follows that $F(\sigma) = Pl(\tilde{\sigma}) = AM(\sigma) = AM(\pi)$. If σ is a mating pair, then we have $F(\sigma) = V = AM(\sigma) = AM(\pi)$. This completes the proof.   □

Again we study the independence of the axioms.

Example 4. (A_1) excluded. Define the function F on G by $F(\pi) = Pl(\tilde{\pi})$, for all profiles π. It is straightforward to check that F satisfies (A), (C) and (A_3). Since $F(v, \tilde{v}) = \{\tilde{v}, v\} \neq V$, for any vertex v, the function F does not satisfy (A_1).

Example 5. (A_3) excluded. Define the function F on G by $F(\pi) = V$, for all profiles π. Obviously, F satisfies axioms (A), (C) and (A_1). Take any two adjacent vertices u and v in G. Then

$$F(u, v) = V \neq \{\tilde{u}, \tilde{v}\} = Pl(\tilde{u}, \tilde{v}). \tag{8}$$

So F does not satisfy (A_3).

Example 6. (C) excluded. Define the function F on G by
$(c1)$: $F(v, \tilde{v}) = V$, for all vertices v in V,
$(c2)$: $F(\pi) = Pl(\tilde{\pi})$, for all profiles π that are not a mating pair.
Clearly, F satisfies (A), (A_1) and (A_3). Take two vertices u and v that are not mates, and let $\pi = (u, \tilde{u}, v, \tilde{v})$. Then, by $(c2)$, we have

$$F(\pi) = Pl(\tilde{\pi}) = \{\tilde{u}, u, \tilde{v}, v\} \neq V = F(u, \tilde{u}) \cap F(v, \tilde{v}). \tag{9}$$

So F does not satisfy *Consistency*.

Also in this case we do not have an example yet that shows the independency of *Anonymity*. Again we leave this as an open problem.

5   Axiomatic Characterization of the Antimedian Function on Complete Graphs

In [13] an extensive study is made of location functions on the complete graph that satisfy the above axioms (A), (B) and (C). So we skip this case here.

As can be expected, due to its nice behavior on K_n, there is also a simple axiomatic characterization of the antimedian function on complete graphs. For the one vertex-graph see above. So let $n > 1$, and let $V = \{v_1, v_2, \ldots, v_n\}$ be the vertex set of K_n. Note that, in writing V in this way, we have chosen a *preferred ordering* of the vertices in V. Consistent with the approach above we seek axioms besides (A) and (C) that involve as few profiles as possible. Recall that $\{\pi\}$ is the set of vertices occurring in π. The set W_π is the set of vertices that occur the least in π. Note that, if $\{\pi\}$ is a proper subset of V, then

$$W_\pi = V - \{\pi\}. \tag{10}$$

Moreover, if π contains all vertices exactly m times, for some $m > 0$, then $W_\pi = V$. Obviously we have $AM(\pi) = W_\pi$. The two axioms we have in mind are

Completeness. $F(v_1, v_2, \ldots, v_n) = V$.

Complement. $F(x) = V - \{x\}$, for each $x \in V$.

In the *Completeness* axiom we have only one profile that contains each element of V once and the elements are in the preferred ordering. The *Complement* axiom involves only profiles containing one element.

Theorem 3. *Let F be a consensus function on K_n with $n > 1$. Then F is the antimedian function if and only if F satisfies (A), (C), Completeness and Complement.*

Proof. Clearly the antimedian function satisfies the four axioms.

Conversely, let F satisfy the four axioms. Take a profile $\pi = (x_1, x_2, \ldots, x_k)$. If $\{\pi\}$ is a proper subset of V, then we can write π as the concatenation of the singleton profiles $(x_1), (x_2), \ldots, (x_k)$. By *Complement*, the intersection of the sets $F(x_1), F(x_2), \ldots, F(x_k)$ equals $W_\pi = V - \{\pi\}$, and by (C) we are done. If all vertices of V occur exactly m times in π with $m > 0$, then, due to *Anonymity*, we can write $\pi = (v_1, v_2, \ldots, v_n)^m$, and we are done by (C) and *Completeness*.

Now let π be any other profile. Then there is a number $m > 0$ such that some but not all vertices occur exactly m times in π whereas the other vertices occur more than m times in π. Due to *Anonymity*, we can write $\pi = \pi'(v_1, v_2, \ldots, v_n)^m$, where π' is a profile such that $W_{\pi'}$ is the set of vertices that occur exactly m times in π. By the above observations and (C), we have $F(\pi) = F(\pi') \cap V = W_{\pi'} = AM(\pi)$.   □

Again in this case we do not yet have an example that shows whether *Anonymity* is independent from the other axioms. The examples below show the independence of the other three axioms.

Example 7. Complement excluded. Let F be defined by $F(\pi) = V$ for all profiles. Then it fails Complement but satisfies trivially the other axioms.

Example 8. (C) excluded. Let F be defined by
$(k1)$ $F(x) = V - \{x\}$, for any $x \in V$,
$(k2)$ $F(\pi) = V$, for any profile π of length at least 2.
Then F fails (C) but trivially satisfies the other axioms.

Example 9. Completeness excluded. Let F be defined by
$(k3)$ $F(\pi) = \{v_1\}$, for any π with $\{\pi\} = V$,
$(k4)$ $F(\pi) = V - \{\pi\}$, for any π with $\{\pi\} \neq V$.
Clearly F satisfies (A) and *Complement*. By $(k3)$ F fails *Completeness*. It remains to check *Consistency*. So let π and ρ be two profiles. If $\{\pi\} = V = \{\rho\}$, then

$$F(\pi) = F(\rho) = F(\pi\rho) = \{v_1\}, \tag{11}$$

and we are done. If $\{\pi\} = V$ and $\{\rho\} \neq V$, then $F(\pi) \cap F(\rho) \neq \emptyset$ only if ρ does not contain v_1. In this case it again follows that

$$F(\pi) = F(\pi) \cap F(\rho) = \{v_1\} = F(\pi\rho). \tag{12}$$

Finally, let $\{\pi\}$ and $\{\rho\}$ both be proper subsets of V. Then $F(\pi) = V - \{\pi\}$ and $F(\rho) = V - \{\rho\}$. These two sets have a non-empty intersection if and only if $\{\pi\} \cup \{\rho\} = \{\pi\rho\}$ is a proper subset of V. Again we have $F(\pi\rho) = F(\pi) \cap F(\rho)$.

Open Problem. (A) independent?

In this case we want to elaborate a little more on our trials to find an example. Note that for independence of (A) the ordering of the elements in a profile is essential. We want to split π into subprofiles to get a grip on $F(\pi)$. The only subprofiles of π that we consider are those containing consecutive elements of π. So, if we say that π' is a subprofile of π, then it is assumed that π' consists of consecutive elements of π. As above, a profile is of length at least 1, but for our purposes here, a subprofile now may be empty. We use the convention that ρ^m is the empty subprofile if $m = 0$. We set $\tau = (v_1, v_2, \ldots, v_n)$ to be the profile containing each vertex once and in the preferred ordering. Any profile π can be written as

$$\pi = \tau^{m_0} \pi_1 \tau^{m_1} \pi_2 \ \cdots \ \pi_{r-1} \tau^{m_{r-1}} \pi_r \tau^{m_r} \tag{13}$$

such that

(i) $\pi_1, \pi_2, \ldots, \pi_r$ are non-empty subprofiles of π that do *not* contain τ,
(ii) $m_1, m_2, \ldots, m_{r-1} > 0$ and $m_0, m_r \geq 0$.

We call this the *standard form* of π. If π does not contain τ, then we take $r = 1$ and $m_0 = m_1 = 0$. If $\pi = \tau^m$ for some $m > 0$, then we take $r = 0$ and $m_0 = m$. Note that, if $m_j > 0$ for some j, then $\{\pi\} = V$. Hence, if $\{\pi\} \neq V$, then π is the standard form of π.

We distinguish two types of profiles.

Type A $\pi = \tau^{m_0} \pi_1 \tau^{m_1} \pi_2 \ \cdots \ \pi_{r-1} \tau^{m_{r-1}} \pi_r \tau^{m_r}$ with $\cup_{1 \leq j \leq n} \{\pi_j\} \neq V$.
Type B $\pi = \tau^{m_0} \pi_1 \tau^{m_1} \pi_2 \ \cdots \ \pi_{r-1} \tau^{m_{r-1}} \pi_r \tau^{m_r}$ with $\cup_{1 \leq j \leq n} \{\pi_j\} = V$.

If π is of type A, then either $\pi = \tau^m$ for some $m > 0$, in which case

$$AM(\pi) = V, \tag{14}$$

or $\pi = \tau^{m_0} \pi_1 \tau^{m_1} \pi_2 \ \cdots \ \pi_{r-1} \tau^{m_{r-1}} \pi_r \tau^{m_r}$ for some $r \geq 1$, in which case

$$AM(\pi) = V - [\cup_{j=1}^{n} \{\pi_j\}] = \cap_{j=1}^{n} [V - \{\pi_j\}]. \tag{15}$$

Let F be any consensus function satisfying *Consistency, Completeness* and *Complement*. Then it follows that we have $F(\pi) = AM(\pi)$, for any profile π of type A. So the only way to differ from AM is on profiles of type B.

Let π and ρ be two profiles of type A. If the concatenated profile $\pi\rho$ is also of type A, then it is straightforward to check that (C) holds. If $\pi\rho$ is of type B, then it is straightforward to check that $F(\pi) \cap F(\rho) = \emptyset$, so this does not affect *Consistency*. Problems may arise when π or ρ are of type B. Let π and ρ be two profiles with $F(\pi) \cap F(\rho) \neq \emptyset$. We write both in standard form:

$$\pi = \tau^{m_0}\pi_1\tau^{m_1}\pi_2 \ \cdots \ \pi_{r-1}\tau^{m_{r-1}}\pi_r\tau^{m_r}, \tag{16}$$

$$\rho = \tau^{n_0}\rho_1\tau^{n_1}\rho_2 \ \cdots \ \rho_{s-1}\tau^{n_{s-1}}\rho_s\tau^{n_s}. \tag{17}$$

Now consider the case that $m_r = 0$ and $\pi_r = (v_1, v_2, \ldots, v_t)$ and $n_0 = 0$ and $\rho_1 = (v_{t+1}, \ldots, v_n)$, for some t with $1 \le t < n$. If we concatenate π and ρ and consider the standard form of $\pi\rho$, then $\pi_r\rho_1 = \tau$ and π_r and ρ_1 'disappear'. So they do not count when we want to determine the type of $\pi\rho$. So $\pi\rho$ might be of type A, whereas at least one of π and ρ is of type B. This makes it difficult to assign values to profiles of type B. On the other hand, one does not expect that we can deduce *Anonymity* from the other three axioms.

6 Concluding Remarks

The median and antimedian functions satisfy *Anonymity* and *Consistency* on any metric space.

On cocktail-party graphs we need two more axioms: in both cases axiom $(A1)$, for the median case also $(A2)$, and for the antimedian case also $(A3)$. All these axioms are natural and intuitively appealing. Except for *Anonymity*, we have shown independence of the axioms.

Also on the complete graphs we have a simple axiomatic characterization of the antimedian function. We need two more axioms here: *Completeness* and *Complement*. Again independence of (A) is an open problem.

References

1. Arrow, K.: Social Choice and Individual Values, 1st edn. Cowles Commission for Research in Economics - Monographs, vol. 12. Wiley, New York (1951)
2. Arrow, K.J., Sen, A.K., Suzumura, K. (eds.): Handbook of Social Choice and Welfare, vol. 1. North Holland, Amsterdam (2002)
3. Arrow, K.J., Sen, A.K., Suzumura, K. (eds.): Handbook of Social Choice and Welfare, vol. 1. North Holland, Amsterdam (2005)
4. Balakrishnan, K., Brešar, B., Changat, M., Klavžar, S., Imrich, W., Kovše, M., Subhamathi, A.R.: On the Remoteness Function in Median Graphs. Discrete Appl. Math. 157, 3679–3688 (2009)
5. Balakrishnan, K., Brešar, B., Changat, M., Klavžar, S., Kovše, M., Subhamathi, A.R.: Computing Median and Antimedian Sets in Median Graphs. Algorithmica 57, 207–216 (2010)
6. Balakrishnan, K., Brešar, B., Changat, M., Klavžar, S., Kovše, M., Subhamathi, A.R.: Simultaneous Embedding of Graphs as Median and Antimedian Subgraphs. Networks 56, 90–94 (2010)
7. Balakrishnan, K., Changat, M., Klavžar, S., Joseph, M., Peterin, I., Prasanth, G.N., Špacapan, S.: Antimedian Graphs. Australas. J. Combin. 41, 159–170 (2008)
8. Balakrishnan, K., Changat, M., Mulder, H.M., Subhamathi, A.R.: Axiomatic Characterization of the Antimedian Function on Paths and Hypercubes. Discrete Math. Algorithm. Appl. 04, 1250054, 20 pages (2012)

9. Deza, M., Laurent, M.: Geometry of Cuts and Metrics. Springer, Heidelberg (1997)
10. Holzman, R.: An Axiomatic Approach to Location on Networks. Math. Oper. Res. 15, 553–563 (1990)
11. Klavžar, S., Mulder, H.M.: Median Graphs- Characterizations, Location Theory and Related Structures. J. Combin. Math. Combin. Comput. 30, 103–127 (1999)
12. McMorris, F.R., Mulder, H.M., Novick, B., Powers, R.C.: Five Axioms for Location Functions on Median Graphs. To appear in Discrete Math. Algorithms Appl.
13. McMorris, F.R., Mulder, H.M., Novick, B., Powers, R.C., Vohra, R.V.: Axiomatic characterization of voting procedures on K_n (manuscript submitted)
14. McMorris, F.R., Mulder, H.M., Ortega, O.: Axiomatic Characterization of the Mean Function on Trees. Discrete Math. Algorithms Applications 2, 313–329 (2010)
15. McMorris, F.R., Mulder, H.M., Ortega, O.: Axiomatic Characterization of the ℓ_p-Function on Trees. Networks 60, 94–102 (2012)
16. McMorris, F.R., Mulder, H.M., Roberts, F.S.: The Median Procedure on Median Graphs. Discrete Appl. Math. 84, 165–181 (1998)
17. McMorris, F.R., Mulder, H.M., Vohra, R.V.: Axiomatic Characterization of Location Functions. In: Kaul, H., Mulder, H.M. (eds.) Advances in Interdisciplinary Applied Discrete Mathematics, Interdisciplinary Mathematical Sciences, vol. 11, pp. 71–91. World Scientific Publishing, Singapore (2010)
18. McMorris, F.R., Roberts, F.S., Wang, C.: The Center Function on Trees. Networks 38, 84–87 (2001)
19. Minieka, E.: Anticenters and Antimedians of a Network. Networks 13, 35–364 (1983)
20. Mulder, H.M.: The Interval Function of a Graph. Math. Centre Tracts, vol. 132. Math. Centre, Amsterdam (1980)
21. Mulder, H.M.: Median Graphs. A Structure Theory. In: Kaul, H., Mulder, H.M. (eds.) Advances in Interdisciplinary Applied Discrete Mathematics. Interdisciplinary Mathematical Sciences, vol. 11, pp. 93–125. World Scientific Publishing, Singapore (2010)
22. Mulder, H.M., Novick, B.A.: An Axiomization of the Median Function on the n-Cube. Discrete Appl. Math. 159, 139–144 (2011)
23. Mulder, H.M., Novick, B.A.: A Tight Axiomatization of the Median Function on Median Graphs. Discrete Appl. Math. 161, 838–846 (2013)
24. Mulder, H.M., Reid, K.B., Pelsmajer, M.J.: Axiomatization of the Center Function on Trees. Australasian J. Combin. 41, 223–226 (2008)
25. Rao, S.B., Vijayakumar, A.: On the Median and the Antimedian of a Cograph. Int. J. Pure Appl. Math. 46, 703–710 (2008)
26. Vohra, R.: An Axiomatic Characterization of Some Locations in Trees. European J. Operational Research 90, 78–84 (1996)
27. Shilpa, M., Changat, M., Narasimha-Shenoi, P.G.: Axiomizatic Characterization of the Center Function on Some Graph Classes (manuscript submitted)

Tree Path Labeling of Hypergraphs – A Generalization of the Consecutive Ones Property

N.S. Narayanaswamy and Anju Srinivasan

Indian Institute of Technology Madras, Chennai - 600036, India
{swamy,asz}@cse.iitm.ac.in

Abstract. Given a set system $\mathcal{F} \subseteq (2^U \setminus \emptyset)$ of a finite set U of cardinality n and a tree T of size n, does there exist at least one bijection $\phi : U \to V(T)$ such that for each $S \in \mathcal{F}$, the set $\{\phi(x) \mid x \in S\}$ is the vertex set of a path in T? Our main result is that the existence of such a bijection from U to $V(T)$ is equivalent to the existence of a function ℓ from $\mathcal{F}$ to the set of all paths in T such that for any *three*, not necessarily distinct, $S_1, S_2, S_3 \in \mathcal{F}$, $|S_1 \cap S_2 \cap S_3| = |\ell(S_1) \cap \ell(S_2) \cap \ell(S_3)|$. ℓ is referred to as a *tree path labeling* of $\mathcal{F}$.

Keywords: consecutive ones property, algorithmic graph theory, hypergraph isomorphism, interval labeling.

1 Introduction

There is an extensive body of results on matrices with the consecutive ones property (COP), comprising of combinatorial characterizations, efficient data structures, and linear time recognition algorithms, see [Rob51, FG65, Tuc72, BL76, Hsu01, Hsu02, McC04, MM96, NS09]. It is also widely applied, in archeology, scheduling, and graph theory [Kou77, ABH98, HT02, Gol04, HL06]. The first polynomial time algorithm was in a 1965 paper where Fulkerson and Gross [FG65] proved the following result:

Theorem 1.1. *Let A and B be 0-1 matrices satisfying $A^t A = B^t B$. Then either both A and B have the consecutive ones property or neither does. Moreover, if A and B have the same number of rows and A has the consecutive ones property, then there is a permutation matrix P such that $B = PA$.*

More than four decades later, we revisited this result, in [NS09], by considering intervals assigned to sets, and focussing on Intersection Cardinality Preserving Interval Assignments (ICPIAs). We considered a natural set system $\mathcal{F}$ on a universe U of cardinality n associated with a 0-1 matrix with n columns. The universe U is set of column indices in the given binary matrix. Each element of the set system $\mathcal{F}$ corresponds to a row in the matrix, and the associated set is the set of column indices with a 1 in that row. The second entity considered was a function ℓ defined on $\mathcal{F}$ such that for each $S \in \mathcal{F}$, $\ell(S)$ is a path in the tree P_n,

S. Ganguly and R. Krishnamurti (Eds.): CALDAM 2015, LNCS 8959, pp. 150–156, 2015.

where P_n is the path on n vertices, $\{1, \ldots, n\}$, with 1 and n as the only two leaves. ℓ is defined to be an ICPIA if for every set $S \in \mathcal{F}$, $|S| = |\ell(S)|$, and for any two $S_1, S_2 \in \mathcal{F}$, $|S_1 \cap S_2| = |\ell(S_1) \cap \ell(S_2)|$. Clearly, if the binary matrix has the COP witnessed by a permutation ϕ of the column indices, then each set S in $\mathcal{F}$ has a natural associated interval (a set of consecutive integers), $\ell(S) = \{\phi(x) | x \in S\}$. Clearly, ℓ is an ICPIA obtained from ϕ. This natural necessary condition was shown to be sufficient in [NS09] by constructing a permutation witnessing the COP from a ICPIA ℓ for $\mathcal{F}$. Phrased in the terminology of Berge and Rado [BR72], the hypergraph $\mathcal{F}$ is isomorphic to a hypergraph whose edges are paths in P_n if and only if there exists an edge permutation ℓ between $\mathcal{F}$ and the set of paths in P_n such that for any two, not necessarily distinct, $S_1, S_2 \in \mathcal{F}$, $|S_1 \cap S_2| = |\ell(S_1) \cap \ell(S_2)|$. Indeed, our result is a rediscovery of the theorem due to Fulkerson and Gross [FG65]. Results by Fournier [Fou80] generalized both the results of Berge and Rado [BR72], and [FG65] by characterizing isomorphism between two hypergraphs by means of equicardinality of certain edge intersections and the exclusion of certain pairs of subhypergraphs.

Our Result. We consider the following generalization of Theorem 1.1 and of the ICPIA [NS09]- given a set system $\mathcal{F}$ of m elements over a universe U of n elements, and a tree T, does there exist a bijection ϕ from U to the vertices of T such that for each $S \in \mathcal{F}$, the set $\{\phi(x) | x \in S\}$ is the vertex set of a path in T? Our result is that such a bijection ϕ exists if and only if there is a tree path labeling ℓ of the elements of $\mathcal{F}$ such that for any *three*, not necessarily distinct, $S_1, S_2, S_3 \in \mathcal{F}$, $|S_1 \cap S_2 \cap S_3| = |\ell(S_1) \cap \ell(S_2) \cap \ell(S_3)|$. We refer to such a path labeling as an *Intersection Cardinality Preserving Path Labeling* (ICPPL). Clearly, given a bijection ϕ such that $\{\phi(x) | x \in S\}$ is the vertex set of a path in T, we get an ICPPL ℓ in which for each $S \in \mathcal{F}$, $\ell(S) = \{\phi(x) | x \in S\}$. We address the sufficiency condition in this paper: Given a path labeling ℓ of $\mathcal{F}$ from a tree T, does there exist a bijection $\phi : U \rightarrow V(T)$ such that $\ell(S) = \{\phi(x) | x \in S\}$? A given path labeling is referred to as a *feasible tree path labeling* if such a ϕ exists. We show that a given tree path labeling is feasible if and only if it is an ICPPL. This characterization is proved constructively by obtaining a relevant bijection ϕ from a given ICPPL in Section 2. For the sake of clarity, we refer to the ordered pair $(\mathcal{F}, \ell)$ as a path labeling, and we mean that it is a path labeling using paths from a fixed tree T. When $\mathcal{F}$ is unambiguous, we refer to ℓ as a path labeling to refer to $(\mathcal{F}, \ell)$. Further, the term labeling is appropriate as each set in $\mathcal{F}$ is assigned a path label by ℓ. Finally, we assume that the support of $\mathcal{F}$ has the same cardinality as the elements of $V(T)$ that are in some image under ℓ. Mainly, this leaves out uninteresting situations where some element of U does not occur in the support of $\mathcal{F}$, and some elements of $V(T)$ do not occur in any image under ℓ.

2   Characterization of Feasible Tree Path Labelings

Consider a tree path labeling $(\mathcal{F}, \ell)$ on a tree T. Let $(\mathcal{F}, \ell)$ be an ICPPL. If T is a path, then $(\mathcal{F}, \ell)$ is referred to as an ICPIA. The following two lemmas are proved in [NS09]. The first one shows that the preservation of pairwise intersection cardinality in a path labeling when T is a path is sufficient to preserve three way intersection cardinality.

Lemma 2.1. *Let T be a path, S_1, S_2, S_3 be 3 sets, and T_1, T_2, T_3 be paths from T, such that $|S_i \cap S_j| = |T_i \cap T_j|, 1 \leq i, j \leq 3$. Then, $|S_1 \cap S_2 \cap S_3| = |T_1 \cap T_2 \cap T_3|$.*

The following lemma shows that a path labeling that preserves pairwise intersection cardinalities is feasible.

Lemma 2.2. *A path labeling $(\mathcal{F}, \ell)$ on a path is feasible iff it is an ICPIA.*

The following lemmas are useful in subsequent inductive arguments.

Lemma 2.3. *If $(\mathcal{F}, \ell)$ is an ICPPL, and $S_1, S_2, S_3 \in \mathcal{F}$, then $|S_1 \cap (S_2 \setminus S_3)| = |\ell(S_1) \cap (\ell(S_2) \setminus \ell(S_3))|$.*

Proof. Let $P_i = \ell(S_i)$, for all $1 \leq i \leq 3$. $|S_1 \cap (S_2 \setminus S_3)| = |(S_1 \cap S_2) \setminus S_3| = |S_1 \cap S_2| - |S_1 \cap S_2 \cap S_3|$. From the definition of an ICPPL, $|S_1 \cap S_2| - |S_1 \cap S_2 \cap S_3| = |P_1 \cap P_2| - |P_1 \cap P_2 \cap P_3| = |(P_1 \cap P_2) \setminus P_3| = |P_1 \cap (P_2 \setminus P_3)|$. Thus lemma is proved. □

Lemma 2.4. *Let $(\mathcal{F}, \ell)$ be an ICPPL, let $\ell(S_i) = P_i, 1 \leq i \leq 4$. Then, $|\cap_{i=1}^{4} S_i| = |\cap_{i=1}^{4} P_i|$.*

Proof. Consider the sets $S_2 \cap S_1$, $S_3 \cap S_1$, and $S_4 \cap S_1$, and let $P_2 \cap P_1$, $P_3 \cap P_1$, and $P_4 \cap P_1$ be their images, respectively. The intersection of two paths in a tree is a path. Further, since ℓ is an ICPPL, it follows that path labeling to $S_2 \cap S_1$, $S_3 \cap S_1$, and $S_4 \cap S_1$ preserves pairwise intersection cardinalities. Also, it is clear that the 3 sets are assigned paths contained within the path P_1. Now by applying Lemma 2.1, it follows that $|\cap_{i=1}^{4} S_i| = |\cap_{i=1}^{4} P_i|$. □

Corollary 2.5. *In an ICPPL let S_1, S_2, S_3, S_4 be sets assigned to paths P_1, P_2, P_3, P_4, respectively. Further, let $P_1 \cap P_2, P_1 \setminus P_2, P_2 \setminus P_1$ be paths. Then the intersection cardinalities $|(S_1 \setminus S_2) \cap S_3 \cap S_4| = |(P_1 \setminus P_2) \cap P_3 \cap P_4|$, and $|(S_2 \setminus S_1) \cap S_3 \cap S_4| = |(P_2 \setminus P_1) \cap P_3 \cap P_4|$ are preserved.*

Proof. It is clear that $|(P_1 \setminus P_2) \cap P_3 \cap P_4| = |\cap_{i=1}^{4} P_i| - |P_2 \cap P_3 \cap P_4|$. Since we have an ICPPL and from Lemma 2.4, the corollary follows.

Lemma 2.6. *Let $(\mathcal{F}, \ell)$ be an ICPPL, and for an $S \in \mathcal{F}$, let $\ell(S) = P$. Let S_{priv} and P_{priv} be the set of elements of S and P such that for all $S' \in \mathcal{F}$ such that $S' \neq S$, $|S' \cap S_{priv}| = 0$ and $|\ell(S') \cap P_{priv}| = 0$, respectively. Then, $|S_{priv}| = |P_{priv}|$.*

Proof. Let S_{two} and P_{two} be the elements of S and P which are in at least one *more* set and path, respectively. Consider the set of paths $\{P_i \cap P \mid 1 \le i \le n\}$ and the sets $\{S_i \cap S \mid 1 \le i \le n\}$. Since ℓ is an ICPPL, it follows that the mapping ℓ', for each $1 \le i \le m$, $\ell'(S_i \cap S) = P_i \cap P$ is an ICPIA in P. Consequently, from Lemma 2.2, it follows that ℓ' is feasible, that is there is a bijection ϕ from S to P such that for each $1 \le i \le m$, the image of $S_i \cap S$ under ϕ is $P_i \cap P$. Further, in ϕ, the elements of S_{two} are mapped to the elements of P_{two}. The proof is of this claim is by contradiction. Let us assume that there is an $x \in S_{two}$ is mapped to a $v \in P_{priv}$ under ϕ. Since $x \in S_{two}$, let S_i be another set that contains x, other than S. Then it means that $\phi(S_i \cap S)$ which is same as $\ell'(S_i \cap S) = P_i \cap P$ contains v. This implies that $v \in P_i \cap P$, contradiction to the fact that $v \in P_{priv}$. Therefore, under the bijection ϕ, the image of the set S_{two} is the set P_{two}. Conseqently, it follows that $|S_{two}| = |P_{two}|$. This shows that $|S_{priv}| = |S \setminus S_{two}| = |P \setminus P_{two}| = |P_{priv}|$. Hence the lemma.

Lemma 2.7. *Let $(\mathcal{F}, \ell)$ be an ICPPL. Let $S_1, S_2 \in \mathcal{F}$ be two sets such that $\ell(S_1)$ and $\ell(S_2)$ share a common leaf. Consider $\mathcal{F}' = \mathcal{F} \setminus \{S_1, S_2\} \cup \{S_1 \cap S_2, S_1 \setminus S_2, S_2 \setminus S_1\}$. Define a labeling ℓ' as follows: $\ell'(S_1 \cap S_2) = \ell(S_1) \cap \ell(S_2), \ell'(S_1 \setminus S_2) = \ell(S_1) \setminus \ell(S_2), \ell'(S_2 \setminus S_1) = \ell(S_2) \setminus \ell(S_1)$, for all other $S \in \mathcal{F}'$, $\ell'(S) = \ell(S)$. The path labeling $(\mathcal{F}', \ell')$ is an ICPPL.*

Proof. We prove the following invariants:

I $\ell'(R)$ is a path in T,   for all $R \in \mathcal{F}'$

II $|R| = |\ell'(R)|$,   for all $R \in \mathcal{F}'$

III $|R \cap R'| = |\ell'(R) \cap \ell(R')|$,   for all $R, R' \in \mathcal{F}'$

IV $|R \cap R' \cap R''| = |\ell'(R) \cap \ell'(R') \cap \ell'(R'')|$,   for all $R, R', R'' \in \mathcal{F}'$

We use the term "new sets" will refer to the $S_1 \cap S_2, S_1 \setminus S_2, S_2 \setminus S_1$ and its images in ℓ'. Let us consider the possible cases for each of the above invariants for $\mathcal{F}'$:

✳ *Invariant I/II*

    a. R *is not a new set.* It is in $\mathcal{F}$. Therefore, by the definition of ℓ' and the fact that ℓ is an ICPPL, it follows that $\ell'(R) = \ell(R)$ and $|\ell'(R)| = |R|$.

    b. R *is a new set.* If R is in $\mathcal{F}'$ and not in $\mathcal{F}$, then it must be one of the new sets added in $\mathcal{F}'$. In this case, it is clear that for each new set, the image under ℓ' is a path since by definition the chosen sets S_1, S_2 are from $\mathcal{F}$, and the paths labels $\ell(S_1), \ell(S_2)$ have a common leaf. Thus invariant I is proved. Moreover, due to the fact that ℓ is an ICPPL and the definition of ℓ' in terms of ℓ, invariant II is indeed true in $\mathcal{F}'$ for any of the new sets. If $R = S_1 \cap S_2, |R| = |S_1 \cap S_2| = |\ell(S_1) \cap \ell(S_2)| = |\ell'(S_1 \cap S_2)| = |\ell'(R)|$. If $R = S_1 \setminus S_2, |R| = |S_1 \setminus S_2| = |S_1| - |S_1 \cap S_2| = |\ell(S_1)| - |\ell(S_1) \cap \ell(S_2)| = |\ell(S_1) \setminus \ell(S_2)| = |\ell'(S_1 \setminus S_2)| = |\ell'(R)|$. Similarly if $R = S_2 \setminus S_1$.

✳ *Invariant III*

    a. R *and* R' *are not new sets.* Again, here $\ell'(R) = \ell(R)$ and $\ell'(R') = \ell(R')$ by definition of ℓ', and since ℓ is an ICPPL, the invariant is true.

b. *Only one, say R, is a new set.* Due to the fact that ℓ is an ICPPL, Lemma 2.3 and definition of ℓ', it follows that invariant III is true no matter which of the new sets R is equal to. If $R = S_1 \cap S_2$, $|R \cap R'| = |S_1 \cap S_2 \cap R'| = |\ell(S_1) \cap \ell(S_2) \cap \ell(R')| = |\ell'(S_1 \cap S_2) \cap \ell'(R')| = |\ell'(R) \cap \ell'(R')|$. If $R = S_1 \setminus S_2$, $|R \cap R'| = |(S_1 \setminus S_2) \cap R'| = |(\ell(S_1) \setminus \ell(S_2)) \cap \ell(R')| = |\ell'(S_1 \cap S_2) \cap \ell'(R')| = |\ell'(R) \cap \ell'(R')|$. Similarly, if $R = S_2 \setminus S_1$.

c. *R and R' are new sets.* By definition, the new sets and their path images under path label ℓ' are disjoint so $|R \cap R'| = |\ell'(R) \cap \ell'(R)| = 0$. Thus this case is proved.

⁕ *Invariant IV*

a. *R, R', and , R'' are not new sets.* The invariant is true because $\ell'(R) = \ell(R), \ell'(R') = \ell(R'), \ell'(R'') = \ell(R'')$, and based on the fact that ℓ is an ICPPL.

b. *Only one, say R, is a new set.* There are three possibilities for R: $R = S_1 \cap S_2$, $R = S_1 \setminus S_2$, $R = S_2 \setminus S_1$. Let the other two sets be S_3 and S_4. Now $|R \cap S_3 \cap S_4| = |S_1 \cap S_2 \cap S_3 \cap S_4|$. Since ℓ is an ICPPL, Lemma 2.4 is applicable and it follows that $|S_1 \cap S_2 \cap S_3 \cap S_4| = |\ell(S_1) \cap \ell(S_2) \cap \ell(S_3) \cap \ell(S_4)|$. By definition of ℓ' it follows that $\ell(S_1) \cap \ell(S_2) = \ell'(S_1 \cap S_2)$, and $\ell'(S_3) = \ell(S_3), \ell'(S_4) = \ell(S_4)$. Therefore, it follows that $|R \cap S_3 \cap S_4| = |\ell'(S_1 \cap S_2) \cap \ell'(S_3) \cap \ell'(S_4)| = |\ell'(R) \cap \ell'(S_3) \cap \ell'(S_4)|$. For the other two cases, the invariant is ensured due to Corollary 2.5.

c. *At least two, say R, R', are new sets.* By definition, the new sets and their path images in path label ℓ' are disjoint so $|R \cap R'| = |\ell'(R) \cap \ell'(R)| = 0 = |R \cap R' \cap R''| = |\ell'(R) \cap \ell'(R) \cap \ell'(R'')|$. Thus the invariant is maintained in this case.

All the above cases prove the lemma. □

Lemma 2.8. *Let $(\mathcal{F}, \ell)$ be an ICPPL in which there is a leaf $v \in T$ such that there is a unique $S_1 \in \mathcal{F}$ such that $v \in \ell(S_1)$. Further let $X = S_1 \setminus \bigcup_{S \in \mathcal{F}, S \neq S_1} S$, and let $x \in X$. Define ℓ' and $\mathcal{F}'$ as follows: For all $S \in \mathcal{F}$ such that $S \neq S_1$, $\ell'(S) = \ell(S)$, $\ell'(\{x\}) = \{v\}$, $\ell'(S_1 \setminus \{x\}) = \ell(S_1) \setminus \{v\}$, and $\mathcal{F}' = (\mathcal{F} \setminus \{S_1\}) \cup \{\{x\}, S_1 \setminus \{x\}\}$. $(\mathcal{F}', \ell')$ is an ICPPL.*

Proof. We show that all the four invariants given in Lemma 2.7 hold for $(\mathcal{F}', \ell')$. Invariant I is obviously true, since the image of a set under ℓ' is a path. The set X is non-empty, and this follows from Lemma 2.6. We remove exactly one element x from one set S in $\mathcal{F}$ and exactly one vertex v which is a leaf from one path $\ell(S)$ in T. This is because x is exclusive to S and v is exclusive to $\ell(S)$. Due to this fact, it is clear that the intersection cardinality equations do not change, i.e., invariants II, III, IV remain true. □

The following lemma formalizes the characterization of a feasible tree path labeling.

Lemma 2.9. *Let $(\mathcal{F}, \ell)$ be an ICPPL of sets in $\mathcal{F}$ using paths from T. Then there exists a bijection $\phi : U \to V(T)$ such that for each $S \in \mathcal{F}$, $\ell(S) = \{\phi(x) | x \in S\}$.*

Proof. The proof is by induction on the cardinality of U. When U has a single element, the base case, ϕ is trivial. Let the statement be true for all ICPPL $(\mathcal{F}, \ell)$ for which $|U| = k \geq 1$. Now, we prove the statement for an ICPPL $(\mathcal{F}, \ell)$ where $|U| = k + 1$. First, iteratively apply Lemma 2.7. When Lemma 2.8 is applicable, let $x \in U$ and $v \in V(T)$ be a leaf that is considered in the lemma. Define $\phi(x) = l$, and let T' be the resulting tree after removing v from T. Let $U' = U \setminus \{x\}$, $(\mathcal{F}', \ell')$ be the ICPPL in the tree T', where $\mathcal{F}' = \{S \setminus \{x\} \mid S \in \mathcal{F}\}$, and for each $S \in \mathcal{F}$, $\ell'(S) = \ell(S) \setminus \{v\}$. Clearly, $|U'| = k$, and by the induction hypothesis let ϕ' be a bijection from U' to $V(T')$ such that for each $S \in \mathcal{F}'$, $\ell(S) = \{\phi'(x) | x \in S\}$. Consider ϕ from U to $V(T)$ by defining it as follows- $\phi(x) = v$ and for all other $y \in U$, $\phi(y) = \phi'(y)$. Clearly, ϕ is a bijection from $\mathcal{F}$ to $V(T)$ such that for each $S \in \mathcal{F}$, $\ell(S) = \{\phi(x) | x \in S\}$. Hence the lemma. $\square$

Theorem 2.10. *A path labeling $(\mathcal{F}, \ell)$ on tree T is a feasible tree path labeling iff it is an ICPPL.*

Proof. From Lemma 2.9, we know that if $(\mathcal{F}, \ell)$ is an ICPPL then $(\mathcal{F}, \ell)$ is feasible. Now consider the case where $(\mathcal{F}, \ell)$ is feasible tree path labeling. Consequently, there exists a bijection $\phi : \mathcal{F} \to V(T)$ such that for each $S \in \mathcal{F}$, $\ell(S) = \{\phi(x) | x \in S\}$. Since ϕ is a bijection, it follows that for each $k \geq 1$, and for all sets $S_1, S_2, \ldots, S_k \subseteq U$, $|\bigcap_{i=1}^{k} S_i| = |\bigcap_{i=1}^{k} \{\phi(x) | x \in S_i\}|$. In particular, it follows that ℓ is an ICPPL. $\square$

References

[ABH98] Atkins, J.E., Boman, E.G., Hendrickson, B.: A spectral algorithm for seriation and the consecutive ones problem. SICOMP: SIAM Journal on Computing 28 (1998)

[BL76] Booth, K.S., Lueker, G.S.: Testing for the consecutive ones property, interval graphs, and graph planarity using PQ-tree algorithms. Journal of Computer and System Sciences 13(3), 335–379 (1976)

[BR72] Berge, C., Rado, R.: Note on isomorphic hypergraphs and whitney's theorem to families of sets. Journal of Combinatorial Theory (B) 13, 226–241 (1972)

[FG65] Fulkerson, D.R., Gross, O.A.: Incidence matrices and interval graphs. Pac. J. Math. 15, 835–855 (1965)

[Fou80] Fournier, J.C.: lsomorphismes dhypergraphes par intersections equicardinales daretes et configurations exclues. Journal of Combinatorial Theory (B) 29, 321–327 (1980)

[Gol04] Golumbic, M.C.: Algorithmic graph theory and perfect graphs, 2nd edn. Annals of Discrete Mathematics, vol. 57. Elsevier Science B.V. (2004)

[HL06] Hochbaum, D.S., Levin, A.: Cyclical scheduling and multi-shift scheduling: Complexity and approximation algorithms. Discrete Optimization 3(4), 327–340 (2006)

[Hsu01] Hsu, W.-L.: PC-Trees vs. PQ-Trees. In: Wang, J. (ed.) COCOON 2001. LNCS, vol. 2108, pp. 207–217. Springer, Heidelberg (2001)

[Hsu02] Hsu, W.-L.: A simple test for the consecutive ones property. J. Algorithms 43(1), 1–16 (2002)

[HT02]   Hochbaum, D., Tucker, A.: Minimax problems with bitonic matrices. NET-
         WORKS: Networks: An International Journal 40 (2002)
[Kou77] Kou, L.T.: Polynomial complete consecutive information retrieval problems.
         SIAM Journal on Computing 6(1), 67–75 (1977)
[McC04] McConnell, R.M.: A certifying algorithm for the consecutive-ones property.
         In: SODA: ACM-SIAM Symposium on Discrete Algorithms (A Conference
         on Theoretical and Experimental Analysis of Discrete Algorithms) (2004)
[MM96]   Meidanis, J., Munuera, E.G.: A theory for the consecutive ones property.
         In: Proceedings of WSP 1996 - Third South American Workshop on String
         Processing, pp. 194–202 (1996)
[NS09]   Narayanaswamy, N.S., Subashini, R.: A new characterization of matrices
         with the consecutive ones property. Discrete Applied Mathematics 157(18),
         3721–3727 (2009)
[Rob51] Robinson, W.S.: A method for chronologically ordering archaeological de-
         posits. American Antiquity 16(4), 293–301 (1951)
[Tuc72] Tucker, A.: A structure theorem for the consecutive 1's property. J.
         Comb. Theory Series B 12, 153–162 (1972)

On a Special Class of Boxicity 2 Graphs

Sujoy Kumar Bhore, Dibyayan Chakraborty, Sandip Das, and Sagnik Sen

Indian Statistical Institute, Kolkata, India

Abstract. We define and study a class of graphs, called 2-stab interval graphs (2SIG), with boxicity 2 which properly contains the class of interval graphs. A 2SIG is an axes-parallel rectangle intersection graph where the rectangles have unit height (that is, length of the side parallel to Y-axis) and intersects either of the two fixed lines, parallel to the X-axis, distance $1 + \epsilon$ ($0 < \epsilon < 1$) apart. Intuitively, 2SIG is a graph obtained by putting some edges between two interval graphs in a particular rule. It turns out that for these kind of graphs, the chromatic number of any of its induced subgraphs is bounded by twice of its (induced subgraph) clique number. This shows that the graph, even though not perfect, is not very far from it. Then we prove similar results for some subclasses of 2SIG and provide efficient algorithm for finding their clique number. We provide a matrix characterization for a subclass of 2SIG graph.

1 Introduction

An *intersection graph* is a graph whose vertices are represented by sets and two vertices are adjacent if their corresponding sets have non-empty intersection. A graph with *boxicity k* is an intersection graph of axes-parallel k dimensional rectangles. The class of graphs with boxicity 1 is better known as *interval graphs* (intersection graph of connected real intervals) while the class of graphs with boxicity 2 is better known as *rectangle intersection graphs* (intersection of axes-parallel rectangles on a plane).

It is known that several questions (for example, recognition, determining clique number, determining chromatic number) that are NP-hard in general becomes polynomial time solvable when restricted to the class of interval graphs while they remain NP-hard for the family of graphs with boxicity k (for $k \geq 2$). The reason for this dichotomy is probably because interval graphs are *perfect* (defined in Section 2) while boxicity 2 graphs are far from being perfect (those questions are polynomial time solvable for perfect graphs as well).

Naturally we are interested in exploring the objects that lie in between, that is, the proper subclasses of graphs with boxicity 2 that contains all interval graphs. Several such graph classes have been defined and studied [11] [6] [2]. In this article, we too define such a graph class and study its different aspects. But we also keep in mind that 'perfectness' is probably the key word here. Therefore, it is not surprising that a well known perfect graph motivates the definition of our graph class.

A *split graph* is a graph in which the vertices can be pertitioned into a clique and an independent set. Splits graphs are well known examples of perfect graphs.

S. Ganguly and R. Krishnamurti (Eds.): CALDAM 2015, LNCS 8959, pp. 157–168, 2015.

Note that a complete graph and an independent set are the two extreme trivial examples of perfect graphs. So when we put edges between these two types of perfect graphs, that is, complete graph and an independent set, then what we obtain is again perfect.

Motivated by this example, we wondered what would happen if we put edges between other kinds of perfect graphs? We use this idea to naturally define a class of intersection graphs containing all interval graphs with boxicity two. We take two interval graphs and put edges in between (following a particular rule) to obtain a class of intersection graphs, not perfect, which have certain property which enables us to call them "nearly perfect" [5]. That is, the chromatic number of each induced subgraph is bounded by a function of its clique number; a linear function in our case.

Let $y = 1$ be the *lower stab line* and $y = 2 + \epsilon$ be the *upper stab line* where $\epsilon \in (0, 1)$ is a constant. Now consider axes-parallel rectangles with unit height (length of the side parallel to Y-axis) that intersects one of the stab lines. A *2-stab interval graph (2SIG)* is a graph G that can be represented as an intersection graph of such rectangles. Such a representation $R(G)$ of G is called a *2-stab* representation (for example, see Fig. 1). A 2SIG may have more than one 2-stab representation.

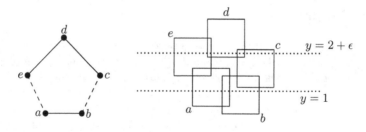

Fig. 1. A representation (left) of a $2SIG$ graph (right)

Notice that, given a representation $R(G)$ of G, each such rectangle intersects exactly one stab line partitioning the vertex set $V(G)$ in two disjoint parts, the *lower partition* V_1 (vertices with corresponding rectangles intersecting the lower stab line) and the *upper partition* V_2 (vertices with corresponding rectangles intersecting the upper stab line). Observe that such a vertex partition depnds on the representation and is not unique. In the remainder of the article, whenever we speak about a 2SIG with a vertex partition $V(G) = V_1 \sqcup V_2$ we will mean the partitions are lower and upper partition due to a representation.

Also note that the induced subgraphs $G[V_1]$ and $G[V_2]$ are interval graphs with intervals corresponding to the projection of their rectangles on X-axis. Hence, indeed, a 2SIG is obtained by putting some edges between two different interval graphs. Also, observe that the definition of 2SIG does not depend on the specific value of the constant ϵ as long as it belongs to the interval $(0, 1)$.

The article is organized in the following manner. In Section 2 we present the necessary definitions and notations. We study the clique number and the chromatic number of 2SIG in Section 3 and justify our claim that 2SIG and some of its subclasses are "nearly perfect" even though not perfect. We provide a matrix characterization for a subclass of 2SIG graph in Section 4. Finally, we conclude the article in Section 5.

2   Preliminaries

The *clique number* $\omega(G)$ of a graph G is the *order* (number of vertices) of the biggest complete subgraph of G. A *k-coloring* of a graph G is an assignment of k colors to the vertices of G such that adjacent vertices receive different colors. The *chromatic number* $\chi(G)$ of a graph G is the minimum k such that G admits a k-coloring. A graph G is *perfect* if $\omega(H) = \chi(H)$ for all induced subgraph H of G.

A graph G is *χ-bounded* if $\chi(H) \leq f(\omega(H))$ for all induced subgraph H of G where f is a bounded integer-valued function [4]. This is what we meant when we used the informal term "nearly perfect".

Recall the definition of 2-stab interval graphs from the previous section. Now by putting more restrictions on our definition of 2SIG we obtain two other interesting subclasses of 2SIG. A *2-stab unit interval graph (2SUIG)* is a 2SIG with a representation where each rectangle is a unit square. The corresponding representation is a *2SUIG representation*. A *2-stab independent interval graph (2SIIG)* is a 2SIG with a representation where the upper partition induces an independent set. The corresponding representation is a *2SIIG representation*.

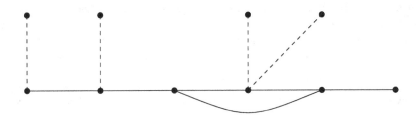

Fig. 2. Example of a bridge triangle free 2SUIG which is also a 2SIIG

Let us fix a representation $R(G)$ of a 2-stab interval graph G with corresponding lower and upper partitions V_1 and V_2, respectively. Then the set of *bridge edges* E_B is the set of edges (denoted by "dashed" edges in the figures) between the vertices of V_1 and V_2 while the set *bridge vertices* V_B is the set of vertices incident to bridge edges (see Fig. 1). For some $v \in V(G)$, the set of *bridge neighbors* $N_B(v)$ is the set of all vertices adjacent to v by a bridge edge. A *bridge triangle* is a triangle (induced K_3) in which exactly two of its edges are bridge edges (note that, no triangle of G can have exactly one or three edges

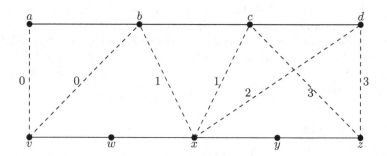

Fig. 3. Example of a 2*SUIG* graph with labeling of bridge edges

from E_B). A *bridge triangle free 2SUIG* is a graph with at least one 2SUIG representation without any bridge triangle (see Fig. 2).

An *orientation* $\overrightarrow{G}$ of a graph G is obtained by replacing its edges with *arcs* (ordered pair of vertices). An orientation $\overrightarrow{G}$ of G is a *transitive orientation* if for each pair of arcs (a,b) and (b,c) we have the arc (a,c) in $\overrightarrow{G}$. We know that the complement of an interval graph admits a transitive orientation [9]. Let $\overrightarrow{I^{\bar{c}}}$ be a transitive orientation of the complement graph of an interval graph I. Let $[u \rightsquigarrow v]$ denote a longest directed path (not necessarily unique) in $\overrightarrow{I^{\bar{c}}}$ from u to v and its length is denoted by l_{uv}.

3   Clique Number and Chromatic Number

Given a 2-stab interval graph G with a vertex partition we tried to figure out what its clique number and chromatic number might be and ended up proving the following bounds. But the drawback of our result is that it does not help us develop any algorithm to compute the exact clique number or chromatic number of G.

Observation 1. *Let* $G = (V_1 \sqcup V_2, E)$ *be a 2SIG graph with a given vertex partition.*

(i) *Then* $max\{\omega(G[V_1]), \omega(G[V_2])\} \leq \omega(G) \leq \omega(G[V_1]) + \omega(G[V_2])$.
(ii) *Then* $max\{\chi(G[V_1]), \chi(G[V_2])\} \leq \chi(G) \leq \chi(G[V_1]) + \chi(G[V_2])$.

Let H be a 2SIG with no bridge edges. Then for H both lower bounds of Observation 1 are tight. Now note that even a complete graph with any vertex partition admits a 2SIG representation. In that case, both the upper bounds of Observation 1 are tight. As the clique number and the chromatic number of an interval graph can be computed in linear time, given a 2SIG with a vertex partition, the lower and upper bounds of Observation 1 can be obtained in linear time as well. As any induced subgraph of a 2SIG is again a 2SIG we have the following result as a direct corollary of the above theorem.

Corollary 1. *Given any 2SIG graph G we have $\chi(H) \leq 2\omega(H)$ for all induced subgraph H of G.*

Proof. Let G be a 2SIG with a vertex partition $V(G) = V_1 \sqcup V_2$. As $G[V_i]$ is an interval graph we have $\omega(G[V_i]) = \chi(G[V_i])$ for all $i \in \{1, 2\}$. Then by Observation 1 we have

$$\chi(G) \leq \chi(G[V_1]) + \chi(G[V_2]) = \omega(G[V_1]) + \omega(G[V_2])$$
$$\leq 2max\{\omega(G[V_1]), \omega(G[V_2])\} \leq 2\omega(G).$$

This completes the proof. □

So in particular 2SIG graphs are χ-bounded which is not surprising as Gyárfás [4] showed that all boxicity 2 graphs are χ-bounded by a quadratic function. We showed that 2SIGs are, in fact, χ-bounded by a linear function. It is known that square intersection graphs are χ-bounded by a linear function [8].

Now we focus on some of the subclasses of 2SIG. First, we show that 2SIIG graphs are χ-bounded by a better function.

Observation 2. *Let $G = (V_1 \sqcup V_2, E)$ be a 2SIIG graph with a given vertex partition. Then $\omega(G) \leq \chi(G) \leq \omega(G) + 1$.*

Moreover, we can enumerate all the maximal cliques and hence, can compute the clique number $\omega(G)$ of G in quadratic time.

Proof. Note that, the lower partition V_1 induces an interval graph and the upper partition V_2 induces an independent set. Hence $\omega(G[V_1]) = \chi(G[V_1])$ while $\omega(G[V_2]) = \chi(G[V_2]) = 1$. Also, Observation 1 implies $\omega(G[V_1]) \leq \omega(G) \leq \omega(G[V_1]) + 1$ and $\chi(G[V_1]) \leq \chi(G) \leq \chi(G[V_1]) + 1$ which implies this result.

We know that it is possible to enumerate all the maximal cliques and to compute the clique number of $G[V_1]$ in linear time [3]. For a vertex v in $G[V_2]$, $N_B(v)$ induces an interval graph. So, the maximal cliques containing v can be enumerated in linear time. Hence we are done. □

Note that it is possible to compute the clique number of a boxicity 2 graph in $|V| log |V|$ time if the intersection model is given. Here we provide a quadratic time solution for the same problem for 2SIIG, a subclass of boxicity 2 graphs, but we do not require the intersection model as our input in this case. It is enough if the vertex partition of the graph is provied. We can prove a similar result for yet another subclass of boxicity 2 graphs, the 2SUIG graphs. The proof is more involved.

Theorem 1. *For any 2SUIG graph G we can enumerate all the maximal cliques and hence, can compute the clique number $\omega(G)$ in polynomial time.*

To prove the above result we need to prove the following lemmas.

Lemma 1. *Let $G = (V_1 \sqcup V_2, E)$ be a 2SUIG graph with a given partition. Then there exist transitive orientations $\overrightarrow{G[V_1]^c}$ and $\overrightarrow{G[V_2]^c}$ such that for every pair of bridge edges u_1v_1 and u_2v_2 with $u_1u_2, v_1v_2 \notin E(G)$ we have the two arcs (u_1, u_2) and (v_1, v_2) in the orientations. Moreover, such orientations can be found in polynomial time.*

Proof. Take $\overrightarrow{G[V_1]^c}$ and $\overrightarrow{G[V_2]^c}$ such that the statement does not hold. The rectangles corresponding to u_1 and v_1 along with their intersection divide the region between the axis parallel lines into two disjoint parts. Hence, the intersection between the rectangles corresponding to u_2 and v_2, cannot be created without introducing redundant intersections. Since, the given graph has representation with the given partition, there must exist $\overrightarrow{G[V_1]^c}$ and $\overrightarrow{G[V_2]^c}$ such that the lemma holds. This can be done in polynomial time as the transitive orientation of the complement of a connected unit interval graph is unique up to reversal [7]. □

So, given a 2SUIG $G = (V_1 \sqcup V_2, E)$ with a partition we can fix transitive orientations $\overrightarrow{G[V_1]^c}$ and $\overrightarrow{G[V_2]^c}$ as in Lemma 1. Now given a bridge vertex $v \in V_i$ a vertex $v' \in V_i$ is its preceeding bridge vertex if each directed path from v' to v does not go through any other bridge vertex. The set of all preceeding bridge vertices of v is denoted by $PBV(v)$.

Now we will assign integer labels to the bridge edges of G. Let $uv \in E_B$ and let $\mathcal{B}(uv)$ be the set of all bridge edges with one vertex incident to it lying completely to the left (that is, strictly less with respect to the transitive orientation) of u or v. We assign the integer label $[e]$ to each edge $e = uv \in E_B$ inductively as follows:

$$[e] = \begin{cases} 0 & \text{if PBV(u) = PBV(v)} = \phi, \\ i+1 & \text{otherwise, where } i = max\{[e'] | e' \in \mathcal{B}(e)\} . \end{cases}$$

Let E_B^i be the bridge edges with label i.

Lemma 2. *In a 2SUIG graph the maximal cliques induced by the vertices incedent to edges of E_B^i can be enumerated in polynomial time.*

Proof. Let $G = (V_1 \sqcup V_2, E)$ be a 2SUIG graph with a given partition. Let G' be the subgraph induced by the vertices incedent to the edges of E_B^i for some fixed index i.. The vertices of G' belonging to the same stab line create a clique (not necessarily maximal in G). Consider the subgraph $G'' \subseteq G'$ containing only edges of E_B^i. Note that, this graph is a bipartite graph. Any maximal bipartite clique in G'' creates a maximal clique in G. All maximal bipartite cliques of G'' can be enumerated in polynomial time [1] (since we can have at most $O(|V(G'')|)$ maximal bipartite cliques). Therefore, the maximal cliques created by the union of the endpoints of E_B^i can be evaluated in polynomial time. □

Now we will show that it is not possible to have bridge edges with different labels in the same maximal clique of a 2SUIG G.

Lemma 3. *Bridge edges with different labels are not part of the same maximal clique in a 2SUIG.*

Proof. Let e, e' be two bridge edges and without loss of generality assume $[e] < [e']$. Then by definition at least one vertex incident to e is not adjacent to one of the vertices incident to e'. Hence we are done. □

Now we are ready to prove our main result.

Proof of Theorem 1: Let $G = (V_1 \sqcup V_2, E)$ be a $2SUIG$ graph with a given partition. We can enumarate all the maximal cliques of G containing at least one bridge edge using Lemma 2 and Lemma 3 in polynomial time. The maximal cliques of $G[V_1]$ and $G[V_2]$ can be enumerated in polynomial time as they are unit interval graphs. □

We could not provide a better χ-bound function for 2SUIG graphs than the one in Observation 2. However, we can provide a better χ-bound function for bridge triangle-free 2SUIG graphs.

Theorem 2. *Let G be a bridge triangle free 2SUIG. Then $\omega(G) \leq \chi(G) \leq \omega(G) + 1$.*

However, we will need to prove some lemmas before proving this result.

Lemma 4. *The bridge vertices of a triangle free 2SUIG graph can be coloured using 2 colors.*

Proof. Let $G = (V_1 \sqcup V_2, E)$ be a triangle free 2SUIG graph with a given partition and representation. Then $G[V_1]$ and $G[V_2]$ are disjoint union of paths. Let $u < v$ if the interval corresponding to u lies in the left of the interval corresponding to v (we compare the starting points) for any $u, v \in V_i$ where $i \in \{1, 2\}$. Furthermore, we say that $e = uv < u'v' = e'$ if $u < u'$ or $v < v'$ where $e, e' \in E_B$, $u, u' \in V_1$ and $v, v' \in V_2$.

We prove the statement using induction on the number of bridge edges. For $i = 1$ the graph is a tree, hence admits a 2-coloring.

Assume that all bridge triangle-free 2SUIG with at most k bridges admits a 3-coloring such that the bridge vertices received only two of the three colors. Let G be a bridge triangle-free 2SUIG with $k + 1$ bridges. Let $e' = u'v'$ be an edge of G such that $e < e'$ for all $e \in E_B$. Delete e' from G to obtain the graph G'. Note that G' admits a 3-coloring where all the bridge edges received only two of the three colors. Let $e'' = uv$ be the edge of G' such that $e < e''$ for all bridge edge e of G'. Suppose they received the colors c_1 and c_2. The subgraph induced by the paths u to u' and v to v' is a cycle. If it is an even cycle then we are done. Otherwise, we can always assign the third color to a non-bridge vertex of the cycle and complete the required coloring.

Note that the proof of our above result has an algorithmic aspect as well and it is not difficult to observe the following result:

Lemma 5. *The chormatic number of a triangle free 2SUIG graph can be decided in polynomial time.*

Now we are ready to prove our main result.

Proof of Theorem 2: Let $G = (V_1 \sqcup V_2, E)$ be a bridge triangle free 2SUIG with a given partition. Note that $G[V_1]$ and $G[V_2]$ are unit interval graphs. We prove the statement using induction on clique number $\omega(G)$. Note that the theorem is true for graphs G with $\omega(G) = 2$ by Lemma 5. Assume that the theorem is true for all bridge triangle free 2SUIG G with $\omega(G) \leq m$. Let $G = (V_1 \sqcup V_2, E)$ be a bridge triangle free 2SUIG with $\omega(G) = m+1$. We delete a maximal independent set from G to obtain the graph G'. Note that $\omega(G') \leq m$ and hence admits a $(m+1)$-coloring by our induction hypothesis. Now we extend this coloring by assigning a new color to the vertices of the deleted maximal independent set to obtain a $(m+2)$-coloring of G. $\qquad\square$

4   Matrix Characterization

A graph G is a *2-stab unit independent interval graph (2SUIIG)* if it admits a 2SUIG representation where the upper partition induces an independent set. The corresponding representation is a *2SUIIG representation*. The corresponding vertex partition $V_1 \sqcup V_2$ is a *canonical partition* if $G[V_1 \cup \{v\}]$ is not a unit interval graph for any $v \in V_2$ (upper partition). It is easy to see that a graph is a *2SUIIG* if and only if it has a canonical partition. For the rest of the section denote the x-coordinate of the bottom-left corner of a unit square representing a vertex v by s_v.

Now we characterize the adjacency matrix of a 2SUIIG. Note that the class of unit interval graphs is a subclass of 2SUIIG. Mertzios [10] showed that a graph is a unit interval graph if and only if its adjacency matrix is a *stair normal interval representation (SNIR) matrix*. We omit the definition of SNIR matrix because of space constraint but one can understand the definition from the example given below (for detailed definition see [10]). Now we will define two more type of matrices.

$$
\begin{bmatrix}
0 & 0 & 0 & 0 & 0 & 0 & 0 & 0 \\
1 & 0 & 0 & 0 & 0 & 0 & 0 & 0 \\
1 & 1 & 0 & 0 & 0 & 0 & 0 & 0 \\
1 & 1 & 1 & 0 & 0 & 0 & 0 & 0 \\
1 & 1 & 1 & 1 & 0 & 0 & 0 & 0 \\
0 & 1 & 1 & 1 & 1 & 0 & 0 & 0 \\
0 & 0 & 0 & 1 & 1 & 1 & 0 & 0 \\
0 & 0 & 0 & 0 & 0 & 0 & 0 & 0
\end{bmatrix}
$$

A *stair* of an SNIR matrix is a lower triangular submatrix with all but the diagonal entries equal to 1.

Definition 1. *The index of the first non-zero element of the ith row of a matrix $\mathcal{A}$ is the First($\mathcal{A}_{i*}$). A proper stab adjacency (PSA) matrix $\mathcal{A} = [\mathcal{A}_{ij}]$ is a matrix with the following properties:*

(i) *Each row has at most two ones and they are consecutive.*

(ii) *For $j < i$ and $\sum_{k} \mathcal{A}_{jk} = 2$ we have First($\mathcal{A}_{i*}$) $\geq$ First($\mathcal{A}_{j*}$).*

(iii) *For $j < i$ and $\sum_{k} \mathcal{A}_{jk} = 1$ we have First($\mathcal{A}_{i*}$) $\geq$ First($\mathcal{A}_{j*}$) -1.*

Definition 2. *An independence stair stab representation (ISSR) matrix*

$$Adj(M) = \left[\begin{array}{c|c} \mathcal{A}' & \mathcal{A}'' \\ \hline \mathcal{A}''^t & 0 \end{array}\right]$$

where the upper half of the matrix is $\mathcal{A}_{m \times (m+n)} = [\mathcal{A}_{ij}]$ with the following properties:

(i) *The submatrix induced by the first m columns is a SNIR matrix.*

(ii) *The submatrix induced by the last n columns is a PSA matrix.*

Note that using the characterization given by Mertzios [10] the SNIR matrix induced by the first m columns $\mathcal{A}$ corresponds to a a unit interval graph I (say). Let $\overrightarrow{I^c}$ be any transitive orientation of its complement.

(iii) *Let $j < i$, $m < k \leq m + n$, $\mathcal{A}_{ik} = 1$ and $\mathcal{A}_{jk} = 1$. Let the rows $\mathcal{A}_{i*}$ and $\mathcal{A}_{j*}$ correspond to the vertices u and v, respectively, of I. Then there is no directed path of length at least two from the v to u in $\overrightarrow{I^c}$.*

(iv) *Let $m < k \leq l \leq m+n$, $\mathcal{A}_{ik} = 1$ and $\mathcal{A}_{jl} = 1$. If the vertices corresponding to the rows $\mathcal{A}_{i*}$ and $\mathcal{A}_{j*}$ are adjacent in I then we have $0 \leq (l - k) \leq 2$.*

Now we are ready to state our main result:

Theorem 3. *A graph is a 2SUIIG graph if and only if it can be represented in ISSR matrix form.*

To prove Theorem 3 we need to prove some lemmas.

Lemma 6. *Let $G = (V_1 \sqcup V_2, E)$ be a 2SUIIG with upper partition V_2. Then for a vertex $v \in V_1$ we have $|N_B(v)| \leq 2$. Moreover, if $N_B(v) = \{x, z\}$ then it is not possible to have a $y \in V_2$ with $s_x < s_y < s_z$.*

Proof. Let $x, y, z \in V_2$ be neighbors of a vertex $v \in V_1$ and $s_x < s_y < s_z$. Note that if a vertex a is adjacent to b then $s_b - 1 \leq s_a \leq s_b + 1$. This implies $s_v \leq s_x + 1 < s_y < s_z - 1 \leq s_v$, a contradiction.

The moreover part can be proved with a similar argument. □

The above lemma gives an upper bound on the number of bridge neighbors of a vertex in V_1.

Lemma 7. *Let $G = (V_1 \sqcup V_2, E)$ be a 2SUIIG with upper partition V_2. For vertices $u, v \in V_1$ with $|N_B(u)| = 2$ and $s_u < s_v$ there exists $w \in N_B(u)$ such that for any $x \in N_B(v)$ we have $s_w \leq s_x$.*

Proof. Let there are vertices $w, x \in V_2$ such that $s_x < s_w$ where $w \in N_B(u)$ and $x \in N_B(v) \setminus N_B(u)$. Then the union of the unit squares corresponding to u and w divides the region enclosed by the stab lines into two disjoint parts. Then the intersection of the unit squares corresponding to v and x cannot be realised. $\square$

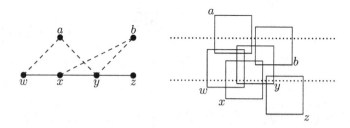

Fig. 4. The scenario dealt with in Lemma 8

Lemma 8. *Let $G = (V_1 \sqcup V_2, E)$ be a 2SUIIG with upper partition V_2. For vertices $x, y \in V_1$ with $|N_B(x)| = 1$ and $s_x < s_y$ we have $s_b - 2 < s_a$ where $b \in N_B(x)$ and $a \in N_B(y)$.*

Proof. Let we have vertices $x, y \in V_1$ with $|N_B(x)| = 1$ and $s_x < s_y$ we have $s_a \leq s_b - 2$ where $b \in N_B(x)$ and $a \in N_B(y)$. Now $s_b < s_x + 1$. If $s_a \leq s_b - 2$ then to realize the intersection between y, a we will have $s_y < s_x$, which is a contradiction. Note that the bound $s_b - 2$ is tight as we can have a situation illustrated in Fig. 4. $\square$

Intuitively, Lemma 7 and Lemma 8 show that in a 2SUIIG representation, the ordering of the intervals corresponding to the vertices in $G[V_1]$ fixes an ordering of the intervals corresponding to the vertices in $G[V_2]$.

Let $G = (V_1 \sqcup V_2, E)$ be a 2SUIIG with upper partition V_2. Assume that $|V_1| = m$ and $|V_2| = n$ We know that it admits a canonical partition. From this canonical partition we will construct a matrix which we will prove is a PSA matrix. Note that $G[V_1]$ is a unit interval graph and hence Mertzios [10] provides a (adjacency) SNIR matrix $\mathcal{A}_{m \times m}$ of $G[V_1]$. This matrix is obtained by putting the vertices of V_1 in a particular order. We will show that, we can obtain a particular order of V_2 such that the biadjacency matrix $\mathcal{A}'_{m \times n}$ of V_1 (taken in the same order as above) and V_2 is a PSA matrix.

Call it $\mathcal{A}_{m \times m}$. $\mathcal{A}'_{m \times n}$ be a zero one matrix. An entry $\mathcal{A}'[i][j]$ is 1 if there is a bridge between i, j. The ordering of the rows of $\mathcal{A}'$ remains same as the ordering of the rows in $\mathcal{A}$. The ordering of the columns of $\mathcal{A}'$ corresponds to an ordering of the vertices of V_2.

We order the vertices of V_2 using the following prescribed rule. Recall that the adjacency matrix $\mathcal{A}$ of $G[V_1]$ produces an interval intersection representation of (may not be unique) the graph. Fix one such representation R. Suppose we have $u, v \in V_1$ and $u', v' \in V_2$ such that $u' \in N_B(u) \setminus N_B(v)$ and $v' \in N_B(v)$. If u, v are not twins (that is, have the same set of neighbors) in $G[V_1]$ and $s_u < s_v$ in R then we want $u' < v'$ in our ordering of V_2. In any other case we take an arbitrary ordering between a pair of vertices $u', v' \in V_2$. This is a well defined algorithm for getting the ordering due to the previously proved lemmas.

Lemma 9. *The matrix $\mathcal{A}'$ is a PSA matrix.*

Proof. Suppose $\mathcal{A}'$ is not PSA matrix. Note that, there cannot be any zero column in it as we have assumed a canonical partition of the graph. Now we will check all the properties mentioned in the definition of the PSA matrix.

Property (i): Follows from Lemma 6.

Property (ii): Follows from Lemma 7.

Property (iii): Follows from Lemma 8.

Actually it is a regulation check using the lemmas proved before keeping in mind the ordering of V_2 obtained by the rule prescribed before this lemma. We skipped the details as it is easy to check.   □

Now we are ready to prove our main theorem.

Proof of Theorem 3: First we will prove the "if" part. Let $G = (V_1 \sqcup V_2, E)$ be a 2SUIIG with upper partition V_2. Then by the discussions above Lemma 9 we obtained an ordering of the vertices of V_1 and V_2. Consider the ordering of $V_1 \sqcup V_2$ by putting the vertices of V_1 in the previously obtained order followed by the vertices of V_2 (in previously obtained order as well). This ordering will give us a matrix of the following type:

$$Adj(G) = \left[\begin{array}{c|c} \mathcal{A} & \mathcal{A}' \\ \hline \mathcal{A}'^t & 0 \end{array} \right]$$

From the previous lemmas we know that the above matrix satisfies the first two properties for being a ISSR matrix. The last two properties are also easy to check using the properties proved before.

For the "only if" part, given a ISSR matrix we can construct a unit interval graph (and generate the intervals corresponding to the vertices) with the SNIR submatrix of it. The intervals corresponding to the rest of the vertices can be generated easily as they are independent following the ordering of the columns of the PSA submatrix.   □.

5 Conclusion

We do not know how difficult it is to determine if a given graph admits a 2SIG representation or not. But we expect it to be NP-hard.

Conjecture 1. Recognising 2SIG is NP-hard.

Also we could not figure out an efficient algorithm for determining the chromatic number of 2SIG or 2SUIG. This are interesting open problems which could be explored in future. Domination number of 2SIG is one of the prospective areas of research as the problem is polynomial time solvable for interval graphs but is NP-hard for boxicity 2 graphs. One can also generalize the concept to define k-stab interval graphs and study its different aspects.

References

1. Alexe, G., Alexe, S., Crama, Y., Foldes, S., Hammer, P.L., Simeone, B.: Consensus algorithms for the generation of all maximal bicliques. Discrete Applied Mathematics 145(1), 11–21 (2004); Graph Optimization IV
2. Brandstädt, A., Spinrad, J.P.: Graph classes: a survey, vol. 3. SIAM (1999)
3. Golumbic, M.C.: Algorithmic Graph Theory and Perfect Graphs, 2nd edn. Annals of Discrete Mathematics. Elsevier Science (2004)
4. Gyárfás, A.: Problems from the world surrounding perfect graphs. Applicationes Mathematicae 19(3-4), 413–441 (1987)
5. Gyárfás, A., Li, Z., Machado, R., Sebo, A., Thomassé, S., Trotignon, N.: Complements of nearly perfect graphs. arXiv preprint arXiv:1304.2862 (2013)
6. Hell, P., Huang, J.: Interval bigraphs and circular arc graphs. Journal of Graph Theory 46(4), 313–327 (2004)
7. Ibarra, L.: The clique-separator graph for chordal graphs. Discrete Applied Mathematics 157(8), 1737–1749 (2009)
8. Kostochka, A.: Coloring intersection graphs of geometric figures with a given clique number. Contemporary Mathematics 342, 127–138 (2004)
9. McConnell, R.M., Spinrad, J.P.: Linear-time modular decomposition and efficient transitive orientation of comparability graphs. In: Proceedings of the Fifth Annual ACM-SIAM Symposium on Discrete Algorithms, pp. 536–545. Society for Industrial and Applied Mathematics (1994)
10. Mertzios, G.B.: A matrix characterization of interval and proper interval graphs. Applied Mathematics Letters 21(4), 332–337 (2008)
11. Zhang, P.: Probe interval graphs and its applications to physical mapping of dna (1994) (unpublished manuscript)

Domination in Some Subclasses of Bipartite Graphs

Arti Pandey and B.S. Panda

Department of Mathematics
Indian Institute of Technology Delhi
Hauz Khas, New Delhi 110016, India
{artipandey,bspanda}@maths.iitd.ac.in

Abstract. A set $D \subseteq V$ is called a dominating set of $G = (V, E)$ if $|N_G[v] \cap D| \geq 1$ for all $v \in V$. The MINIMUM DOMINATION problem is to find a dominating set of minimum cardinality of the input graph. In this paper, we study the MINIMUM DOMINATION problem for star-convex bipartite graphs, circular-convex bipartite graphs and triad-convex bipartite graphs. It is known that the MINIMUM DOMINATION PROBLEM for a graph with n vertices can be approximated within $\ln n$. However, we show that for any $\epsilon > 0$, the MINIMUM DOMINATION problem does not admit a $(1 - \epsilon) \ln n$-approximation algorithm for star-convex bipartite graphs with n vertices unless NP $\subseteq$ DTIME($n^{O(\log \log n)}$). On the positive side, we propose polynomial time algorithms for computing a minimum dominating set of circular-convex bipartite graphs and triad-convex bipartite graphs, by polynomially reducing the MINIMUM DOMINATION problem for these graph classes to the MINIMUM DOMINATION problem for convex bipartite graphs.

Keywords: Domination, Convex bipartite graphs, Graph classes, NP-completeness.

1 Introduction

A vertex v of a graph $G = (V, E)$ is said to *dominate* a vertex w if either $v = w$ or $vw \in E$. A set of vertices D is a *dominating set* of G if every vertex of G is dominated by at least one vertex of D. The *domination number* of a graph G, denoted by $\gamma(G)$, is the cardinality of a minimum dominating set of G. The MINIMUM DOMINATION problem is to find a dominating set of minimum cardinality of the input graph G. Given a positive integer k and a graph G, the DOMINATION DECISION problem is to decide whether G has a dominating set of cardinality at most k. The concept of domination and its variations are widely studied as can be seen in [1, 2].

The DOMINATION DECISION problem is known to be NP-complete for general graphs [3], and remains NP-complete even for bipartite graphs [4] and for chordal bipartite graphs [5]. However, the MINIMUM DOMINATION problem is polynomial time solvable for convex bipartite graphs [6].

S. Ganguly and R. Krishnamurti (Eds.): CALDAM 2015, LNCS 8959, pp. 169–180, 2015.

In this paper we study the domination problem for three subclasses of bipartite graphs: circular-convex bipartite graphs, triad-convex bipartite graphs, and star-convex bipartite graphs. The class of circular-convex bipartite graphs was introduced by Liang and Blum [7] and has been studied recently by researchers (see [7, 8, 9, 10]). The triad-convex bipartite graphs and star-convex bipartite graphs are studied in [11, 9, 12]. The contributions of our paper are summarized as follows.

1. We prove that the MINIMUM DOMINATION problem for a star-convex bipartite graph G with n vertices does not admit a $(1 - \epsilon) \ln n$-approximation algorithm for any $\epsilon > 0$ unless NP $\subseteq$ DTIME($n^{O(\log \log n)}$). We also prove that the MINIMUM DOMINATION problem is linear-time solvable for bounded degree star-convex bipartite graphs.
2. We propose a polynomial time algorithm to solve the MINIMUM DOMINATION problem for circular-convex bipartite graphs. Since, the class of convex bipartite graphs is a subclass of circular-convex bipartite graphs, our algorithmic result for circular-convex bipartite graphs reduces the complexity gap between bipartite graphs and convex bipartite graphs.
3. We propose a polynomial time algorithm to solve the MINIMUM DOMINATION problem for triad-convex bipartite graphs.

2   Preliminaries

For a graph $G = (V, E)$, the sets $N_G(v) = \{u \in V(G) | uv \in E\}$ and $N_G[v] = N_G(v) \cup \{v\}$ denote the *open neighborhood* and *closed neighborhood* of a vertex v, respectively. For a set $S \subseteq V$, the sets $N_G(S) = \cup_{u \in S} N_G(u)$ and $N_G[S] = N_G(S) \cup S$ denote the *open neighborhood* and the *closed neighborhood* of S, respectively. The *degree* of a vertex v is $|N_G(v)|$ and is denoted by $d_G(v)$. If $d_G(v) = 1$, then v is called a *pendant vertex*. A vertex adjacent to some pendant vertex is called a *support vertex*. For $S \subseteq V$, let $G[S]$ denote the subgraph induced by G on S. A graph $G = (V, E)$ is said to be *bipartite* if $V(G)$ can be partitioned into two disjoint sets X and Y such that every edge of G joins a vertex in X to another vertex in Y. Such a partition (X, Y) of V is called a *bipartition*. A bipartite graph with bipartition (X, Y) of V is denoted by $G = (X, Y, E)$. Let n and m denote the number of vertices and number of edges of G, respectively. A bipartite graph $G = (X, Y, E)$ is called *circular-convex bipartite graph* if a circular ordering can be defined on the vertices of X, such that for every vertex y in Y, the open neighborhood of y is a circular arc in this ordering. A tree with exactly one non-pendant vertex is a *star*. A bipartite graph $G = (X, Y, E)$ is called a *tree-convex bipartite graph*, if a tree $T = (X, E_X)$ can be defined, such that for every vertex y in Y, the neighborhood of y induces a subtree of T. Tree-convex bipartite graphs are recognizable in linear time, and the associated tree T can also be constructed in linear-time [13]. For T a path or a star, G is called *convex-bipartite graphs* or *star-convex bipartite graphs*. For T a *triad*, that is, three paths with a common end-vertex, G is called a *triad-convex bipartite graph.* The following result is known regarding the complexity of the MINIMUM DOMINATION problem in convex bipartite graphs.

Theorem 1. *[14] A minimum dominating set of convex bipartite graphs can be computed in $O(n^2)$ time.*

Let D be a minimum dominating set of G. Let p be a pendant vertex of G, and s be the support vertex adjacent to p. Then either p or s must belong to D. If $s \notin D$, then clearly $p \in D$. Now $(D \setminus \{p\}) \cup \{s\}$ is a minimum dominating set containing s. Given a minimum dominating set of G, using the above process we can construct a minimum dominating set of G containing all the support vertices of G. Hence, in view of Theorem 1, we have the following theorem.

Theorem 2. *A minimum dominating set of a convex bipartite graph G that contains all the support vertices of G can be computed in $O(n^2)$ time.*

3   Star-convex Bipartite Graphs

In this section, we prove hardness results for the MINIMUM DOMINATION problem in star-convex bipartite graphs.

We give below a necessary and sufficient condition for a bipartite graph to be a star-convex bipartite graph. This will be useful in the polynomial reduction.

Lemma 1. *A bipartite graph $G = (X, Y, E)$ is a star-convex bipartite graph if and only if there exists a vertex x in X such that every vertex y in Y is either a pendant vertex or is adjacent to x.*

Proof. The proof is straightforward and hence is omitted.   □

To show the hardness results for the MINIMUM DOMINATION problem in star convex bipartite graphs, we need a well studied problem, named the MIN SET COVER problem.

Let S be any non-empty set and F be a family of subsets of S. For the set system (S, F), a set $C \subseteq F$ is called a *cover* of S, if every element of S belongs to at least one element of C. The MIN SET COVER problem is to find a minimum cardinality cover of S for a given set system (S, F). For a given positive integer k and a set system (S, F), the DECIDE SET COVER problem is to decide whether S has a cover of cardinality at most k. The following results are known.

Theorem 3. *[15] The DECIDE SET COVER problem is NP-complete.*

Theorem 4. *[16] The MIN SET COVER problem for input instance (S, F) does not admit a $(1 - \epsilon) \ln |S|$-approximation algorithm for any $\epsilon > 0$ unless $NP \subseteq DTIME(|S|^{O(\log \log |S|)})$. Furthermore, this inapproximability result holds for the case when the size of the input collection F is no more than the size of the set S.*

Now we are ready to prove the following result:

Theorem 5. *The* MINIMUM DOMINATION *problem for a star-convex bipartite graph G with n vertices does not admit a $(1 - \epsilon)\ln n$-approximation algorithm for any $\epsilon > 0$ unless $NP \subseteq DTIME(n^{O(\log \log n)})$.*

Proof. To prove the theorem, we provide an approximation preserving reduction from the MIN SET COVER problem to the MINIMUM DOMINATION problem for star-convex bipartite graphs.

Let the set system (S, F), where $S = \{S_1, S_2, \ldots, S_p\}$ and $F = \{C_1, C_2, \ldots, C_q\}$, $q \leq p$ be an instance of the MIN SET COVER problem. Now we construct in polynomial time the star-convex bipartite graph $G = (X, Y, E)$ as follows.

1. For each element S_i in the set S, add a vertex x_i in partite set X of G.
2. For each set C_j in the collection F, add a vertex c_j in partite set Y of G.
3. Add a vertex x_{p+1} in X, a vertex y_{q+1} in Y, and set of edges $\{x_{p+1}y_j \mid 1 \leq j \leq q+1\}$ in E.
4. If an element S_i belongs to set C_j, then add an edge between vertices x_i and c_j in graph G.

Formally, $X = \{x_1, x_2, \ldots, x_p, x_{p+1}\}$, $Y = \{y_1, y_2, \ldots, y_q, y_{q+1}\}$, and $E = \{x_iy_j \mid S_i \in C_j\} \cup \{x_{p+1}c_j \mid 1 \leq j \leq q+1\}$. Clearly G is a star-convex bipartite graph with n vertices, where $n = p + q + 2$. Now first we need to prove the following claim:

Claim. S has a cover of cardinality k if and only if G has a dominating set of cardinality $k + 1$.

Proof. Let C be a cover of S of cardinality k. Then the set $D = \{x_{p+1}\} \cup \{c_j \mid C_j \in C\}$ is a dominating set of cardinality $k + 1$.

Conversely suppose that D is a dominating set of G of cardinality $k + 1$. Then either x_{p+1} or c_{q+1} must belong to D. Define $D = D \setminus \{x_{p+1}, c_{q+1}\}$. Then the cardinality of the modified set D is at most k, and D dominates all the vertices of the set $X \setminus \{x_{p+1}\}$. Every vertex $x \in X$ is either dominated by itself or one of its neighbor. If $x \in X \setminus \{x_{p+1}\}$ dominates itself, then remove x from D and add one of its neighbor in D. Let us call the resultant set D'. Then $D' \subseteq Y \setminus \{y_{q+1}\}$ and dominates all the vertices of the set $X \setminus \{x_{p+1}\}$. Now define $C = \{C_j \in F \mid y_j \in D'\}$. Then C is a cover of S of cardinality at most k. Hence S has a cover of cardinality k. □

By the above claim, if D^* is a minimum dominating set of G and C^* is a minimum cover of S for the set system (S, F), then $|D^*| = |C^*| + 1$. Now we assume that the MINIMUM DOMINATION problem can be approximated with ratio α, where $\alpha = (1 - \epsilon)\ln n$ for some fixed $\epsilon > 0$, by using an approximation algorithm A. Let l be a fixed positive integer. Then the algorithm APPROX-SET-COVER constructs a solution of the MIN SET COVER problem.

Clearly, the algorithm APPROX-SET-COVER is a polynomial time algorithm. If the cardinality of a minimum cover of S is at most l, then it can be computed in polynomial time. So, here we will analyze the case, where the cardinality of minimum cover of S is greater than l. Let C^* denotes a minimum

---

Algorithm 1. APPROX-SET-COVER(S,F)

Input: A set S and a collection F of subsets of S.
Output: A cover of S.
begin
    if *there exists a cover C of S of cardinality $\leq l$* **then**
        | $C_a = C$;
    else
        Construct the graph G;
        Compute a dominating set D of G using the approximation algorithm A;
        Construct a cover C of S from dominating set D (as illustrated in the proof of
        the claim);
        $C_a = C$;
    end
    return C_a;
end

---

cover of S and D^* denotes a minimum dominating set of G. So $|C^*| > l$. If C_a is a cover of S computed by the algorithm APPROX-SET-COVER, then,
$$|C_a| \leq |D| \leq \alpha|D^*| = \alpha(1 + |C^*|) = \alpha(1 + \tfrac{1}{|C^*|})|C^*| < \alpha(1 + \tfrac{1}{l})|C^*|.$$

Since ϵ is fixed, there always exist a positive integer l such that $\frac{1}{l} < \epsilon$. So $|C_a| < \alpha(1+\epsilon)|C^*| = (1-\epsilon^2)\ln n|C^*| = (1-\epsilon')\ln n|C^*| \approx (1-\epsilon')\ln p|C^*|$ (since $\ln n \approx \ln p$ for sufficiently large values of p). Hence $|C_a| < (1 - \epsilon')\ln p|C^*|$, and therefore the algorithm APPROX-SET-COVER approximates set cover within ratio $(1 - \epsilon')\ln p$ for some $\epsilon > 0$.

By Theorem 4, if the MIN SET COVER problem can be approximated within $(1 - \epsilon')\ln p$, then NP $\subseteq$ DTIME $(p^{O(\log \log p)})$. It follows that if the MINIMUM DOMINATION problem can be approximated within $(1 - \epsilon)\ln n$ for any $\epsilon > 0$, then NP $\subseteq$ DTIME $(n^{O(\log \log n)})$. Hence, the MINIMUM DOMINATION problem can not be approximated within $(1 - \epsilon)\ln n$ unless NP $\subseteq$ DTIME $(n^{O(\log \log n)})$.
        □

The following corollary follows from the claim of Theorem 5, Theorem 3, and the fact that the DOMINATION DECISION problem is in NP for star-convex bipartite graphs.

Corollary 1. *The* DOMINATION DECISION *problem is NP-complete for star-convex bipartite graphs.*

The MINIMUM DOMINATION problem is APX-complete even for bipartite graphs with maximum degree 3. But for bounded degree star-convex bipartite graphs, we can prove the following result.

Theorem 6. *The* MINIMUM DOMINATION *problem is linear-time solvable for bounded degree star-convex bipartite graphs.*

Proof. The proof is omitted due to space constraint.
        □

4 Circular-convex Bipartite Graphs

In this section we propose a polynomial time algorithm to compute a minimum dominating set in a circular-convex bipartite graph G. A bipartite graph $G = (X, Y, E)$ with $|X| = n_1$ and $|Y| = n_2$ is called a circular-convex bipartite graph if there exists a circular ordering $\prec$ on X, say $x_1 \prec x_2 \prec \ldots \prec x_{n_1} \prec x_{(n_1+1)} = x_1$, such that for every vertex y in Y, either $N_G(y) = \{x_i, x_{i+1}, \ldots, x_j\}$ or $N_G(y) = \{x_j, x_{j+1}, \ldots, x_{n_1}, x_1, \ldots, x_i\}$ for $1 \leq i \leq j \leq n_1$. A sequence of consecutive vertices in the clock-wise direction in the circular ordering $\prec$ on X is called a *circular arc* and the first vertex and the last vertex in the circular arc are called left end point and right end point of the circular arc, respectively.

The idea behind our algorithm is as follows. We construct a collection of subsets of $V(G)$, say $\mathbb{F}$, satisfying the following properties:

• Cardinality of $\mathbb{F}$ is polynomial in n, where n is the number of vertices of G.
• For at least one set, say F in $\mathbb{F}$, there exists a minimum dominating set of G that contains F.

Thus, our problem of finding a minimum dominating set, say D^*, of G is reduced to the problem of finding a minimum dominating set among all dominating sets containing F of G, say D_F, for each $F \in \mathbb{F}$. Then $|D^*| = min\{|D_F| \mid F \in \mathbb{F}\}$. Note that a minimum dominating set D_F containing the set F is not necessarily a minimum dominating set of G. Now corresponding to each set $F \in \mathbb{F}$, we construct a convex bipartite graph G_F, from the circular-convex bipartite graph G. We compute the minimum dominating set, say D, of G_F, and then we show that the set D_F can be computed from D in constant time.

First, we prove the following lemma.

Lemma 2. *Let $G = (X, Y, E)$ be a circular-convex bipartite graph, and D^* be a minimum dominating set of G. Then, for a vertex $x_i \in D^* \cap X$, $|N_G(x_i) \cap D^*| \leq 2$.*

Proof. Since for every vertex $y \in N_G(x_i)$, $N_G(y)$ is a circular arc in X containing the vertex x_i, hence $N_G(N_G(x_i))$ is also a circular arc in X containing the vertex x_i. Now suppose that x_c, x_d be two end points of this arc. Then there exist some $y_j, y_k \in N_G(x_i)$ such that $x_c \in N_G(y_j)$ and $x_d \in N_G(y_k)$. Then the vertices y_j, y_k dominates all the vertices of circular arc $N_G(N_G(x_i))$. Hence, if for a minimum dominating set D^* of G, $|N_G(x_i) \cap D^*| \geq 3$, then we can define a set $D' = (D^* \setminus N_G(x_i)) \cup \{y_j, y_k\}$ such that D' is also a dominating set of G and $|D'| < |D^*|$, which is a contradiction. □

Let D^* be a minimum dominating set of G. If D^* does not contain any vertex from X, then $D^* = Y$. Define $S_0 = Y$. If D^* contains at least one vertex say x_i from X, then we have the following cases:

1. None of the neighbor of x_i belongs to D^*.
2. Exactly one neighbor of x_i belongs to D^*.
3. Exactly two neighbors of x_i belong to D^*.

However, neither we know which vertex x_i belongs to D^* nor we know which neighbor of x_i belongs to D^*. Let D_{x_i} denotes the minimum dominating set of G containing $x_i \in X$, but not containing any neighbor of x_i. Define $S_1 = \{D_{x_i} \mid x_i \in X\}$. Let $D_{x_i}^{y_j}$ denotes the minimum dominating set of G containing x_i and exactly one neighbor y_j of x_i. Define $S_2 = \{D_{x_i}^{y_j} \mid x_i \in X, y_j \in N_G(x_i)\}$. Let $D_{x_i}^{y_j, y_k}$ denotes the minimum dominating set of G containing x_i and exactly two neighbors y_j, y_k of x_i. Define $S_3 = \{D_{x_i}^{y_j, y_k} \mid x_i \in X, y_j, y_k \in N_G(x_i)\}$. Now define $S = S_0 \cup S_1 \cup S_2 \cup S_3$. Then the following lemma is straightforward.

Lemma 3. *The minimum cardinality set in S is a minimum dominating set of G, and $|S| = O(|X||Y|^2)$.*

Construction of Convex Bipartite Graphs

As discussed above, we need to consider three cases depending on the cardinality of the set $N_G(x_i) \cap D^*$ by assuming that $x_i \in D^*$. In each of the three cases, corresponding to the circular-convex bipartite graph G, we define a convex bipartite graph.

CASE 1: No Neighbor of x_i Belong to D^*

Construction 1: Define a graph G_{x_i} by deleting the vertex x_i and all its neighbors. Clearly, $G_{x_i} = (X_{x_i}, Y_{x_i}, E_{x_i})$ is a convex bipartite graph.

Lemma 4. *If D is a minimum dominating set of G_{x_i}, then $D \cup \{x_i\}$ is a dominating set of G. Moreover, if there exists a minimum dominating set, say D^* of G such that $x_i \in D^*$, and $N_G(x_i) \cap D^* = \emptyset$, then $|D^*| = |D| + 1$.*

Proof. The proof is easy and hence is omitted.   □

CASE 2: Exactly One Neighbor of x_i, Say y_j, Belongs to D^*

Construction 2: Define a graph $G_{x_i}^{y_j}$ in the following way:

Remove vertices in $N_G(x_i) \setminus \{y_j\}$. Split x_i into x_i', x_i''. Split y_j into y_j', y_j''. Note that $N_G(y_j)$ is a circular arc in X containing x. Let x_a and x_b denote the end points of this arc. Join y_j' with the vertices of this arc present on one side of x_i. Join y_j'' with the vertices of this arc present on another side of x_i. Join x_i' with y_j', and x_i'' with y_j''. Take four new vertices p', p'', q', q'', join p' with x_i', p'' with x_i'', q' with y_j', and q'' with y_j''. Then $G_{x_i}^{y_j} = (X_{x_i}^{y_j}, Y_{x_i}^{y_j}, E_{x_i}^{y_j})$ is a convex bipartite graph.

Lemma 5. *If D is a minimum dominating set of $G_{x_i}^{y_j}$ containing all the support vertices of $G_{x_i}^{y_j}$, that is, $\{x_i', x_i'', y_j', y_j''\} \subseteq D$, then $(D \setminus \{x_i', x_i'', y_j', y_j''\}) \cup \{x_i, y_j\}$ is a dominating set of G. Moreover, if there exists a minimum dominating set, say D^* of G such that $x_i \in D^*$, and $N_G(x_i) \cap D^* = \{y_j\}$, then $|D^*| = |D| - 2$.*

Proof. The proof is omitted due to space constraint.   □

CASE 3: Exactly Two Neighbors of x_i say y_j, y_k Belong to D^*

Construction 3: Define a graph $G_{x_i}^{y_j, y_k}$ in the following way:

Remove vertices in $N_G(x_i) \setminus \{y_j, y_k\}$. Split x_i into x_i', x_i''. Note that $N_G(\{y_j, y_k\})$ is a circular arc in X containing x. Let x_a and x_b denote the end points of this arc. Join y_j with the vertices of this arc present on one side of x_i. Join y_k

with the vertices of this arc present on another side of x_i. Join x_i' with y_j, and x_i'' with y_k. Take four new vertices p', p'', q', q'', join p' with x_i', p'' with x_i'', q' with y_j, and q'' with y_k.

Then $G_{x_i}^{y_j,y_k} = (X_{x_i}^{y_j,y_k}, Y_{x_i}^{y_j,y_k}, E_{x_i}^{y_j,y_k})$ is a convex bipartite graph, since if we define a path $P = q', x_i', x_{i+1}, \ldots, x_a, \ldots, x_b, \ldots, x_i'', q''$, then for every y in $Y_{x_i}^{y_j,y_k}$, $N_{G_{x_i}^{y_j,y_k}}(y)$ induces a subpath (subtree) in P.

Lemma 6. *If D is a minimum dominating set of $G_{x_i}^{y_j,y_k}$ containing all the support vertices of $G_{x_i}^{y_j,y_k}$, that is, $\{x_i', x_i'', y_j, y_k\} \subseteq D$, then $(D \setminus \{x_i', x_i''\}) \cup \{x_i\}$ is a dominating set of G. Moreover, if there exists a minimum dominating set, say D^* of G such that $x_i \in D^*$, and $N_G(x_i) \cap D^* = \{y_j, y_k\}$, then $|D^*| = |D| - 1$.*

Proof. The proof is omitted due to space constraint.   □

Now we are ready to present the detailed algorithm for finding a minimum dominating set in a circular-convex bipartite graph G.

Algorithm 2. DOM-CIRCULAR-CONVEX()

Input: A circular-convex bipartite graph $G = (X, Y, E)$, where $X = \{x_1, x_2, \ldots, x_{n_1}\}$ and $Y = \{y_1, y_2, \ldots, y_{n_2}\}$.
Output: A minimum dominating set D of the graph G.
$S = \emptyset$;
for $i=1$ *to* n_1 **do**
  Construct G_{x_i} using Construction 1;
  Find minimum dominating set D of G_{x_i};
  Update $D = D \cup \{x_i\}$; $S = S \cup \{D\}$;
end
for $i=1$ *to* n_1 **do**
  foreach $y_j \in N_G(x_i)$ **do**
    Construct $G_{x_i}^{y_j}$ using Construction 2;
    Find a minimum dominating set D containing all support vertices of $G_{x_i}^{y_j}$;
    Update $D = (D \cap (X \cup Y)) \cup \{x_i, y_j\}$; $S = S \cup \{D\}$;
  end
end
for $i=1$ *to* n_1 **do**
  foreach $y_j, y_k \in N_G(x_i)$ **do**
    Construct $G_{x_i}^{y_j,y_k}$ using Construction 3;
    Find a minimum dominating set D containing all support vertices of $G_{x_i}^{y_j,y_k}$;
    Update $D = (D \cap (X \cup Y)) \cup \{x_i, y_j, y_k\}$; $S = S \cup \{D\}$;
  end
end
Find the minimum cardinality set in S, say D^*;
return D^*.

The proof of correctness of the above algorithm follows from Lemma 2, Lemma 4, Lemma 5 and Lemma 6, and the running time of the algorithm is $O(n^5)$ which follows from Lemma 3 and Theorem 2. Hence we have the following theorem.

Theorem 7. *A minimum dominating set of circular-convex bipartite graph can be computed in $O(n^5)$ time.*

5   Triad-Convex Bipartite Graphs

In this section, we propose a polynomial time algorithm to compute a minimum dominating set in triad-convex bipartite graphs. For this, we make a polynomial reduction from triad-convex bipartite graphs to convex bipartite graphs, the graph class for which the MINIMUM DOMINATION problem is already shown to be polynomial time solvable.

Let $G = (X, Y, E)$ be a triad-convex bipartite graph with a triad $T = (X, E_X)$ defined on X, such that for every vertex $y \in Y$, $T[N_G(y)]$ is a subtree of T. Let $X = \{x_0\} \cup X_1 \cup X_2 \cup X_3$, such that for each i, $1 \leq i \leq 3$, $X_i = \{x_{i,1}, x_{i,2}, \ldots, x_{i,n_i}\}$. Also suppose that $x_{i,0} = x_0$ for all i, $1 \leq i \leq 3$. For each i, $1 \leq i \leq 3$, let $P_i = T[\{x_0\} \cup X_i]$ is a path, and $E(P_i) = \{x_{i,k}x_{i,k+1} \mid 0 \leq k < n_i\}$. Note that x_0 is a common vertex in all the three paths P_1, P_2, and P_3.

Let D be a minimum dominating set of G. If $x_0 \notin D$ and $D \cap X_1 \neq \emptyset$, define $a = min\{j \mid x_{1,j} \in D\}$. Hence $x_{1,a}$ is the vertex of D nearest to x_0 in path P_1. Similarly, if $x_0 \notin D$ and $D \cap X_2 \neq \emptyset$, define $b = min\{j \mid x_{2,j} \in D\}$. Similarly, if $x_0 \notin D$ and $D \cap X_3 \neq \emptyset$, define $c = min\{j \mid x_{3,j} \in D\}$.

Now before giving the reduction from triad-convex bipartite graphs to convex bipartite graphs, we prove the following lemma:

Lemma 7. *There exists a minimum dominating set D of G such that $|N_G(x_0) \cap D| \leq 3$.*

Proof. Let D be a minimum dominating set of G. If $|N_G(x_0) \cap D| \leq 3$, then D is the required set, and we are done. So, now suppose that $|N_G(x_0) \cap D| > 3$.

Let $X^0 = N_G(N_G(x_0))$. Then every vertex present in $N_G(x_0) \cap D$ either dominates itself or dominates some vertices of the set X^0. For each i, $1 \leq i \leq 3$, find $y_i \in N_G(x_0)$ such that $N_G(y_i) \cap X_i = X^0 \cap X_i$. Note that such a vertex y_i must exist. Then y_i dominates the set $X^0 \cap X_i$. Now define $D^* = (D \setminus N_G(x_0)) \cup \{x_0, y_1, y_2, y_3\}$. Note that D^* is also a dominating set of G of cardinality at most the cardinality of the set D, and hence a minimum dominating set of G. Also $|N_G(x_0) \cap D^*| \leq 3$. Hence D^* is the required set.   □

Now let D be a minimum dominating set of G for which $|N(x_0) \cap D| \leq 3$. Define $D \cap N_G(x_0) = Y^0$. For each i, $1 \leq i \leq 3$, let y_i be a vertex in Y^0 such that $N_G(y_i) \cap X_i = N_G(Y^0) \cap X_i$. Hence $D \cap N_G(x_0) = \{y_1, y_2, y_3\}$. If $|D \cap N_G(x_0)| < 3$, then some of the elements in the set $\{y_1, y_2, y_3\}$ may be same. Now we discuss the following three cases:

CASE 1: $x_0 \in D$ **and** $N_G(x_0) \cap D = \emptyset$.
Construction 4: Define $G_0 = G \setminus (\{x_0\} \cup N_G(x_0))$.
Then $G_0 = (X_0, Y_0, E_0)$, where $X_0 = X \setminus \{x_0\}$, $Y_0 = Y \setminus N_G(x_0)$, and $E_0 = \{e = xy \in E(G) \mid x \in X_0, y \in Y_0\}$.

It is easy to observe that each component of G_0 is a convex bipartite graph. Let $G_1, G_2, \ldots, G_k$ be the k components of G_0. Let $D_1, D_2, \ldots, D_k$ be minimum dominating sets of graphs $G_1, G_2, \ldots, G_k$, respectively.

Lemma 8. *Let D be a minimum dominating set of G_0, then $D \cup \{x_0\}$ is a dominating set of G. Moreover, if there exists a minimum dominating set, say D^* of G such that $x_0 \in D^*$, and $N_G(x_0) \cap D^* = \emptyset$, then $|D^*| = |D| + 1$.*

Proof. The proof is easy and hence is omitted. $\Box$

CASE 2: $x_0 \in D$ and $1 \le |N_G(x_0) \cap D| \le 3$.
Construction 5: Define the graph $G_0^{y_1}$ in the following way:

Let $X' = X_1 \cup \{x_0\}$, $Y' = \{y_1\} \cup \{y \in Y \mid N_G(y) \subseteq X_1\}$, and $G' = G[X' \cup Y']$. Now add two vertices f_1, g_1 in G'. Also add the edges $f_1 y_1, g_1 x_0$ in G'. Call the resultant graph as $G_0^{y_1}$.

Formally, $G_0^{y_1} = (X_0^{y_1}, Y_0^{y_1}, E_0^{y_1})$ where $X_0^{y_1} = \{x_0, f_1\} \cup X_1$, $Y_0^{y_1} = \{y_1, g_1\} \cup \{y \in Y \mid N_G(y) \subseteq X_1\}$, and $E_0^{y_1} = \{f_1 y_1, g_1 x_0\} \cup \{e = xy \in E \mid x \in X_0^{y_1}, y \in Y_0^{y_1}\}$.

Similarly, define the graphs $G_0^{y_2}$ and $G_0^{y_3}$. It is easy to observe that the graphs $G_0^{y_1}$, $G_0^{y_2}$, $G_0^{y_3}$ defined above, are all convex bipartite graphs. Let $D_0^{y_1}$ be a minimum dominating set of $G_0^{y_1}$ containing all the support vertices, that is $\{x_0, y_1\} \subseteq D_0^{y_1}$, $D_0^{y_2}$ be a minimum dominating set of $G_0^{y_2}$ containing all the support vertices, that is $\{x_0, y_2\} \subseteq D_0^{y_2}$, and $D_0^{y_3}$ be a minimum dominating set of $G_0^{y_3}$ containing all the support vertices, that is $\{x_0, y_3\} \subseteq D_0^{y_3}$, then we prove the following statement.

Lemma 9. *The set $D = D_0^{y_1} \cup D_0^{y_2} \cup D_0^{y_3}$ is a dominating set of G. Moreover, if there exist a minimum dominating set D^* of G such that $\{x_0, y_1, y_2, y_3\} \subseteq D^*$, then D is also a minimum dominating set of G.*

Proof. The proof is omitted due to space constraint. $\Box$

CASE 3: $x_0 \notin D$ and $1 \le |N_G(x_0) \cap D| \le 3$.
Construction 6: If $X_1 \cap D = \emptyset$, then define $G^{y_1} = (X^{y_1}, Y^{y_1}, E^{y_1})$ where $X^{y_1} = \{f_1\}$, $Y^{y_1} = \{y_1\} \cup \{y \in Y \mid N_G(y) \subseteq X_1\}$, and $E^{y_1} = \{f_1 y_1\}$.

Similarly, define G^{y_2} and G^{y_3}.

If $X_1 \cap D \ne \emptyset$, then define $G_a^{y_1} = (X_a^{y_1}, Y_a^{y_1}, E_a^{y_1})$ where $X_a^{y_1} = \{f_1\} \cup \{x_{1,p} \mid p \ge a\}$, $Y_a^{y_1} = \{y_1, g_1\} \cup \{y \in Y \mid N_G(y) \subseteq X_1\}$, and $E_a^{y_1} = \{f_1 y_1, g_1 x_{1,p}\} \cup \{e = xy \in E \mid x \in X_a^{y_1}, y \in Y_a^{y_1}\}$.

Similarly, define $G_b^{y_2}$ and $G_c^{y_3}$.

It is easy to observe that graphs G^{y_1}, G^{y_2}, G^{y_3}, $G_a^{y_1}$, $G_b^{y_2}$, $G_c^{y_3}$ defined above, are all convex bipartite graphs. Let D^{y_1}, D^{y_2}, D^{y_3}, $D_a^{y_1}$, $D_b^{y_2}$, $D_c^{y_3}$, denote the minimum dominating sets containing all the support vertices of G^{y_1}, G^{y_2}, G^{y_3}, $G_a^{y_1}$, $G_b^{y_2}$, $G_c^{y_3}$, respectively. Then we prove the following statement.

Lemma 10. *Let $C = \{(A \cup B \cup C) \mid A \in \{D^{y_1}, D_a^{y_1}\}, B \in \{D^{y_2}, D_b^{y_2}\}, C \in \{D^{y_3}, D_c^{y_3}\}\}$. Then every set $D \in C$ is a dominating set of G. Moreover, if there exists a dominating set, say D^*, of G satisfying the following conditions:*
(i) $x_0 \notin D^*$ and $1 \le |N_G(x_0) \cap D^*| \le 3$,
(ii) either $X_1 \cap D^* = \emptyset$ or $x_{1,a}$ is the vertex of D^* nearest to x_0 in path P_1,
(iii) either $X_2 \cap D^* = \emptyset$ or $x_{2,b}$ is the vertex of D^* nearest to x_0 in path P_2,

(iv) either $X_3 \cap D^* = \emptyset$ or $x_{3,c}$ is the vertex of D^* nearest to x_0 in path P_3, then $|D^*| = |D'|$ for some $D' \in \mathcal{C}$.

Proof. The proof is omitted due to space constraint.   □

Now we are ready to present the detailed algorithm for finding a minimum dominating set in a triad-convex bipartite graph G.

Algorithm 3. DOM-TRIAD-CONVEX(G)

Input: A triad-convex bipartite graph $G = (X, Y, E)$ with a triad $T = (X, E_X)$, where $X = \{x_0\} \cup X_1 \cup X_2 \cup X_3$, and $|X_1| = n_1$, $|X_2| = n_2$, $|X_3| = n_3$.
Output: A minimum dominating set D of the graph G.
$S = \emptyset$;
Find $\mathbb{P}$, the set of all non-empty subsets of $N_G(x_0)$ of cardinality at most 3.
Construct G_0 using Construction 4;
Find minimum dominating set D_0 of G_0;
$S = S \cup \{D_0\}$;
foreach *set $P \in \mathbb{P}$* **do**
  Find a vertex $y_1 \in P$ such that $N_G(y_1) \cap X_1 = N_G(P) \cap X_1$;
  Find a vertex $y_2 \in P$ such that $N_G(y_2) \cap X_2 = N_G(P) \cap X_2$;
  Find a vertex $y_3 \in P$ such that $N_G(y_3) \cap X_3 = N_G(P) \cap X_3$;
  Construct $G_0^{y_1}, G_0^{y_2}, G_0^{y_3}$ using Construction 5;
  Find minimum dominating sets $D_0^{y_1}, D_0^{y_2}, D_0^{y_3}$ containing all support vertices of
  $G_0^{y_1}, G_0^{y_2}, G_0^{y_3}$, respectively;
  $S = S \cup \{D_0^{y_1} \cup D_0^{y_2} \cup D_0^{y_3}\}$;
  Construct $G^{y_1}, G^{y_2}, G^{y_3}$ using Construction 6;
  Find minimum dominating sets $D^{y_1}, D^{y_2}, D^{y_3}$ containing all support vertices of
  $G^{y_1}, G^{y_2}, G^{y_3}$, respectively;
  $S = S \cup \{D^{y_1} \cup D^{y_2} \cup D^{y_3}\}$;
  for *r=1 to n_1* **do**
    for *s=1 to n_2* **do**
      for *t=1 to n_3* **do**
        Construct $G_r^{y_1}, G_s^{y_2}, G_t^{y_3}$ using Construction 6;
        Find minimum dominating sets $D_r^{y_1}, D_s^{y_2}, D_t^{y_3}$ containing all support
        vertices of $G_r^{y_1}, G_s^{y_2}, G_t^{y_3}$, respectively;
        $S = S \cup \{(A \cup B \cup C) \mid A \in \{D^{y_1}, D_r^{y_1}\}, B \in \{D^{y_2}, D_s^{y_2}\}, C \in \{D^{y_3}, D_s^{y_3}\}\}$;
      end
    end
  end
end
Find a minimum cardinality set in S, say D^*;
return D^*.

The proof of correctness of the above algorithm follows from Lemma 8, Lemma 9 and Lemma 10, and running time of the algorithm is $O(n^8)$ which follows from Theorem 2. Hence we have the following theorem.

Theorem 8. *The algorithm DOM-TRIAD-CONVEX(G) can be implemented in $O(n^8)$ time.*

6   Conclusion

In this paper, we proved an approximation hardness result of the MINIMUM
DOMINATION problem for star-convex bipartite graphs. On the positive side, we
showed that the MINIMUM DOMINATION problem is polynomial time solvable
for circular-convex bipartite graphs and triad-convex bipartite graphs. It will be
an interesting problem to propose algorithms with better time complexity for
the MINIMUM DOMINATION problem for circular-convex bipartite graphs and
triad-convex bipartite graphs.

References

[1] Haynes, T.W., Hedetniemi, S.T., Slater, P.J.: Domination in graphs: Advanced
topics, vol. 209. Marcel Dekker Inc., New York (1998)

[2] Haynes, T.W., Hedetniemi, S.T., Slater, P.J.: Fundamentals of Domination in
Graphs, vol. 208. Marcel Dekker Inc., New York (1998)

[3] Garey, M.R., Johnson, D.S.: Computers and Interactability: a guide to the theory
of NP-completeness. W.H. Freeman and Co., San Francisco (1979)

[4] Bertossi, A.A.: Dominating sets for split and bipartite graphs. Inform. Process.
Lett. 19, 37–40 (1984)

[5] Müller, H., Brandstädt, A.: The NP-completeness of steiner tree and dominating
set for chordal bipartite graphs. Theoret. Comput. Sci. 53, 257–265 (1987)

[6] Damaschke, P., Müller, H., Kratsch, D.: Domination in convex and chordal bipartite graphs. Inform. Process. Lett. 36, 231–236 (1990)

[7] Liang, Y.D., Blum, N.: Circular convex bipartite graphs: maximum matching and
hamiltonial circuits. Inform. Process. Lett. 56, 215–219 (1995)

[8] Lu, M., Liu, T., Xu, K.: Independent domination: Reductions from circular- and
triad-convex bipartite graphs to convex bipartite graphs. In: Fellows, M., Tan,
X., Zhu, B. (eds.) FAW-AAIM 2013. LNCS, vol. 7924, pp. 142–152. Springer,
Heidelberg (2013)

[9] Lu, Z., Liu, T., Xu, K.: Tractable connected domination for restricted bipartite graphs (Extended abstract). In: Du, D.-Z., Zhang, G. (eds.) COCOON 2013.
LNCS, vol. 7936, pp. 721–728. Springer, Heidelberg (2013)

[10] Liu, T., Lu, M., Lu, Z., Xu, K.: Circular convex bipartite graphs: Feedback vertex
sets. Theoret. Comput. Sci. (2014), doi: 10.1016/j.tcs.2014.05.001

[11] Song, Y., Liu, T., Xu, K.: Independent domination on tree convex bipartite graphs.
In: Snoeyink, J., Lu, P., Su, K., Wang, L. (eds.) AAIM 2012 and FAW 2012. LNCS,
vol. 7285, pp. 129–138. Springer, Heidelberg (2012)

[12] Liu, W.J.T., Wang, C., Xu, K.: Feedback vertex sets on restricted bipartite graphs.
Theoret. Comput. Sci. 507, 41–51 (2013)

[13] Zhang, Y., Bao, F.S.: A review of tree convex sets test. Computational Intelli-
gence 28, 358–372 (2012)

[14] Bang-Jensen, J., Huang, J., MacGillivray, G., Yeo, A.: Domination in convex bipartite and convex-round graphs. Recent Trends in Computational Math. and its
Applications. International Journal of Mathematical Sciences 5 (2006)

[15] Karp, R.M.: Reducibility among combinatorial problems. In: Miller, R.E.,
Thatcher, J.W. (eds.) Complexity of Computer Computations, pp. 85–103.
Plenum, New York (1972)

[16] Feige, U.: A threshold of $\ln n$ for approximating set cover. J. ACM 45, 634–652
(1998)

Parameterized Analogues of Probabilistic Computation

Ankit Chauhan and B.V. Raghavendra Rao

Department of Computer Science and Engineering,
Indian Institute of Technology Madras,
Chennai, India
{ankitch,bvrr}@cse.iitm.ac.in

Abstract. We study structural aspects of randomized parameterized computation. We introduce a new class W[P]-PFPT as a natural parameterized analogue of PP. Our definition uses the machine based characterization of the parameterized complexity class W[P] obtained by Chen et.al [TCS 2005]. We translate most of the structural properties and characterizations of the class PP to the new class W[P]-PFPT.

We study a parameterization of the polynomial identity testing problem based on the degree of the polynomial computed by the arithmetic circuit. We obtain a parameterized analogue of the well known Schwartz-Zippel lemma [Schwartz, JACM 80 and Zippel, EUROSAM 79].

Additionally, we introduce a parameterized variant of permanent, and prove its $\#W[1]$ completeness.

1 Introduction

Parameterized Complexity Theory provides a formal framework for finer complexity analysis of problems by allowing a parameter along with the input. It was pioneered by Downey and Fellows [11,10] two decades ago. Since then, it has revolutionized algorithmic research [20], and led to the development of several important algorithmic techniques.

Fixed Parameter Tractability (FPT) forms the central notion of tractability in Parameterized Complexity Theory. Here, any problem that is decidable in deterministic time $f(k)\text{poly}(n)$ is deemed to be tractable, where k is the parameter and f any computable function. Several NP hard problems including the vertex cover problem are known to be tractable under this notion [14].

The W-hierarchy serves as the basis for all intractable problems in the parameterized world. W[1], the smallest member of W-hierarchy, consists of problems that are FPT equivalent to the p-clique problem [14]. The limit of W-hierarchy, W[P] encapsulates all problems solvable in non-deterministic $f(k)\text{poly}(n)$ time using at most $g(k)\log n$ non-deterministic bits [7,14], where f and g are arbitrary computable functions.

There have been significant efforts towards understanding the structure of parameterized complexity classes in the last two decades. Specifically exact characterizations of the W-hierarchy and other related hierarchies are known [12]. (See also [14,11].)

Apart from non-deterministic computation, probabilistic computation serves as one of the crucial building blocks of Complexity Theory. Probabilistic complexity classes have been well studied in the literature and has been an active area of research for more

S. Ganguly and R. Krishnamurti (Eds.): CALDAM 2015, LNCS 8959, pp. 181–192, 2015.
© Springer International Publishing Switzerland 2015

than three decades. There are a significant number of parameterized algorithms that use randomization [11, Chapter 8]. Hence, development of randomized complexity classes in the parameterized framework is necessary to understand the use of randomization in the parameterized setting.

Müller [19,18] was the first to do a systematic development and study of parameterized randomization. He defined bounded error probabilistic parameterized classes such as W[P]-BPFPT and W[1]-BPFPT. Further, he obtained amplification results and conditions for derandomization of these classes. Further, Müller [19] studied several parameterizations of the well known polynomial identity testing problem (ACIT) and obtained several hardness results as well has upper bounds in terms of the newly defined randomized classes.

We continue the line of research initiated by Müller [19] and study a parameterized variant of probabilistic computation with unbounded error and establish a relationship with the corresponding parameterized counting class.

It should be noted that almost all of the randomized FPT algorithms use randomness of the same magnitude as their running times. However, such an algorithm cannot be visualized as a non-deterministic algorithm with $f(k) \log n$ random bits, where $f(k)$ is an arbitrary computable function. This is in stark contrast to the classical setting, where every randomized algorithm with bounded error probability can also be seen as a non-deterministic algorithm with the same time bound. So it is desirable to have randomized FPT algorithms that use at most $O(f(k) \log n)$ random bits instead of $f(k)\text{poly}(n)$ random bits. As a first step towards this we obtain such an algorithm for a suitable parameterization of ACIT.

Finally, following the recent developments in the parameterized complexity theory of counting problems [5,8,9], we develop a parameterized variants of the problems of computing permanent and determinant of a matrix.

Our Results. We make an attempt at understanding the relations between counting and probabilistic classes. We focus on a probabilistic analogue of the class W[P]. Using the notion of k-restricted Turing machines [7], we introduce W[P]-PFPT as a parameterized variant of the probabilistic polynomial time (PP). As in the classical complexity setting, we establish a close connection between W[P]-PFPT and the counting class #W[P] (Theorem 2). Further, we show that W[P]-PFPT is closed under complementation and symmetric differences. (Theorem 3 and Lemma 3.)

We consider the polynomial identity testing problems (ACIT) with the *syntactic degree* (See Section 2 for a definition) as a parameter. Using the construction of hitting set generators by Shpilka and Volkovich [22], we obtain what can be called as a parameterized analogue of the celebrated Schwartz-Zippel Lemma [21,24]. (Theorem 4.)

Finally, we introduce a parameterized variant of the permanent function p-perm and prove that it characterizes the class #W[1]. (Theorem 6.) Analogously, a variant of the determinant function (p-det) and show that it is Fixed Parameter Tractable (Theorem 5).

2 Preliminaries

We include some of the definitions from Parameterized Complexity theory and Complexity theory here. For Parameterized Complexity, the notations in [11,14] are followed. Definitions of complexity classes can be found in e.g., [13,4].

A *parameterized* language is a set $P \subseteq \Sigma^* \times \mathbb{N}$, where Σ is a finite alphabet. If $(x, k) \in \Sigma^* \times \mathbb{N}$ is an input instance of a parameterized language, then x is referred to as the *input* and k as the *parameter*.

A *parameterized counting* problem is a pair (f, k), where $f : \Sigma^* \to \mathbb{N}$ is a counting function and k is the parameter and Σ is a finite alphabet. For notational convenience, we will denote a parameterized counting problem as a function $f : \Sigma^* \times \mathbb{N} \to \mathbb{N}$, where the second argument to f is considered as the parameter.

A parameterized language $P \subseteq \Sigma^* \times \mathbb{N}$ is said to be *fixed-parameter tractable* if there is an algorithm that given a pair $(x, k) \in \Sigma^* \times \mathbb{N}$, decides if $(x, k) \in P$ in at most $O(f(k)|x|^c)$ steps, where $f : \mathbb{N} \to \mathbb{N}$ is a computable function and $c \in \mathbb{N}$ is a constant.

Definition 1. FPT *denotes the class of all parameterized languages that are fixed parameterized tractable.*

A parameterized language L is said to be in RFPT (Randomized FPT) if there is a $f(k)\mathrm{poly}(n)$ time bounded randomized machine accepting L with bounded one-sided error probability. See [14,11] for more details.

Definition 2. *A k-restricted machine is a non-deterministic $g(k)\mathrm{poly}(n)$ time bounded Random Access Machine (RAM) that uses at most $f(k)$ non-deterministic words, where f and g are arbitrary computable functions. Here we assume that the word size is $O(\log n)$, where n is the length of the input.*

A k-restricted Turing machine is a non-deterministic $g(k)\mathrm{poly}(n)$ time Turing machine that makes at most $f(k) \log n$ non-deterministic moves, where f and g are arbitrary computable functions.

Definition 3. *A* tail *non-deterministic machine is a k-restricted machine in which all non-deterministic steps are among last $f(k)$ steps.*

W[P] is the class of all parameterized problems (Q, k) that can be decided by a k-restricted non-deterministic machine (for more details see chapter 3 in [14]). W[1] is the class of all parameterized problems (Q, k) that can be decided by *tail* non-deterministic machine (for more details see [7]).

For a non-deterministic machine M, let $\#\mathrm{acc}_M(x, k)$ and $\#\mathrm{rej}_M(x, k)$ respectively denote the number of accepting and rejecting paths of M on input (x, k). Define $\mathrm{gap}_M(x, k) = \#\mathrm{acc}_M(x, k) - \#\mathrm{rej}_M(x, k)$.

Definition 4. *[14] A parameterized counting function (f, k) over the alphabet Σ is in $\#W[P]$ if there is a k-restricted non-deterministic machine M such that $f(x, k) = \#\mathrm{acc}_M(x, k)$.*

Definition 5. *A probabilistic k-restricted machine is a probabilistic $g(k)\mathrm{poly}(n)$ time bounded RAM that make at most $f(k)$ probabilistic moves, where f and g are some computable functions. Here we assume that one probabilistic move involves choosing a random word of $O(\log n)$ bits.*

A language L is said to be in W[P]-RFPT [19] if there is a k-restricted probabilistic machine such that $(x, k) \in L \implies Pr[M \text{ accepts } (x, k)] \geq 2/3$; and $x \notin L \implies$

$Pr[M$ rejects $(x, k)] = 0$. It should be noted that for a langauage in RFPT the number of random bits can be a big as $f(k)\text{poly}(n)$ whereas for languages in W[P]-RFPT it is bounded by $O(f(k) \log n)$.

An *arithmetic circuit* C is a directed acyclic graph with labelling on the vertices as follows. Nodes of in-degree zero are called *input* gates and are labelled from $\{-1, 1\} \cup \{x_1, \ldots, x_n\}$ where $x_1, \ldots, x_n$ are the input variables. The remaining gates are labelled $\times$ or $+$. An arithmetic circuit has exactly one gate of zero out-degree called the *output* gate. Every gate v in an arithmetic circuit can naturally be associated with a polynomial $p_v \in \mathbb{Z}[x_1, \ldots, x_n]$, where the polynomials associated at input nodes are either constants or variables. If $v = v_1 + v_2$ then $p_v = p_{v_1} + p_{v_2}$ and if $v = v_1 \times v_2$ then $p_v = p_{v_1} \times p_{v_2}$. The polynomial computed by the circuit C is the polynomial associated with its only output gate and is denoted by p_C. The size of an arithmetic circuit is the number of gates in it and is denoted by $\text{size}(C)$.

We associate a number called the *syntactic degree* (syntdeg)[1] with every gate of an arithmetic circuit C. For a leaf node v, $\text{syntdeg}(v) = 1$. If $v = v_1 + v_2$ then $\text{syntdeg}(v) = \max\{\text{syntdeg}(v_1), \text{syntdeg}(v_2)\}$ and if $v = v_1 \times v_2$ then $\deg(v) = \text{syntdeg}(v_1) + \text{syntdeg}(v_2)$. It should be noted that the degree of the polynomial computed by a circuit is bounded by its syntactic degree.

Remark 1. the parameter $d_\times$ introduced in [19] is closely related to syntdeg, in fact $\text{syntdeg} \leq 2^{d_\times} \leq 2^{\text{syntdeg}}$.

In [3], Alon obtained a characterization for multivariate polynomials that are not identically zero known as the Combinatorial Nullstellensatz:

Proposition 1 (Combinatorial Nullstellensatz, [3]). *Let $P \in \mathbb{K}[x_1, \ldots, x_n]$ be a polynomial where for every $i \in [n]$, the degree of x_i is bounded by t. Let $S \subseteq \mathbb{K}$ be a finite set of size at least $t + 1$, and $A = S^n$. Then $P \equiv 0 \iff P(a) = 0, \forall a \in A$.*

3 Probabilistic Computation

In this section, we develop a parameterized analogue of the classical complexity class PP. Our definition of W[P]-PFPT is based on k-restricted probabilistic Turing machines.

Throughout this section unless otherwise stated, $f(k)$ denotes an arbitrary computable function, and $P(n, k) = f(k) \log n$. For an input x, we denote $n = |x|$.

Definition 6. *A parameterized language L is said to be in the class W[P]-PFPT if there is a k-restricted probabilistic Turing machine M such that for any $(x, k) \in \Sigma^* \times \mathbb{N}$ we have,*

$$(x, k) \in L \Rightarrow \Pr[M \text{ accepts } (x, k)] > \frac{1}{2}$$

$$(x, k) \notin L \Rightarrow \Pr[M \text{ accepts } (x, k)] \leq \frac{1}{2}$$

where the probabilities are over the random choices made by M.

---

[1] Syntactic degree is also known as the formal degree [16] and is a standard parameter for arithmetic circuits.

Without loss of generality, we assume $\Sigma = \{0,1\}$.

In the classical setting, PP is known to have several characterizations based on, 1) difference between two #P functions [15], 2) difference between the number of accepting and rejecting paths of a polynomial time bounded non-deterministic Turing machine [15], 3) logics based on majority quantifiers [17] and 4) large fan-in circuits with threshold gates [2]. We observe that all of the characterizations except (3) hold for W[P]-PFPT. However, it is not clear if the majority quantifier logical characterization of PP [17] translates to the parameterized setting.

Definition 7 (Diff-FPT, Gap-FPT). *A parameterized function $f : \Sigma^* \times \mathbb{N} \to \mathbb{Z}$ is said to be in* Diff-FPT *if there are two functions $g, h \in $ #W[P] *such that $f(x,k) = g(x,k) - h(x,k)$.*

f is said to be in Gap-FPT *if there is a k-restricted TM M such that $f(x,k) = $ #acc$_M(x,k) - $#rej$_M(x,k)$, $\forall (x,k) \in \Sigma^* \times \mathbb{N}$.*

Firstly, we observe that the two classes Gap-FPT and Diff-FPT coincide.

Lemma 1. Gap-FPT $=$ Diff-FPT

Proof. To show Gap-FPT $\subseteq$ Diff-FPT: Let $f \in$ Gap-FPT, then there is a k-restricted M with $f(x,k) = $ #acc$_M(x,k) - $#rej$_M(x,k)$. Let M' be a new machine that simulates M on input (x,k) and accepts if and only if M rejects (x,k). Then we have $f(x,k) = $ #acc$_M(x,k) - $#acc$_{M'}(x,k)$. For the converse inclusion, let $f \in$ Diff-FPT, and M_1, M_2 be such that $f(x,k) = $ #acc$_{M_1}(x,k) - $#acc$_{M_2}(x,k)$. Let M be a new machine: on input (x,k), Guess a non-deterministic bit $b \in \{0,1\}$. Run M_1 if $b = 0$, M_2 otherwise. If $b = 0$ and M_1 accepts then accept. If $b = 1$, and M_2 accepts then reject. In all other cases guess another non-deterministic bit b' and accept if $b' = 1$ and reject otherwise. Then #acc$_M(x,k) - $#rej$_M(x,k) = $ #acc$_{M_1}(x,k) - $#acc$_{M_2}(x,k) = f(x,k)$

Lemma 2. Gap-FPT *is closed under taking p-bounded summations and products, i.e., if $g_1, \ldots, g_{t(k)} \in$ Gap-FPT, then so are $g_1 + g_2 \cdots + g_{t(k)}$ and $g_1 \times g_2 \times \cdots \times g_{t(k)}$, where t is any computable function.*

Proof. The arguments here are straightforward adaptations of proofs from classical complexity. We include it here for completeness. For summation, we can construct a new machine M that first guesses $i \in [1, t(k)]$ and and runs the k-restricted machine for g_i on (x,k).

For product, we will show for the case when $t(k) = 2$. Let $f_1, f_2 \in$ Gap-FPT. Let M_1 and M_2 as the k-restricted machines such that $f_i(x,k) = $ #acc$_{M_i}(x,k) - $#rej$_{M_i}(x,k)$, $1 \leq i \leq 2$. Let $\overline{M_i}$ be the machine that flips the answers of M_i. Let M be k-restricted machine defined as follows: On input (x,k) first simulate M_1 on (x,k). If M_1 accepts then run M_2 on (x,k) and accept if and only if M_2 does so. If M_1 rejects then run $\overline{M_2}$ on (x,k) and accept if and only if $\overline{M_2}$ does so. It can be seen that $f_1(x,k)f_2(x,k) = $ #acc$_M(x,k) - $#rej$_M(x,k)$.

The above argument can be generalized to the case $t(k) \geq 2$.

Theorem 1. *Let L be a parameterized language. The following are equivalent:*

1. *L* $\in$ W[P]-PFPT.
2. *There is a k−restricted Turing machine M such that,*
   $(x,k) \in L \iff$ #accept$_M(x,k)$ − #reject$_M(x,k) > 0$.
3. *There is a function f* $\in$ Gap-FPT *such that* $(x,k) \in L \iff f(x,k) > 0$
4. *There is a B* $\in$ FPT *and* $P(n,k) = f(k)\log n$ *such that* $(x,k) \in L$ *if and only if*
   $|\{y \in \{0,1\}^{P(n,k)} \mid (x,y,k) \in B\}| \geq 2^{P(n,k)-1} + 1.$

Proof (Theorem 1). $(1 \Rightarrow 2)$ Let $L \in$ W[P]-PFPT. Let M be a k-restricted probabilistic machine for L. Then,

$$(x,k) \in L \Rightarrow Pr[\text{M accept}(x,k)] > \frac{1}{2} \Rightarrow \frac{\#\text{accept}_M(x,k)}{\#\text{accept}_M(x,k) + \#\text{reject}_M(x,k)} > \frac{1}{2}$$
$$\Rightarrow \#\text{accept}_M(x,k) - \#\text{reject}_M(x,k) > 0$$

$(2 \Rightarrow 3)$ This directly follows from the definition of Gap-FPT.

$(3 \Rightarrow 4)$ Let $f \in$ Gap-FPT with $(x,k) \in L \iff f(x,k) > 0$, and M be a k-restricted machine with $f(x,k) = \text{gap}_M(x,k)$. Let $P(n,k)$ be the number of non-deterministic bits used by M on an input of length n with parameter k. Then $\text{gap}_M(x,k) > 0 \implies \#\text{acc}_M(x,k) > 2^{P(n,k)}/2 = 2^{P(n,k)-1}$. Let

$$B = \{\langle x,y,k \rangle \mid M \text{ on the non-deterministic path defined by } y \text{ accepts } x.\}$$

Clearly, $B \in$ FPT and $\#\text{acc}_M(x,k) = |\{y \in \{0,1\}^{P(n,k)} \mid \langle x,y,k \rangle \in B\}|$. Thus $(x,k) \in L \implies |\{y \in \{0,1\}^{P(n,k)} \mid \langle x,y,k \rangle \in B\}| > 2^{P(n,k)-1}$.

$(4 \Rightarrow 1)$ Let L as given in 4. Let M be k-restricted machine that on input (x,k) guesses a string $y \in \{0,1\}^{P(n,k)}$ and accepts if and only if $\langle x,y,k \rangle \in B$. Then we have $x \in L \iff \#\text{acc}_M(x,k) > 2^{P(n,k)-1} \iff Pr[M \text{ accepts } (x,k)] > 1/2$.

Similar to the case of PP, we observe that an FPT machine with oracle access to a function in $\#W[P]$ is equivalent to an FPT machine with a language in W[P]-PFPT as an oracle.

Theorem 2. FPT$^{\#W[P]}$ = FPT$^{W[P]\text{-PFPT}}$

Proof (Theorem 2). We show the containment in both the directions. We start with the easier direction, i.e., we show FPT$^{W[P]\text{-PFPT}} \subseteq$ FPT$^{\#W[P]}$.

Let $L \in$ FPT$^{W[P]\text{-PFPT}}$ and M be a deterministic oracle Turing machine that runs in time $f(k)\text{poly}(n)$ and $A \in$ W[P]-PFPT with $L = L(M^A)$. We need to show that $L \in FPT^{\#W[P]}$. By Theorem 1, there are two parameterized functions $g,h : \{0,1\}^* \times k \to \mathbb{N}$ with $g,h \in \#W[P]$ such that

$$(x,k) \in A \iff g(x,k) - h(x,k) > 0. \tag{1}$$

Let $\gamma : \{0,1\}^* \times k \to \mathbb{N}$ where $\gamma(0x,k) = g(x,k)$, and $\gamma(1x,k) = h(x,k), \forall x \in \{0,1\}^*$. On strings of length 0 and 1, γ can be defined arbitrarily. We have $\gamma \in \#W[P]$, since on input $y = ax$, with $a \in \{0,1\}$, the machine would run the machine for g if $a = 0$ and machine for h if $a = 1$.

We can simulate a query (y, k') made by the machine M to A by two queries to the function γ: (1) Query $(0y, k')$ and (2) $(1y, k')$, compute the difference of the values obtained and use (1) to decide the membership of (y, k') in A. Thus we can conclude $L \in \mathsf{FPT}^{\#\mathsf{W[P]}}$.

For the reverse containment, given a Turing machine M, let L_M be the language defined as : $L_M = \{((x, k, y) \in \Sigma^* \times \mathbb{N} \times \mathbb{N} \mid \#\mathrm{acc}_M(x, k) > y\}$

Claim. Let M be a k-restricted Turing machine, then $L_M \in \mathsf{W[P]}$-PFPT.

Proof (Claim). Let M' be a Turing machine runs in FPT time computing function $t(x, k, y)$, that on input (x, k, y), produces exactly y accepting paths, where y is represented in binary, and $y \in [0, 2^{P(n,k)}]$. Clearly, M' is a k-restricted Turing machine, since it needs to use only $P(n, k)$ many non-deterministic bits. Thus the function $t(x, k, y) = y$ is in $\#\mathsf{W[P]}$. Let $f_M(x, k, y) = \#\mathrm{acc}_M(x, k) - y$. Then by Lemma 1 f_M is in $\mathsf{gapW[P]}$ and the claim now follows from Theorem 1. $\square$

Let $L \in \mathsf{FPT}^{\#\mathsf{W[P]}}$, then there is a deterministic oracle Turing machine M' that runs FPT time, and a function $g \in \#\mathsf{W[P]}$ such that $L = M'^g$. Let M be a k-restricted Turing machine that uses at most $f(k) \log n$ non-deterministic steps such that $g(x, k) = \#\mathrm{acc}_M(x, k)$. We use the standard binary search technique to show that $g(x, k)$ can be computed using $O(kn)$ many queries to the language L_M.

Input (x, k), oracle access to L_M. **Output** $g(x, k)$.
1. Initialize $p = P(|x|, k)$, $y = 2^p$.
2. Repeat steps 3 & 4 until $p \geq 0$
3. Query (x, k, y) to the oracle; If YES, then set $b_p = 1$ and $y = y + 2^{p-1}$; Else set $b_p = 0$.
4. Set $p = p - 1$
5. Return $a = \mathrm{binary}(b_p b_{p-1} \ldots b_0)$.

In the above $\mathrm{binary}(b_p b_{p-1} \ldots b_0) = \sum_{i=0}^{p} 2^i b_i$. Clearly, the algorithm above runs in time $f(k)\mathrm{poly}(n)$, and hence computing g can be done in FPT with oracle access to $L_M \in \mathsf{W[P]}$-PFPT. This concludes the inclusion in the converse direction. $\square$

Theorem 3. $\mathsf{W[P]}$-PFPT *is closed under complementation.*

Proof can be found in the complete version of paper [6].

Lemma 3. $\mathsf{W[P]}$-PFPT *is closed under symmetric difference.*

Proof. The proof essentially follows the ideas in the classical setting [2]. Let $L_1, L_2 \in \mathsf{W[P]}$-$\mathsf{PFPT}$. By Theorem 1, we have $B_1, B_2 \in \mathsf{FPT}$ and $P(n, k) = f(k) \log n$ such that for any $x \in \{0, 1\}^*$, $k \in \mathbb{N}$ and $i \in \{1, 2\}$,

$$(x, k) \in L_i \iff |\{y_i \in \{0, 1\}^{P(n,k)} | (x, k, y_i) \in B_i\}| \geq 2^{P(n,k)-1} + 1$$

Using a construction similar to the one used in the proof of Theorem 3, we get parameterized languages B_1' and B_2', and a function $P'(n, k) = f'(k) \log n$ with the following property for $1 \leq i \leq 2$:

$$(x, k) \in L_i \implies |\{y_i \in \{0, 1\}^{P'(n,k)} | (x, k, y_i) \in B_i'\}| \geq 2^{P'(n,k)-1} + 1; \text{ and}$$
$$(x, k) \notin L_i \implies |\{y_i \in \{0, 1\}^{P'(n,k)} | (x, k, y_i) \in B_i'\}| \leq 2^{P'(n,k)-1} - 1.$$

Let $a_1(x), a_2(x) \in \mathbb{Z}$ such that $|\{y \in \{0,1\}^{P(n,k)} \mid \langle x,y \rangle \in B'_i\}| = 2^{P(n,k)-1} + a_i(x)$ for $1 \leq i \leq 2$. Thus $|\{y \in \{0,1\}^{P(n,k)} \mid \langle x,y \rangle \notin B'_i\}| = 2^{P(n,k)-1} - a_i(x)$. For $x \in \Sigma^*$, let

$$
\begin{aligned}
\ell(x,k) &\triangleq |S(x,k)| \\
&= |(2^{P'(n,k)-1} + a_1)(2^{P'(n,k)-1} - a_2) \\
&\quad + (2^{P'(n,k)-1} - a_1)(2^{P'(n,k)-1} + a_2)| \\
&= (2^{2P'(n,k)-1} - a_1 a_2),
\end{aligned}
$$

where $S(x,k) = \{\langle y_1, y_2 \rangle \mid (< x, y_1 > \in B_1 \wedge \langle x, y_2 \rangle \notin B_2) \vee (\langle x, y_1 \rangle \notin B_1 \wedge \langle x, y_2 \rangle \in B_2)\}$. Now, if $x \in L_1 \triangle L_2$ then either $(a_1 \geq 1$ and $a_2 < 0)$ or $(a_1 < 0$ and $a_2 \geq 1)$ then $\ell > 2^{2P(n,k)-1}$ and if $x \notin L_1 \triangle L_2$ then either both a_1 and a_2 are greater than equal to 1 or both are less than 1, and in both the cases $\ell \leq 2^{2P(n,k)-1}$. Let M' be a k-restricted Turing machine that on input (x,k) guesses two strings y_1 and y_2 of length $P'(n,k)$ each, and queries (x,k,y_i) to B'_i, $1 \leq i \leq 2$, accepts if and only if exactly one of the oracle answers is YES. It can be seen that $\#\mathrm{acc}_{M'}(x,k) = \ell(x,k)$. We conclude $L_1 \triangle L_2 \in \mathsf{W[P]\text{-}PFPT}$.

4   Polynomial Identity Testing

Müller [19] studied the Arithmetic Circuit Identity Testing (ACIT) problem with various parameters and obtained upper bounds as well as hardness results for each of the parameters considered. However none of the parameters considered in [19] seem adequate for developing a complexity theory for the parameterized probabilistic and counting classes along the lines of classical complexity classes.

Recall that, in ACIT we are given an arithmetic circuit C as an input and the task is to test if the polynomial computed by C is identically zero. We consider the degree of the polynomial computed by C as a parameter.

Problem 1 (p-acit). *Input*: Arithmetic circuit C, $\mathsf{syntdeg}(C) \leq k$.
Parameter: k.
Task: Test if the polynomial computed by C is identically zero.

Our main objective now is to show that p-acit $\in \mathsf{W[P]\text{-}RFPT}$. However, it should be noted that this does not follow directly from the Schwartz-Zippel Lemma, since it would require $O(n \log k)$ random bits. So the challenge here is to reduce the number of random bits required to $f(k) \log n$. Towards this, we use a mapping defined by Shpilka and Volkovich [22] that reduces the number of variables from n to $2k$. Then we apply Alon's Combinatorial Nullstellensatz [3] to obtain what can be treated as a parameterized version of the Schwartz-Zippel lemma.

We begin with a few observations on polynomials of degree at most k. Let S be any finite subset of $\mathbb{K}$ that includes $0 \in \mathbb{K}$ and let $W_n^k(S)$ denote the set of all vectors in S^n with at most k non zero entries i.e, the set of all vectors of Hamming weight at most k.

Lemma 4. *Let f be an n-variate polynomial of degree at most k. Then*

$$ f \equiv 0 \iff \forall a \in W_n^k(S) \ f(a) = 0, $$

where $S \subset \mathbb{K}$ has at least $k+1$ elements.

Proof. For simplicity, we denote $W_n^k(S)$ by W_n^k. The proof is by induction on n. For the base case, suppose $n \leq k$. Since individual degrees of each variable is bounded by k, by Proposition 1, we have $f \equiv 0 \iff f(a) = 0 \,\forall\, a \in S^n$, for an S with $|S| \geq k$.

For the induction step, let $n > k$, and $f(a) = 0 \,\forall\, a \in W_n^k$. For $i \in \{1, \dots, n\}$, let $f_i = f|_{x_i=0}$, i.e., f substituted with $x_i = 0$. Note that each of the f_i is a degree k polynomial on at most $n - 1$ variables, and $\forall a \in W_{n-1}^k$ $f_i(a) = 0$. By the induction hypothesis, we have $f_i \equiv 0$, and hence x_i divides f. Repeating the argument for all $i \in [1, \dots, n]$, we have $x_1 x_2 \cdots x_n$ divides f, and hence $\deg(f) \geq n > k$, a contradiction since $\deg(f) = k < n$. Thus we conclude $\forall a \in W_n^k$ $f(a) = 0 \implies f \equiv 0$. The converse direction is trivially true.

We need a function introduced by Shpilka and Volkovich [22], that gives a map $G_k : \mathbb{K}[x_1, \dots, x_n] \to \mathbb{K}[y_1, \dots, y_{2k}]$ and serves as a non-identity preserving for a large class of polynomials. We observe that G_k also functions as a non-identity preserving map for the class of all n variate polynomials of degree at most k. We begin with the definition of the generator G_k.

Definition 8 (Shpilka-Volkovich Hitting Set Generator, [22]). *Let $a_1, \dots, a_n$ be distinct elements in $\mathbb{K}$. Let $G_k^i \in \mathbb{K}[y_1, \dots, y_k, z_1, \dots, z_k]$ be the polynomial defined as follows:*

$$G_k^i(y_1, \dots, y_k, z_1, \dots, z_k) = \sum_{j=1}^{k} L_i(y_i) z_i, \quad \text{where } L_i(x) = \frac{\prod_{j \neq i}(x - a_j)}{\prod_{j \neq i}(a_i - a_j)}.$$

The generator G_k is defined as $G_k \overset{\triangle}{=} (G_k^1, \dots, G_k^n)$.

Lemma 5. *For any finite set $S \subset \mathbb{K}$, then $W_n^k(S) \subseteq \{(G_k^1(a), \dots, G_k^n(a)) \mid a \in (S \cup \{a_1, \dots, a_n\})^{2k}\}$.*

Proof. The proof essentially follows the arguments in [22]. We include a sketch here for the sake of completeness. Note that,

$$L_i(\alpha) = \begin{cases} 0 & \alpha = a_j, \text{ if } j \neq i \\ 1 & \text{if } \alpha = a_i. \end{cases}$$

Thus if we set $y_\ell = a_i$, then the image of G_k^i contains z_i as a summand. By ensuring that $y_j, i \neq j$ gets some $a_\ell, i \neq \ell$, we get $G_k^i = z_i$. In this way we can obtain all vectors of Hamming weight k, by setting y_i's and z_i's accordingly.   □

Combining Lemma 5 with Lemma 4 we have:

Lemma 6. *Let f be a polynomial of degree at most k. Then $f \equiv 0 \iff f(G_k) \equiv 0$.*

Theorem 4. p-acit *is in* W[P]-RFPT

Proof. By Lemma 6 p-acit reduces to testing identity of $2k$-variate polynomials of degree $O(nk)$ (since the polynomials L_i have degree n). Now applying the Schwartz-Zippel lemma [21,24], we obtain a randomized algorithm that uses $O(2k \log(nk))$ random bits and runs in time polynomial in n and k.

5 Parameterized Permanent vs Determinant

The determinant (det) permanent (perm) functions are defined as

$$\det(A) = \sum_{\sigma \in S_n} \prod_{i=1}^{n} \text{sgn}(\sigma) a_{i,\sigma(i)} \tag{2}$$

$$\text{perm}(A) = \sum_{\sigma \in S_n} \prod_{i=1}^{n} a_{i,\sigma(i)}, \tag{3}$$

where $A = (a_{i,j})_{1 \le i,j \le n} \in \mathbb{N}^{n \times n}$, S_n is the set of all permutations on n symbols and sgn is the sign function for permutations. It is known that, given an integer matrix A, computing A can be done in polynomial time (e.g., Gaussian Elimination method). In his celebrated paper, Valiant [23] showed that computing perm of an integer even for 0 or 1 matrix is complete for #P. Though there are several natural counting problems that characterize #$W[1]$, it is desirable to have a parameterized variant of permanent so that we get access to the algebraic properties of the permanent function.

Naturally, we expect any parameterized variant of permanent to be a function of degree k in n^2 variables, where k is the parameter. One way to achieve this would be to restrict the summation given in (3) that move exactly k-elements. Formally, a permutation $\sigma \in S_n$ is said to be a k-permutation, if $|\{i \mid \sigma(i) \ne i\}| = k$. Let $S_{n,k}$ denote the set of all k-permutations on n symbols.

Definition 9. *Let k be a parameter. The parameterized determinant (p-det) and permanent (p-perm) functions of a matrix $A \in \mathbb{Z}^{n \times n}$ are defined as follows:*

$$\text{p-det}(A, k) = \sum_{\sigma \in S_{n,k}} \prod_{i \ne \sigma(i)} \text{sgn}(\sigma) a_{i\sigma(i)}$$

$$\text{p-perm}(A, k) = \sum_{\sigma \in S_{n,k}} \prod_{i \ne \sigma(i)} a_{i\sigma(i)},$$

where k is a parameter. By abusing the notation, we also let p-perm denote the problem of computing p-perm of an $n \times n$ matrix, where k is the parameter.

Quite expectedly, p-det is FPT and p-perm can be shown to be #$W[1]$ complete under fpt-reductions. We start with the tractability of p-det.

Theorem 5. p-det *on integer matrices is fixed parameter tractable.*

Proof. Let $A \in \mathbb{Z}^{n \times n}$, and k be the parameter. Let A' be the matrix obtained from A by replacing the diagonal entries in A by zeroes. Clearly p-det$(A, k) =$ p-det(A', k). Let x be a formal variable. Then $\det(xA')$ is a univariate polynomial of degree bounded by n, and the coefficient of x^k in $\det(xA')$ is equal to p-det(A, k). The value p-det(A, k) be recovered using the standard interpolation of univariate polynomials.   □

Theorem 6. p-perm *on matrices in $\mathbb{N}^{n \times n}$ is #$W[1]$ complete. The hardness holds even in the case of 0-1 matrices.*

Proof. It is known that counting k-matchings in a bipartite graph is complete for #W[1] even in the weighted case [9]. We prove a parameter preserving equivalence between p-perm and the problem of counting k-matchings in a bipartite graph which completes the proof. For the upper bound, we give a reduction from p-perm to counting k-matchings in a bipartite graph. For a given matrix $A \in \mathbb{N}^{n \times n}$ define the matrix A' by setting the diagonal entries of A to zero, i.e., $A'[i, j] = A[i, j]$ if $i \neq j$ and $A'[i, i] = 0$, $1 \leq i \leq n$. Note that p-perm(A) = p-perm(A'). Every k-permutation of $\{1, \ldots, n\}$ corresponds to a matching of size k in the bipartite graph G' with A' is the bipartite adjacency matrix. Thus p-perm(A') =the sum of weights of k-matchings in G'.

For the hardness we give a reduction in the reverse direction, i.e., a parameter preserving reduction from counting the number of k-matchings in a bipartite graph $G = (U, V, E)$ to computing p-perm of an integer matrix.

Let $G = (U, V, E)$ be a bipartite graph. Without loss of generality, assume that $U = V = \{1, \ldots, n\}$. Construct a new bipartite graph $G' = (U', V', E')$ with $U' = V' = \{1, \ldots, 2n\}$. For every edge of the form $(i, j) \in E, i \neq j$, G' contains the edge $(i, j) \in E'$. For edges of the form $(i, i) \in E$, G' contains the edge $(i, n + i) \in E'$. Note that the vertices $n + 1, \ldots, 2n$ in U' are isolated vertices.

Note that the number of matchings in G and those in G' of a given size k are equal. Let A' be the bipartite adjacency matrix of G'. Every k-permutation of $[2n]$ that contributes a non-zero value to p-perm(A') corresponds to a matching of size k in G'. Moreover, none of the k-matchings in G' will have an edge of the form $(i, i), i \in [2n]$. Thus, p-perm(A', k) = #matchings of size k in G'. This completes the proof.   □

6 Conclusions

We have studied parameterized variants of probabilistic computation. We hope that our definition of W[P]-PFPT leads to further developments in the structural aspects of probabilistic and counting complexities in the parameterized world. Further, W[P]-PFPT might be useful in defining a parameterized variant of the Counting Hierarchy (CH) which could in turn have implications to parameterized complexity of numerical and algebraic computation [1]. Though definition of a parameterized CH based on W[P]-PFPT is straightforward, the usefulness of such a definition would rely on W[P]-PFPT being closed under intersection, which is not known currently.

Further, we believe any fixed parameter tractable randomized algorithm should naturally place the problem in W[P]. One way to achieve this is to obtain randomized FPT algorithms that use at most $O(f(k) \log n)$ random bits. As a first step towards this direction, we introduce a natural parameterization to the polynomial identity testing for which we obtain such an algorithm. We hope our observations will lead to further development of randomness efficient parameterized algorithms.

Acknowledgements. We thank anonymous reviewers for their comments on an earlier version of this paper which helped in improving the presentation of the article.

References

1. Allender, E., Bürgisser, P., Kjeldgaard-Pedersen, J., Miltersen, P.B.: On the complexity of numerical analysis. SIAM J. Comput. 38(5), 1987–2006 (2009)
2. Allender, E., Wagner, K.W.: Counting hierarchies: Polynomial time and constant. Bulletin of the EATCS 40, 182–194 (1990)
3. Alon, N.: Combinatorial nullstellensatz. Combinatorics, Problem and Computing 8 (1999)
4. Arora, S., Barak, B.: Computational Complexity: A Modern approach. Cambridge Univeristy Press (2009)
5. Bläser, M., Curticapean, R.: Weighted counting of k-matchings is #W[1]-hard. In: IPEC, pp. 171–181 (2012)
6. Chauhan, A., Rao, B.V.R.: Parameterized analogues of probabilistic computation. CoRR, abs/1409.7790 (2014)
7. Chen, Y., Flum, J., Grohe, M.: Machine-based methods in parameterized complexity theory. Theor. Comput. Sci. 339(2-3), 167–199 (2005)
8. Curticapean, R.: Counting matchings of size k is $\sharp$W[1]-hard. In: Fomin, F.V., Freivalds, R., Kwiatkowska, M., Peleg, D. (eds.) ICALP 2013, Part I. LNCS, vol. 7965, pp. 352–363. Springer, Heidelberg (2013)
9. Curticapean, R., Marx, D.: Complexity of counting subgraphs: only the boundedness of the vertex-cover number counts. CoRR, abs/1407.2929 (2014); to Appear in FOCS 2014
10. Downey, R.G., Fellows, M.R.: Fixed-parameter intractability. In: Structure in Complexity Theory Conference, pp. 36–49 (1992)
11. Downey, R.G., Fellows, M.R.: Parameterized Complexity. Springer (1997)
12. Downey, R.G., Fellows, M.R., Regan, K.W.: Parameterized circuit complexity and the W hierarchy. Theor. Comput. Sci. 191(1-2), 97–115 (1998)
13. Du, D.-Z., Ko, K.-I.: Theory of Computational Complexity. Springer (2000)
14. Flum, J., Grohe, M.: Parameterized Complexity Theory. Springer (2008)
15. Fortnow, L.: Counting complexity. In: Hemaspaandra, L., Selman, A. (eds.) Complexity Theory Retrospective II, pp. 81–107 (1997)
16. Kayal, N., Saha, C., Saptharishi, R.: A super-polynomial lower bound for regular arithmetic formulas. In: STOC, pp. 146–153 (2014)
17. Kontinen, J.: A logical characterization of the counting hierarchy. ACM Trans. Comput. Log. 10(1) (2009)
18. Müller, M.: Parameterized derandomization. In: Grohe, M., Niedermeier, R. (eds.) IWPEC 2008. LNCS, vol. 5018, pp. 148–159. Springer, Heidelberg (2008)
19. M üller, M.: Parameterized Randomization. PhD thesis, Albert-Ludwigs-Universität Freiburg im Breisgau (2008)
20. Niedermeier, R.: Invitation to Fixed-Parameter Algorithms. Oxford Lecture Series in Mathematics and Its Applications, vol. 31. Oxford University Press (2006)
21. Schwartz, J.T.: Fast probabilistic algorithms for verification of polynomial identities. J. ACM 27(4), 701–717 (1980)
22. Shpilka, A., Volkovich, I.: Improved polynomial identity testing for read-once formulas. In: Dinur, I., Jansen, K., Naor, J., Rolim, J. (eds.) APPROX and RANDOM 2009. LNCS, vol. 5687, pp. 700–713. Springer, Heidelberg (2009)
23. Valiant, L.G.: The complexity of computing the permanent. Theor. Comput. Sci. 8, 189–201 (1979)
24. Zippel, R.: Probabilistic algorithms for sparse polynomials. In: Ng, K.W. (ed.) EUROSAM 1979 and ISSAC 1979. LNCS, vol. 72, pp. 216–226. Springer, Heidelberg (1979)

Algebraic Expressions of Rhomboidal Graphs

Mark Korenblit

Holon Institute of Technology, Israel
korenblit@hit.ac.il

Abstract. The paper investigates relationship between algebraic expressions and graphs. We consider rhomboidal non-series-parallel graphs, specifically, a digraph called a full square rhomboid. Our intention is to simplify the expressions of full square rhomboids. We describe two decomposition methods for generating expressions of rhomboidal graphs and carry out their comparative analysis.

1 Introduction

A *graph* $G = (V, E)$ consists of a *vertex set* V and an *edge set* E, where each edge corresponds to a pair (v, w) of vertices. A graph $G^2 = (V, E')$ is a *square of a graph* $G = (V, E)$ if $E' = \{(u, w) : (u, w) \in E \vee ((u, v) \in E \wedge (v, w) \in E)$ for some $v \in V\}$. A two-terminal directed acyclic graph (*st-dag* in [2]) has only one source and only one sink.

We consider a *labeled graph* in which each edge has a unique label. Each path between the source and the sink (a *spanning path*) in an st-dag can be represented by a product of all edge labels of the path. We define the sum of edge label products corresponding to all possible spanning paths of an st-dag G as the *canonical expression* of G. An algebraic expression is called an *st-dag expression* (a *factoring of an st-dag* in [2]) if it is algebraically equivalent to the canonical expression of an st-dag. An st-dag expression consists of literals (edge labels), and the operators $+$ (disjoint union) and $\cdot$ (concatenation, also denoted by juxtaposition). An expression of an st-dag G will be denoted by $Ex(G)$ and the total number of literals in the expression will be defined as its *complexity*.

A *series-parallel graph* is defined recursively so that a single edge is a series-parallel graph and a graph obtained by a *parallel* or a *series composition* of series-parallel graphs is series-parallel [2]. A series-parallel graph expression has a representation in which each literal appears only once [2], [12]. This representation is the shortest for a series-parallel graph expression. For example, the canonical expression of the series-parallel graph presented in Figure 1 is $abd + abe + acd + ace + fe + fd$ and it can be reduced to $(a(b + c) + f)(d + e)$.

A *Fibonacci graph* [9] has vertices $\{1, 2, 3, \ldots, n\}$ and edges $\{(v, v + 1) \mid v = 1, 2, \ldots, n - 1\} \cup \{(v, v + 2) \mid v = 1, 2, \ldots, n - 2\}$. As shown in [5], an st-dag is series-parallel if and only if it does not contain a subgraph which is a homeomorph of the *forbidden subgraph* positioned between vertices 1 and 4 of the Fibonacci graph illustrated in Figure 2. Thus a Fibonacci graph gives a generic example of non-series-parallel graphs.

S. Ganguly and R. Krishnamurti (Eds.): CALDAM 2015, LNCS 8959, pp. 193–204, 2015.
© Springer International Publishing Switzerland 2015

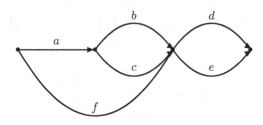

Fig. 1. A series-parallel graph

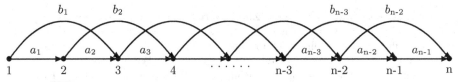

Fig. 2. A Fibonacci graph

Mutual relations between graphs and expressions are discussed in a number of works. Specifically, [16], [17], and [22] consider the correspondence between series-parallel graphs and read-once functions. A Boolean function is defined as *read-once* if it may be computed by a formula in which every variable appears exactly once (*read-once formula*) [7]. Hence, a series-parallel graph expression is reduced to the representation that can be considered as a read-once formula (Boolean operations are replaced by arithmetic ones).

Problems related to computations on graphs have applications in different areas. Specifically, many network problems (e.g., flow [23], scheduling [6], reliability [21] problems) which are either intractable or have complicated solutions in the general case are solvable for series-parallel graphs.

An expression of a homeomorph of the forbidden subgraph belonging to any non-series-parallel st-dag has no representation in which each literal appears once. For example, consider the subgraph positioned between vertices 1 and 4 of the Fibonacci graph shown in Figure 2. Possible shortest representations of its expression are $a_1 (a_2 a_3 + b_2) + b_1 a_3$ or $(a_1 a_2 + b_1) a_3 + a_1 b_2$. For this reason, an expression of a non-series-parallel st-dag can not be represented as a read-once formula. However, for arbitrary functions, which are not read-once, generating the optimum factored form is NP-complete [24].

The problem of factoring Boolean functions into compact formulae is one of the basic operations in algorithmic logic synthesis since the complexity of a logic circuit and computation time depend on the number of literals. Some algorithms developed in order to obtain good factored forms are described in [15], [8].

A symbolic approach to scheduling of a robotic line simulated by a Fibonacci graph is considered in [14]. The method uses the max-algebra tools and allows the shortest-path problem to be interpreted as the computation of the st-dag expression. The complexity of this problem is determined by the complexity of the st-dag expression.

A method for automated composition of algebraic expressions in complex business process modeling based on acyclic directed graph reductions is introduced in

[19]. The method transforms business step dependencies into digraphs and finally generates algebraic expressions.

Thus expressions with a minimum (or, at least, a polynomial) complexity may be considered as a key to generating efficient algorithms on distributed systems.

In [12] we presented an algorithm, which generates the expression of $O\left(n^2\right)$ complexity for an n-vertex Fibonacci graph. In [13] we discussed a *square rhomboid* (Figure 3). This graph looks like a planar approximation of the square of a *rhomboid* [11] (obtained by consecutive series compositions of *rhomb* st-dags) in which edges labeled by a, b, c (Figure 3) are absent. Geometrically, a square rhomboid (SR for brevity) is a "gluing" of two Fibonacci graphs, i.e., it is the next harder one in a sequence of increasingly non-series-parallel graphs.

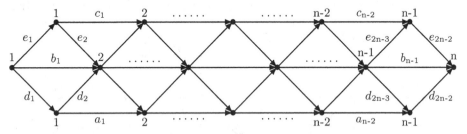

Fig. 3. A square rhomboid of size n

In this paper we investigate a more complicated graph called a *full square rhomboid* (FSR) which is a real square of a rhomboid and, in addition to all edges of an SR, has edges labeled by f and g (Figure 4). Our intention is to generate and simplify the expressions of full square rhomboids.

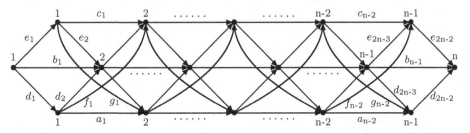

Fig. 4. A full square rhomboid of size n

The set of vertices of N-vertex SR and FSR consists of $\frac{N+2}{3}$ middle (*basic*), $\frac{N-1}{3}$ *upper*, and $\frac{N-1}{3}$ *lower* vertices. Upper and lower vertices numbered x will be denoted in formulae by $\overline{x}$ and $\underline{x}$, respectively. SR and FSR including n basic vertices will be denoted by $SR(n)$ and $FSR(n)$, respectively, and will be called an SR and an FSR of size n.

Being complicated graphs of a regular structure, rhomboidal graphs may be applied for simulating advanced homogeneous systems, e.g., production lines or telecommunication networks. At that square rhomboids being planar graphs are more appropriate for structures in which crossing edges are undesirable whereas full square rhomboids are suitable for more general applications.

2   Generating Expressions for Square Rhomboids

The expressions of square rhomboids are generated using two-vertex decomposition method (2-VDM) and one-vertex decomposition method (1-VDM) [13]. Both methods are based on revealing subgraphs in the initial graph. The resulting expression is produced by a special composition of subexpressions describing these subgraphs.

2-VDM is applied as follows. For a non-trivial SR subgraph with a source p and a sink q we choose two *decomposition vertices* one of which belongs to the upper group and the other one belongs to the lower group. These vertices have the same number i chosen as $\frac{q+p-1}{2}$ ($\lceil\frac{q+p-1}{2}\rceil$ or $\lfloor\frac{q+p-1}{2}\rfloor$). We conditionally split each SR through its decomposition vertices (see the example in Figure 5).

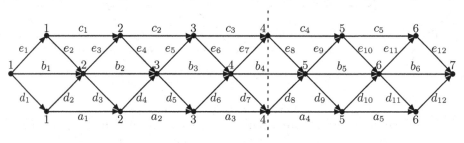

Fig. 5. Decomposition of a square rhomboid by 2-VDM

Two kinds of subgraphs are revealed in the graph in the course of decomposition. The first of them is an SR with a fewer number of vertices than the initial SR. The second one is an SR supplemented by two additional edges at one of four sides. Possible varieties of this st-dag (we call it a *single-leaf square rhomboid* and denote by $\widehat{SR}$) are four subgraphs of an SR in Figure 5 positioned between vertices 1 and $\overline{4}$, $\overline{4}$ and 8, 1 and $\underline{4}$, $\underline{4}$ and 8. Let $\widehat{SR}(n)$ (an $\widehat{SR}$ of size n) denote an $\widehat{SR}$ including n basic vertices.

We denote by $E(p,q)$ a subexpression related to an SR subgraph with a source p and a sink q. We denote by $E(p,\overline{q})$, $E(\overline{p},q)$, $E(p,\underline{q})$, $E(\underline{p},q)$ subexpressions related to $\widehat{SR}$ subgraphs with a source p and a sink $\overline{q}$, a source $\overline{p}$ and a sink q, a source p and a sink $\underline{q}$, and a source $\underline{p}$ and a sink q.

One can see that any path from vertex 1 to vertex 8 in Figure 5 passes either through one of decomposition vertices ($\overline{4}$ or $\underline{4}$) or through edge b_4. Therefore, in the general case a current subgraph is decomposed into six new subgraphs and

$$E(p,q) \leftarrow E(p,i)b_iE(i+1,q) + E(p,\overline{i})E(\overline{i},q) + E(p,\underline{i})E(\underline{i},q). \qquad (1)$$

Subgraphs described by subexpressions $E(p,i)$ and $E(i+1,q)$ include all paths from vertex p to vertex q passing through edge b_i. Subgraphs described by subexpressions $E(p,\overline{i})$ and $E(\overline{i},q)$ include all paths from vertex p to vertex q passing via vertex $\bar{i}$. Subgraphs described by subexpressions $E(p, \underline{i})$ and $E(\underline{i}, q)$ include all paths from vertex p to vertex q passing via vertex $\underline{i}$.

An $\widehat{SR}$ subgraph is decomposed into six new subgraphs in the same way as an SR (see the example in Figure 6). Two decomposition vertices (one from the upper and one from the lower group of vertices) with the same absolute ordinal numbers are selected in the $\widehat{SR}$. These vertices are chosen so that the location of the split is in the middle of the subgraph.

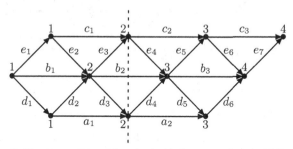

Fig. 6. Decomposition of a single-leaf square rhomboid by 2-VDM

Three kinds of subgraphs are revealed in an $\widehat{SR}$ in the course of decomposition. The first and the second of them are an SR and an $\widehat{SR}$, respectively. The third one is an SR supplemented by two additional pairs of edges (one pair is on the left and another one is on the right). Possible varieties of this st-dag (we call it a *dipterous square rhomboid* and denote it by $\widehat{\widehat{SR}}$) are illustrated in Figure 7(a) (a *parallelogram* $\widehat{\widehat{SR}}$ *graph*) and Figure 7(b) (a *trapezoidal* $\widehat{\widehat{SR}}$ *graph*). Let $\widehat{\widehat{SR}}(n)$ (an $\widehat{\widehat{SR}}$ of size n) denote an $\widehat{\widehat{SR}}$ including n basic vertices.

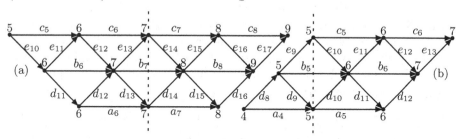

Fig. 7. Decomposition of dipterous square rhomboids by 2-VDM

An $\widehat{\widehat{SR}}$ subgraph is decomposed into six new subgraphs in the same way as an SR and an $\widehat{SR}$ (see examples in Figure 7(a, b)). The number i of the upper and the lower decomposition vertices for a current $\widehat{\widehat{SR}}$ subgraph positioned between vertices p and q, is chosen as $\frac{q+p}{2}$ ($\lceil \frac{q+p}{2} \rceil$ or $\lfloor \frac{q+p}{2} \rfloor$). In the course of decomposition, two kinds of subgraphs are revealed in an $\widehat{\widehat{SR}}$. They are an $\widehat{SR}$ and an $\widehat{\widehat{SR}}$.

1-VDM consists in splitting a non-trivial SR with a source p and a sink q through one decomposition vertex i located in the basic group of the subgraph. The number i is chosen as $\frac{q+p}{2}$ ($\lceil\frac{q+p}{2}\rceil$ or $\lfloor\frac{q+p}{2}\rfloor$) - see the example in Figure 8.

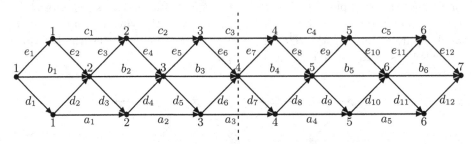

Fig. 8. Decomposition of a square rhomboid by 1-VDM

As for 2-VDM, two SR subgraphs and four $\widehat{SR}$ subgraphs are revealed in the course of decomposition. Any path from vertex 1 to vertex 7 in Figure 8 passes through decomposition vertex 4 or through edge c_3 or through edge a_3. Therefore, in the general case a current subgraph is decomposed into six new subgraphs and

$$E(p,q) \leftarrow E(p,i)E(i,q) + E(p,\overline{i-1})c_{i-1}E(\overline{i},q) + E(p,\underline{i-1})a_{i-1}E(\underline{i},q). \quad (2)$$

Subgraphs described by subexpressions $E(p,i)$ and $E(i,q)$ include all paths from vertex p to vertex q passing through vertex i. Subgraphs described by subexpressions $E(p,\overline{i-1})$ and $E(\overline{i},q)$ include all paths from vertex p to vertex q passing via edge c_{i-1}. Subgraphs described by subexpressions $E(p,\underline{i-1})$ and $E(\underline{i},q)$ include all paths from vertex p to vertex q passing via edge a_{i-1}.

An $\widehat{SR}$ subgraph is decomposed by 1-VDM through a decomposition vertex selected in its basic group into six new subgraphs in a similar way to 2-VDM. The decomposition vertex is chosen so that the location of the split is in the middle of the subgraph. The decomposition also gives one SR subgraph, three $\widehat{SR}$ subgraphs, and two $\widehat{\widehat{SR}}$ subgraphs.

Finally, an $\widehat{\widehat{SR}}$ subgraph is also decomposed into two $\widehat{SR}$ subgraphs and four $\widehat{\widehat{SR}}$ subgraphs. The number i of the decomposition vertex in the basic group for a current $\widehat{\widehat{SR}}$ subgraph is chosen as $\frac{q+p+1}{2}$ ($\lceil\frac{q+p+1}{2}\rceil$ or $\lfloor\frac{q+p+1}{2}\rfloor$).

Thus by the master theorem [4], the total number of literals $T(n)$ in expressions $Ex(SR(n))$ derived both by 2-VDM and 1-VDM is $O\left(n^{\log_2 6}\right)$.

However, numerically 1-VDM is significantly more efficient than 2-VDM [13]. Specifically, for $n = 50$, the expression $Ex(SR(n))$ derived by the best algorithm based on 2-VDM contains 43585 literals while the expression $Ex(SR(n))$ derived by 1-VDM contains only 28741 literals.

3   Generating Expressions for Full Square Rhomboids

Now, we attempt to apply 2-VDM and 1-VDM to a full square rhomboid.

Analogously to graphs mentioned in the previous section, we define *single-leaf full square rhomboid* of size n denoted by $\widehat{FSR}(n)$ and *dipterous full square rhomboids (trapezoidal* and *parallelogram*) of size n denoted by $\widehat{\widehat{FSR}}(n)$. These graphs, in addition to all edges in corresponding $\widehat{SR}$ and $\widehat{\widehat{SR}}$ graphs, have edges labeled by f and g (as in Figure 4).

We denote by $E(p,q)$ a subexpression related to an FSR subgraph with a source p and a sink q. We denote by $E(p,\overline{q})$, $E(\overline{p},q)$, $E(p,\underline{q})$, $E(\underline{p},q)$ subexpressions related to $\widehat{FSR}$ subgraphs with a source p and a sink $\overline{q}$, a source $\overline{p}$ and a sink q, a source p and a sink $\underline{q}$, and a source $\underline{p}$ and a sink q. We denote by $E(\overline{p},\overline{q})$, $E(\overline{p},\underline{q})$, $E(\underline{p},\overline{q})$, $E(\underline{p},\underline{q})$ subexpressions related to $\widehat{\widehat{FSR}}$ subgraphs with a source $\overline{p}$ and a sink $\overline{q}$, a source $\overline{p}$ and a sink $\underline{q}$, a source $\underline{p}$ and a sink $\overline{q}$, and a source $\underline{p}$ and a sink $\underline{q}$, respectively.

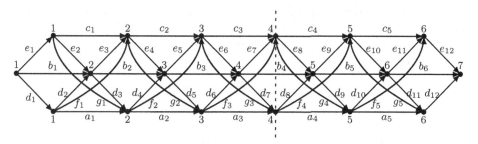

Fig. 9. Decomposition of a full square rhomboid by 2-VDM

Figure 9 illustrates the example of a decomposition of an FSR by 2-VDM. Appearance of new edges does not change the essence of the splitting procedure because these edges (labeled by f and g) do not cross the "splitting line" that passes between vertices $\overline{4}$ and $\underline{4}$. In all revealed subgraphs, edges f_i and g_i are together with pairs of edges d_{2i} and e_{2i+1}, and e_{2i} and d_{2i+1}, respectively. Therefore, in the general case a current FSR subgraph is decomposed into two FSR subgraphs and four $\widehat{FSR}$ subgraphs, and its expression is the same as in statement (1).

$\widehat{FSR}$ and $\widehat{\widehat{FSR}}$ subgraphs are also decomposed by 2-VDM into six new subgraphs in a similar way to $\widehat{SR}$ and $\widehat{\widehat{SR}}$ subgraphs. Thus the complexity of the expression $Ex(FSR(n))$ derived by 2-VDM is also $O\left(n^{\log_2 6}\right)$.

Now, consider decomposition of an FSR by 1-VDM (see the example in Figure 10). One can see that edges f_3 and g_3 cross the "splitting line" that passes through vertex 4. Hence, any path from vertex 1 to vertex 7 passes through decomposition vertex 4 or through one of the following edges: c_3, a_3, g_3, f_3.

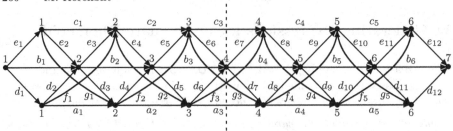

Fig. 10. Decomposition of a full square rhomboid by 1-VDM

Therefore, in the general case

$$E(p,q) \leftarrow E(p,i)E(i,q) + E(p,\overline{i-1})c_{i-1}E(\overline{i},q) + E(p,\underline{i-1})a_{i-1}E(\underline{i},q)+ \quad (3)$$
$$E(p,\overline{i-1})g_{i-1}E(\underline{i},q) + E(p,\underline{i-1})f_{i-1}E(\overline{i},q).$$

Additional parts $E(p,\overline{i-1})g_{i-1}E(\underline{i},q)$ and $E(p,\underline{i-1})f_{i-1}E(\overline{i},q)$ which are absent in statement (2), describe all paths from vertex p to vertex q passing via edges g_{i-1} and f_{i-1}, respectively.

Hence, the expression of a current subgraph of size n derived by 1-VDM includes ten subexpressions related to subgraphs of size $n' \approx \frac{n}{2}$. This expression can be simplified by putting subexpressions which appear twice, outside the brackets. Finally, statement (3) may be presented as

$$E(p,q) \leftarrow E(p,i)E(i,q) + E(p,\overline{i-1})\left(c_{i-1}E(\overline{i},q) + g_{i-1}E(\underline{i},q)\right)+ \quad (4)$$
$$E(p,\underline{i-1})\left(a_{i-1}E(\underline{i},q) + f_{i-1}E(\overline{i},q)\right),$$

i.e., the resulting expression for FSR consists of eight subexpressions related to subgraphs of size $n' \approx \frac{n}{2}$ and four additional literals. The expressions for $\widehat{FSR}$ and $\widehat{\widehat{FSR}}$ are constructed in the same way. Thus by the master theorem, the complexity of the expression $Ex(FSR(n))$ derived by 1-VDM is $O\left(n^3\right)$.

Therefore, 2-VDM that numerically is less efficient than 1-VDM for square rhomboids is significantly more efficient for full square rhomboids. For this reason we compute $Ex(FSR)$ by the following recursive relations based on 2-VDM:

1. $E(p,p) = 1$
2. $E(p,\overline{p}) = e_{2p-1}$
3. $E(p,\underline{p}) = d_{2p-1}$
4. $E(\overline{p},p+1) = e_{2p}$
5. $E(\underline{p},p+1) = d_{2p}$
6. $E(\overline{p},\overline{p+1}) = c_p + e_{2p}e_{2p+1}$
7. $E(\overline{p},\underline{p+1}) = e_{2p}d_{2p+1} + g_p$
8. $E(\underline{p},\overline{p+1}) = d_{2p}e_{2p+1} + f_p$
9. $E(\underline{p},\underline{p+1}) = a_p + d_{2p}d_{2p+1}$
10. $E(p,\overline{q})=E(p,i)b_iE(i+1,\overline{q}) + E(p,\overline{i})E(\overline{i},\overline{q}) + E(p,\underline{i})E(\underline{i},\overline{q}),\ i=\left\lfloor\frac{q+p}{2}\right\rfloor\ (q>p)$
11. $E(p,\underline{q})=E(p,i)b_iE(i+1,\underline{q}) + E(p,\overline{i})E(\overline{i},\underline{q}) + E(p,\underline{i})E(\underline{i},\underline{q}),\ i=\left\lfloor\frac{q+p}{2}\right\rfloor\ (q>p)$

12. $E(\overline{p}, q) = E(\overline{p}, i)b_i E(i + 1, q) + E(p, \overline{i})E(\overline{i}, q) + E(\overline{p}, \underline{i})E(\underline{i}, q)$, $i = \lfloor \frac{q+p}{2} \rfloor$
    $(q > p + 1)$

13. $E(\underline{p}, q) = E(\underline{p}, i)b_i E(i + 1, q) + E(\underline{p}, \overline{i})E(\overline{i}, q) + E(\underline{p}, \underline{i})E(\underline{i}, q)$, $i = \lfloor \frac{q+p}{2} \rfloor$
    $(q > p + 1)$

14. $E(\overline{p}, \overline{q}) = E(\overline{p}, i)b_i E(i+1, \overline{q}) + E(\overline{p}, \overline{i})E(\overline{i}, \overline{q}) + E(\overline{p}, \underline{i})E(\underline{i}, \overline{q})$, $i = \frac{q+p}{2}$ $(q > p+1)$

15. $E(\overline{p}, \underline{q}) = E(\overline{p}, i)b_i E(i+1, \underline{q}) + E(\overline{p}, \overline{i})E(\overline{i}, \underline{q}) + E(\overline{p}, \underline{i})E(\underline{i}, \underline{q})$, $i = \frac{q+p}{2}$ $(q > p+1)$

16. $E(\underline{p}, \overline{q}) = E(\underline{p}, i)b_i E(i+1, \overline{q}) + E(\underline{p}, \overline{i})E(\overline{i}, \overline{q}) + E(\underline{p}, \underline{i})E(\underline{i}, \overline{q})$, $i = \frac{q+p}{2}$ $(q > p+1)$

17. $E(\underline{p}, \underline{q}) = E(\underline{p}, i)b_i E(i+1, \underline{q}) + E(\underline{p}, \overline{i})E(\overline{i}, \underline{q}) + E(\underline{p}, \underline{i})E(\underline{i}, \underline{q})$, $i = \frac{q+p}{2}$ $(q > p+1)$

18. $E(p, q) = E(p, i)b_i E(i+1, q) + E(p, \overline{i})E(\overline{i}, q) + E(p, \underline{i})E(\underline{i}, q)$, $i = \frac{q+p-1}{2}$ $(q > p)$.

The following lemma results from relations 6 – 9 and 14 – 17.

Lemma 1. *Complexities of expressions* $Ex\left(\text{trapezoidal } \overset{\frown}{\widehat{FSR}}(n)\right)$ *and* $Ex\left(\text{parallelogram } \overset{\frown}{\widehat{FSR}}(n)\right)$ *derived by 2-VDM are equal.*

The following proposition results from Lemma 1 and relations 1 – 18.

Proposition 1. *The total number of literals $T(n)$ in the expression $Ex(FSR(n))$ derived by 2-VDM is defined recursively as follows:*

1) $T(1) = 0$; *2)* $\widehat{T}(1) = 1$; *3)* $\widehat{\widehat{T}}(1) = 3$

4) $T(n) = T\left(\lceil \frac{n}{2} \rceil\right) + T\left(\lfloor \frac{n}{2} \rfloor\right) + 2\widehat{T}\left(\lceil \frac{n}{2} \rceil\right) + 2\widehat{T}\left(\lfloor \frac{n}{2} \rfloor\right) + 1$ $(n > 1)$

5) $\widehat{T}(n) = T\left(\lceil \frac{n}{2} \rceil\right) + \widehat{T}\left(\lfloor \frac{n}{2} \rfloor\right) + 2\widehat{T}\left(\lceil \frac{n}{2} \rceil\right) + 2\widehat{T}\left(\lfloor \frac{n}{2} \rfloor\right) + 1$ $(n > 1)$

6) $\widehat{\widehat{T}}(n) = \widehat{T}\left(\lceil \frac{n}{2} \rceil\right) + \widehat{T}\left(\lfloor \frac{n}{2} \rfloor\right) + 2\widehat{T}\left(\lceil \frac{n}{2} \rceil\right) + 2\widehat{T}\left(\lfloor \frac{n}{2} \rfloor\right) + 1$ $(n > 1)$,

where $\widehat{T}(n)$ and $\widehat{\widehat{T}}(n)$ are the total numbers of literals in $Ex(\widehat{FSR}(n))$ and $Ex\left(\widehat{\widehat{FSR}}(n)\right)$, respectively.

It is of interest to obtain exact formulae describing complexity of the expression $Ex(FSR(n))$ derived by 2-VDM. We attempt to do it for n that is a power of two, i.e., $n = 2^k$ for some positive integer $k \geq 1$. Formulae (4 – 6) of Proposition 1 are presented in this case as

$$\begin{cases} T(n) = 2T\left(\frac{n}{2}\right) + 4\widehat{T}\left(\frac{n}{2}\right) + 1 \\ \widehat{T}(n) = T\left(\frac{n}{2}\right) + 3\widehat{T}\left(\frac{n}{2}\right) + 2\widehat{\widehat{T}}\left(\frac{n}{2}\right) + 1 \\ \widehat{\widehat{T}}(n) = 2\widehat{T}\left(\frac{n}{2}\right) + 4\widehat{\widehat{T}}\left(\frac{n}{2}\right) + 1, \end{cases} \tag{5}$$

respectively. The following explicit formulae for simultaneous recurrences (5) are obtained by the method for linear recurrence relations solving [20]:

$$T(n) = \frac{89}{45}n^{\log_2 6} - \frac{20}{9}n^{\log_2 3} - \frac{1}{5}$$

$$\widehat{T}(n) = \frac{89}{45}n^{\log_2 6} - \frac{5}{9}n^{\log_2 3} - \frac{1}{5}$$

$$\widehat{\widehat{T}}(n) = \frac{89}{45}n^{\log_2 6} + \frac{10}{9}n^{\log_2 3} - \frac{1}{5}.$$

Table 1. Complexities for 1-VDM, 2-VDM, and modified 2-VDM

n	$T(n)$, 1-VDM	$\widehat{T}(n)$, 1-VDM	$\widehat{\widehat{T}}(n)$, 1-VDM	$T(n)$, 2-VDM	$\widehat{T}(n)$, 2-VDM	$\widehat{\widehat{T}}(n)$, 2-VDM	$T(n)$, 2-VDM mod.	$\widehat{T}(n)$, 2-VDM mod.	$\widehat{\widehat{T}}(n)$, 2-VDM mod.
1	0	1	3	0	1	3	0	1	3
2	5	10	15	5	10	15	5	10	15
3	20	29	42	28	33	48	20	29	42
4	53	66	85	51	66	81	51	66	81
5	104	123	152	120	135	170	104	119	152
6	175	204	243	189	224	259	157	192	227
7	284	323	388	278	313	358	262	297	342
8	409	474	531	367	412	457	367	412	457
9	608	665	760	574	619	704	526	571	652
10	793	888	975	781	866	951	685	766	847
20	6325	6800	7061	5027	5282	5537	4435	4678	4921
30	21351	22326	23301	13077	13472	13867	12789	13184	13579
40	50905	53280	54063	31183	31948	32713	27583	28312	29041
50	98935	100690	102445	54493	55518	56543	48445	49470	50495
60	171195	176070	178995	80043	81228	82413	78315	79500	80685

4   A Combined Method for Generating Expressions of Full Square Rhomboids

Despite on the asymptotic advantage of 2-VDM for full square rhomboids, expressions constructed by 1-VDM are shorter for some small values of n. One can see (Table 1) that complexities for 1-VDM (three leftmost columns) are smaller than corresponding complexities for 2-VDM (three middle columns) when $n = 3$ and, as a result, when $n = 5$ and $n = 6$. Expressions of graphs with these sizes are included by expressions of graphs with larger sizes.

We modify 2-VDM through generating expressions of graphs with size 3 by 1-VDM and obtain the following new values: $T(5) = 104$, $\widehat{T}(5) = 119$, $\widehat{\widehat{T}}(5) = 154$, $T(6) = 157$, $\widehat{T}(6) = 192$, $\widehat{\widehat{T}}(6) = 227$. Hence, all new values except $\widehat{\widehat{T}}(5)$ are not greater than corresponding values for 1-VDM presented in Table 1. For this reason, we additionally improve 2-VDM and derive by 1-VDM the expression $Ex\left(\widehat{\widehat{FSR}}(5)\right)$ as well. The final complexities for modified 2-VDM are presented in three rightmost columns of Table 1. For all n except $n = 3$ and $n = 5$ they are determined in accordance with Proposition 1. In addition, we use the following formulae which result from statement (4) and from analogous relations for computing expressions $Ex\left(\widehat{FSR}\right)$ and $Ex\left(\widehat{\widehat{FSR}}\right)$ by 1-VDM:

$$T(n) = T\left(\left\lfloor \tfrac{n}{2} \right\rfloor + 1\right) + T\left(\left\lceil \tfrac{n}{2} \right\rceil\right) + 2\widehat{T}\left(\left\lfloor \tfrac{n}{2} \right\rfloor\right) + 4\widehat{T}\left(\left\lceil \tfrac{n}{2} \right\rceil - 1\right) + 4 \quad (n = 3)$$

$$\widehat{T}(n) = T\left(\left\lfloor \tfrac{n}{2} \right\rfloor + 1\right) + \widehat{T}\left(\left\lceil \tfrac{n}{2} \right\rceil\right) + 4\widehat{T}\left(\left\lfloor \tfrac{n}{2} \right\rfloor\right) + 2\widehat{\widehat{T}}\left(\left\lceil \tfrac{n}{2} \right\rceil - 1\right) + 4 \quad (n = 3)$$

$$\widehat{\widehat{T}}(n) = \widehat{T}\left(\left\lfloor \tfrac{n}{2} \right\rfloor + 1\right) + \widehat{T}\left(\left\lceil \tfrac{n}{2} \right\rceil\right) + 2\widehat{T}\left(\left\lfloor \tfrac{n}{2} \right\rfloor\right) + 4\widehat{\widehat{T}}\left(\left\lceil \tfrac{n}{2} \right\rceil - 1\right) + 4 \quad (n = 3, 5).$$

5   Conclusions and Future Work

Various non-series-parallel graphs (Fibonacci graphs, square rhomboids, etc.) have expressions with polynomial complexity despite their relatively complex structure. Full square rhomboids fill up the family of these graphs. The existence of a decomposition method for a graph G is a sufficient condition for the existence of such expression for G. Its complexity depends, in particular, on the number of revealed subgraphs in each recursive step of the decomposition procedure. Different decomposition methods may be applied to the same class of graphs and one of the methods may be more efficient for one class and less efficient for another one.

Note that expressions with polynomial complexity are not obtained for every st-dag of a regular structure. For example, applying a decomposition method to a directed grid graph is an open problem. While a rhomboid is obtained by consecutive series compositions of *rhomb* st-dags whose underlying graphs are C_4 graphs, a directed grid graph can be considered to be a "gluing" of a lot of rhombs. Its special case, a directed *ladder graph* [18] is not investigated from this perspective as well. At last, generating an expression with polynomial complexity is problematic for a *complete st-dag* [11].

An undirected graph in which every subgraph has a vertex of degree at most k is called k-*inductive* [10]. For instance, trees are 1-inductive graphs, and planar graphs are 5-inductive. *Random scale-free networks* [1] demonstrate important practical examples of k-inductive graphs. As follows from [3], a graph G is k-inductive if and only if the edges of G can be oriented to form a directed acyclic graph with out-degree of its vertices at most k. Thus underlying graphs of Fibonacci graphs are 2-inductive while underlying graphs of square and full square rhomboids are 3-inductive.

We intend to extend the presented decomposition technique to a class of st-dags whose underlying graphs are k-inductive.

Acknowledgment. The author thanks Vadim E. Levit for helpful advises.

References

1. Barabási, A.-L., Albert, R.: Emergence of Scaling in Random Networks. Science 286(5439), 509–512 (1999)
2. Bein, W.W., Kamburowski, J., Stallmann, M.F.M.: Optimal Reduction of Two-Terminal Directed Acyclic Graphs. SIAM Journal of Computing 21(6), 1112–1129 (1992)
3. Chrobak, M., Eppstein, D.: Planar Orientations with Low Out-Degree and Compaction of Adjacency Matrices. Theoretical Computer Science 86(2), 243–266 (1991)

4. Cormen, T.H., Leiseron, C.E., Rivest, R.L.: Introduction to Algorithms. The MIT Press, Cambridge (2001)
5. Duffin, R.J.: Topology of Series-Parallel Networks. Journal of Mathematical Analysis and Applications 10, 303–318 (1965)
6. Finta, L., Liu, Z., Milis, I., Bampis, E.: Scheduling UET-UCT Series-Parallel Graphs on Two Processors. Theoretical Computer Science 162(2), 323–340 (1996)
7. Golumbic, M.C., Gurvich, V.: Read-Once Functions. In: Crama, Y., Hammer, P.L. (eds.) Boolean Functions: Theory, Algorithms and Applications, pp. 519–560. Cambridge University Press, New York (2011)
8. Golumbic, M.C., Mintz, A., Rotics, U.: Factoring and Recognition of Read-Once Functions Using Cographs and Normality and the Readability of Functions Associated with Partial k-Trees. Discrete Applied Mathematics 154(10), 1465–1477 (2006)
9. Golumbic, M.C., Perl, Y.: Generalized Fibonacci Maximum Path Graphs. Discrete Mathematics 28, 237–245 (1979)
10. Irani, S.: Coloring Inductive Graphs On-Line. Algorithmica 11(1), 53–72 (1994)
11. Korenblit, M.: Efficient Computations on Networks, Ph.D. Thesis, Bar-Ilan University, Israel (2004)
12. Korenblit, M., Levit, V.E.: On Algebraic Expressions of Series-Parallel and Fibonacci Graphs. In: Calude, C.S., Dinneen, M.J., Vajnovszki, V. (eds.) DMTCS 2003. LNCS, vol. 2731, pp. 215–224. Springer, Heidelberg (2003)
13. Korenblit, M., Levit, V.E.: A One-Vertex Decomposition Algorithm for Generating Algebraic Expressions of Square Rhomboids. In: Fellows, M., Tan, X., Zhu, B. (eds.) FAW-AAIM 2013. LNCS, vol. 7924, pp. 94–105. Springer, Heidelberg (2013)
14. Levit, V.E., Korenblit, M.: A Symbolic Approach to Scheduling of Robotic Lines. In: Intelligent Scheduling of Robots and Flexible Manufacturing Systems, pp. 113–125. The Center for Technological Education Holon, Israel (1996)
15. Mintz, A., Golumbic, M.C.: Factoring Boolean Functions Using Graph Partitioning. Discrete Applied Mathematics 149(1-3), 131–153 (2005)
16. Mundici, D.: Functions Computed by Monotone Boolean Formulas with no Repeated Variables. Theoretical Computer Science 66, 113–114 (1989)
17. Mundici, D.: Solution of Rota's Problem on the Order of Series-Parallel Networks. Advances in Applied Mathematics 12, 455–463 (1991)
18. Noy, M., Ribó, A.: Recursively Constructible Families of Graphs. Advances in Applied Mathematics 32, 350–363 (2004)
19. Oikawa, M.K., Ferreira, J.E., Malkowski, S., Pu, C.: Towards Algorithmic Generation of Business Processes: From Business Step Dependencies to Process Algebra Expressions. In: Dayal, U., Eder, J., Koehler, J., Reijers, H.A. (eds.) BPM 2009. LNCS, vol. 5701, pp. 80–96. Springer, Heidelberg (2009)
20. Rosen, K.H. (ed.): Handbook of Discrete and Combinatorial Mathematics. CRC Press, Boca Raton (2000)
21. Satyanarayana, A., Wood, R.K.: A Linear Time Algorithm for Computing K-Terminal Reliability in Series-Parallel Networks. SIAM Journal of Computing 14(4), 818–832 (1985)
22. Savicky, P., Woods, A.R.: The Number of Boolean Functions Computed by Formulas of a Given Size. Random Structures and Algorithms 13, 349–382 (1998)
23. Tamir, A.: A Strongly Polynomial Algorithm for Minimum Convex Separable Quadratic Cost Flow Problems on Two-Terminal Series-Parallel Networks. Mathematical Programming 59, 117–132 (1993)
24. Wang, A.R.R.: Algorithms for Multilevel Logic Optimization, Ph.D. Thesis, University of California, Berkeley (1989)

Solving Hamiltonian Cycle by an EPT Algorithm for a Non-sparse Parameter

Sigve Hortemo Sæther[*]

University of Bergen

Abstract. Many hard graph problems, such as Hamiltonian Cycle, become FPT when parameterized by treewidth, a parameter that is bounded only on sparse graphs. When parameterized by the more general parameter clique-width, Hamiltonian Cycle becomes W[1]-hard, as shown by Fomin et al. [5]. Sæther and Telle address this problem in their paper [14] by introducing a new parameter, split-matching-width, which lies between treewidth and clique-width in terms of generality. They show that even though graphs of restricted split-matching-width might be dense, solving problems such as Hamiltonian Cycle can be done in FPT time.

Recently, it was shown that Hamiltonian Cycle parameterized by treewidth is in EPT [1,6], meaning it can be solved in $n^{\mathcal{O}(1)}2^{\mathcal{O}(k)}$-time. In this paper, using tools from [6], we show that also parameterized by split-matching-width Hamiltonian Cycle is EPT. To the best of our knowledge, this is the first EPT algorithm for any "globally constrained" graph problem parameterized by a non-trivial and non-sparse structural parameter. To accomplish this, we also give an algorithm constructing a branch decomposition approximating the minimum split-matching-width to within a constant factor. Combined, these results show that the algorithms in [14] for Edge Dominating Set, Chromatic Number and Max Cut all can be improved. We also show that for Hamiltonian Cycle and Max Cut the resulting algorithms are asymptotically optimal under the Exponential Time Hypothesis.

1 Introduction

The problem of finding a Hamiltonian Cycle in a graph - a simple cycle covering all the vertices of the graph - is NP-complete [10]. One way to handle an NP-hard problem is by investigating its parameterized complexity, for various choices of parameter. Unlike a lot of other NP-hard graph problems, Hamiltonian Cycle does not have a natural parameter, since the solution size is the number of vertices in the input graph. Instead, we may look at structural parameterizations of the input graph, for instance its treewidth or clique-width.

A lot of NP-hard graph problems become fixed parameter tractable (FPT, solvable in $f(k)n^{\mathcal{O}(1)}$-time for parameter-value k) when parameterized by treewidth. Many examples of problems that can be checked locally, e.g., Independent Set, Vertex Cover, Dominating Set and so on, are even EPT when

---

[*] Supported by the Norwegian Research Council.

parameterized by treewidth, meaning that the problems can be solved in time $2^{\mathcal{O}(k)}n^{\mathcal{O}(1)}$ [4] (also referred to as having a single exponential algorithm). When parameterized by clique-width, hardly any of these problems are known to be EPT. For instance Dominating Set has recently been shown solvable in time $2^{\mathcal{O}(k \log k)}n^{\mathcal{O}(1)}$ for clique-width k [11], but this is still not EPT.

For problems that have a global constraint, like Steiner Tree, Hamiltonian Cycle and Feedback Vertex Set, EPT algorithms parameterized by treewidth were for a long time not known. For example, the asymptotically best algorithm for Hamiltonian Cycle was for a long time the folklore $n^{\mathcal{O}(1)}k^{\mathcal{O}(k)}$ time algorithm, resulting in a belief that graph problems with a global requirement may not have EPT algorithms. Recently, however, a breakthrough paper by Cygan et al. [3] gave a randomized EPT algorithm for Hamiltonian Cycle, and other problems with global constraints, when parameterized by treewidth. Shortly after this, Bodlaender et al. [1] and then Fomin et al. [6], also found deterministic EPT algorithms for Hamiltonian Cycle parameterized by treewidth. Both the papers [1,6] are general, in the sense that they provide a framework for solving many problems. Graph classes of bounded treewidth are all sparse, so one may wonder if using either of these new frameworks will help in finding similar EPT results for globally constrained problems like Hamiltonian Cycle for a parameter bounded also on non-sparse graph classes. The classical structural graph parameter bounded also on some non-sparse graphs is clique-width. Unfortunately, it is unlikely that such a result exists for clique-width, as Hamiltonian Cycle has been shown to be W-hard when parameterized by clique-width [5]. So, we must focus on a non-sparse parameter which is less general than clique-width. Examples of some such parameters are modular-width, shrub-depth, neighbourhood diversity, twin-cover, and the newly introduced split-matching-width (see Figure 1).

In the recent paper [7] Gajarský et al. give an FPT algorithm (but not EPT) for Hamiltonian Cycle parameterized by modular-width, and show W-hardness when parameterized by shrub-depth. Split-matching-width is a new parameter introduced by Sæther and Telle [14] for which Hamiltonian Cycle is FPT [14]. Unlike modular-width, split-matching-width generalizes treewidth, so it is a good candidate for applying the framework used for treewidth.

In this paper, we will show that using the framework of [6] we can solve Hamiltonian Cycle in time $2^{\mathcal{O}(k)}n^{\mathcal{O}(1)}$ for parameter k being split-matching-width. The approach will be similar to that of [14] in the sense that it consists of two parts; (1) given a graph G, finding a branch decomposition of low split-matching-width, and then (2) solving Hamiltonian Cycle on G with a runtime depending on the split-matching-width of the computed branch decomposition. We will in this paper improve on the results from [14] by showing the following two theorems that when combined results in an EPT algorithm for Hamiltonian Cycle parameterized by split-matching-width.

Theorem 1. *Given a graph G of split-matching-width less than k, in $n^{\mathcal{O}(1)}2^{\mathcal{O}(k)}$ time we can find a branch decomposition of split-matching-width less than $16k$.*

Theorem 2. *Given a graph G and a branch decomposition of split-matching-width k, we can decide if G has a Hamiltonian Cycle in time $n^{\mathcal{O}(1)}2^{\mathcal{O}(k)}$.*

Another result of Theorem 1 is that we can improve the runtime of the algorithms for solving Edge Dominating Set, Chromatic Number, and Max Cut parameterized by split-matching-width described in [14]. In fact, under the Exponential Time Hypothesis the asymptotic runtimes for Max Cut, Hamiltonian Cycle, and Edge Dominating Set become optimal [8,9][1]. (I.e., no $n^{\mathcal{O}(1)}2^{o(k)}$ algorithm exists.)

This paper is organized as follows: In Section 2, we give the necessary definitions and background needed for the rest of the paper and in Section 3 we prove Theorem 2. In Section 4 we prove Theorem 1, and then we end this paper in Section 5 where we give a short summary.

Due to space restrictions, certain proofs have been left out. These proofs can be found in [13]. The symbol $\star$ denotes that the proof can be found [13].

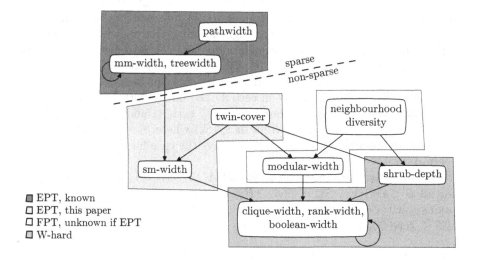

Fig. 1. A small overview of how certain graph width parameters relate to each other. A directed path from parameter a to b means that there exists a function f so that for all graphs G, $b(G) \leq f(a(G))$. The arrows related to sm-width are from [14], for the others see Gajarský et al. [7]. The colors depict the complexity of Hamiltonian Cycle parameterized with the respective parameters.

2   Preliminaries and Terminology

Graph and Set Preliminaries. We work on simple undirected graphs $G = (V, E)$ and denote the set of vertices and set of edges of a graph G by $V(G)$ and $E(G)$, respectively. We use n to denote the number of vertices of the graph in question. For an edge between vertices u and v, we simply write uv. For a path P, by writing uPv we mean that the endpoints of P are u and v. For a graph G and subset $A \subseteq V(G)$, we denote by $G[A]$ the subgraph of G *induced* by A. That is,

---

[1] See the full version of the paper [13] for the lower bound on Edge Dominating Set.

the vertex set $V(G[A])$ of $G[A]$ is A and the edge set is $E(G[A]) = \{uv \in E(G) : u, v \in A\}$. For disjoint sets $A, B \subseteq V(G)$, we denote by $G[A, B]$ the bipartite subgraph of G induced by the pair (A, B). That is $V(G[A, B]) = A \cup B$ and $E(G[A, B]) = \{uv \in E(G) : u \in A, v \in B\}$. For a set of vertices $S \subseteq V(G)$, we denote by $N_G(S)$ all the vertices in $V(G) \setminus S$ adjacent to S. We omit the subscript G in $N_G(S)$ when it is clear from context. For a single vertex v, we write $N_G(v)$ instead of $N_G(\{v\})$. To *contract* an edge uv means to replace the vertices u and v by a new vertex v_{uv} adjacent to exactly the same vertices as u and v combined. For a set $A \subseteq V(G)$, when $V(G)$ is clear from context, we write $\overline{A}$ to mean the set $V(G) \setminus A$. For a graph G and subsets $A, B, C \subseteq V(G)$, we say that C separates A and B if there are no paths from $A \setminus C$ to $B \setminus C$ in $G[\overline{C}]$.

Hamiltonian Cycle and Certificates. A Hamiltonian Cycle in a graph G is a simple cycle in G of size $|V(G)|$. In this paper we will solve Hamiltonian Cycle using dynamic programming. The framework we use will be the same as in [14]. That is, a *certificate* is a set of edges forming vertex disjoint paths or a Hamiltonian cycle, and a witness is a Hamiltonian cycle. We denote by $\text{cert}(X)$ all the certificates C so that $C \subseteq E(G[X])$. For two disjoint sets A and B and certificates $s_a \in \text{cert}(A)$ and $s_b \in \text{cert}(B)$, we define the operation $\oplus(s_a, s_b)$ to be the set of all the certificates s' in $\text{cert}(A \cup B)$ on the form $s' = s_b \cup s_b \cup E'$ for some set $E' \subseteq E(G[A, B])$. For sets $S_A \subseteq \text{cert}(A)$ and $S_B \subseteq \text{cert}(B)$ we denote by $\oplus(S_A, S_B)$ the union of all $\oplus(S_a, S_b)$ where $S_a \in S_A$ and $S_b \in S_B$. From these definitions, we see that it is always the case for disjoint sets A and B, $\text{cert}(A \cup B) = \oplus(\text{cert}(A), \text{cert}(B))$.

For two sets $A_1, A_2 \subseteq \text{cert}(A)$, we say that A_1 *preserves* A_2, denoted $A_1 \preceq_A A_2$, if for all $B \in \text{cert}(V \setminus A)$, if $\oplus(A_1, \{B\})$ contains a witness, then $\oplus(A_2, \{B\})$ contains a witness. We note that if $A_1 \preceq_A A_2$ then for any C disjoint from A, and $X \subseteq \text{cert}(C)$, we have $\oplus(A_1, X) \preceq_{A \cup C} \oplus(A_2, X)$.

Splits and Split Decompositions. A *split* of a connected graph G is a bi-partition (A, B) of the vertices $V(G)$ where $|A|, |B| \geq 2$ and for all $a \in N(B)$, $N(a) \cap B = N(A)$. That is, all vertices in A adjacent to B have the same neighbourhood in B. Notice that this property holds if and only if also for all $b \in N(A)$, $N(b) \cap A = N(B)$. A bi-partition (A, B) where either A or B consists of at most one vertex is said to be a *trivial split*.

A graph G having a split (A, B) can be *decomposed* into smaller graphs G_A and G_B where G_A is the graph G with all the vertices of B replaced by a single vertex v, called a *marker*, adjacent to the same vertices in G_A as B is adjacent to in G. G_B is in the same way the graph G where we replace the vertices A by the marker vertex v so that $N_{G_B}(v) = N_G(A)$. A graph without a split is called a *prime graph*. Since all graphs of three or less vertices trivially do not contain any splits, we say that a prime graph on more than three vertices is a *non-trivial* prime graph.

A *split decomposition* of a graph G is a recursive decomposition of G so that all of the obtained graphs are prime. For a split decomposition of G into $G_1, G_2, \ldots, G_k$, a *split decomposition tree* is a tree T where each vertex corresponds to a prime graph and we have an edge between two vertices if and only if the prime graphs they correspond to share a marker. That is, the edge set of the tree is $E(T) = \{v_i v_j : v_i, v_j \in V(T) \text{ and } V(G_i) \cap V(G_j) \neq \emptyset\}$. See Figure 2 for an example. Given a split decomposition of graph G with prime graphs $G_1, G_2, \ldots, G_k$, we define $\text{tot}(v : G_i)$ recursively to be $\{v\}$ if $v \in V(G)$, and otherwise to be $\bigcup_{u \in V(G_j) \setminus \{v\}} \text{tot}(u : G_j)$ for the graph $G_j \neq G_i$ containing the marker v in the split decomposition. Another way of saying this latter part by the use of the split decomposition tree T is: if v is not in $V(G)$, then $\text{tot}(v : G_i)$ is defined to be the vertices of $V(G)$ residing in the prime graphs of the connected component in $T[V(T) \setminus \{G_i\}]$ where v is also located. For a set $V' \subseteq V(G_i)$, we define $\text{tot}(V' : G_i)$ to be the union of $\text{tot}(v : G_i)$ for all $v \in V'$. We define the *active set* of a vertex $v \in G_i$, denoted $\text{act}(v : G_i)$ to be the vertices of $\text{tot}(v : G_i)$ that are contributing to the neighborhood of v in G_i. That is, $\text{act}(v : G_i)$ is defined as $N(V(G) \setminus \text{tot}(v : G_i))$. Note that if G has a split decomposition into prime graphs $G_1, \ldots, G_k$, then for any marker v there are exactly two prime graphs G_i and G_j containing v, and we have $\text{tot}(v : G_i) \cup \text{tot}(v : G_j) = V(G)$. When the prime graph G_i is clear from context, we denote $\text{tot}(X : G_i)$ and $\text{act}(X : G_i)$ simply as $\text{tot}(X)$ and $\text{act}(X)$. See Figure 2 for an example of $\text{tot}()$ and $\text{act}()$. For a prime graph G' and vertex $v \in V(G')$, when we say the *weight* of v, we mean the cardinality of $\text{act}(v)$.

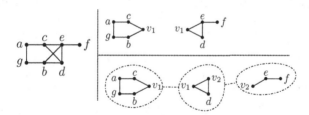

Fig. 2. A graph G on vertices a, b, c, d, e, f, g on the left, and a the graph G decomposed through the split $(\{a, b, c, g\}.\{d, e, f\})$ to the upper right. Notice the introduction of marker v_1. To the lower right we have a split decomposition tree of G with markers v_1 and v_2. In the prime graph with vertices $\{v_1, d, v_2\}$ we have $\text{tot}(v_1) = \{a, b, c, g\}$ and $\text{act}(v_1) = \{b, c\}$. The weight of v_2 in the rightmost prime graph drawn in the split decomposition tree is three while in the middle prime graph it is one.

Split-matching-width. A branch decomposition of a set X is a description of recursively dividing X into smaller and smaller sets until only single element sets are left. Formally, a branch decomposition of X is a pair (T, δ) where T is a subcubic tree (a tree of max degree three) and δ is a bijection from the leaves in T to the elements in X. In this paper, when we say a branch decomposition of a graph G, we will mean a branch decomposition of $V(G)$.

A cut of X is a bi-partition of X. Each edge e of a branch decomposition (an edge in the tree) describes a cut/bi-partition of X, namely the cut (A, B) where

A are all the elements that are mapped to from δ by the leaves on one side of e and B are the elements mapped by δ from the leaves of the other side of e. For a set X, a *cut function* $f : 2^X \to \mathbb{N}$ is a symmetric $(f(A) = f(\overline{A}))$ function on subsets of X. For a branch decomposition (T, δ) of G and a cut function f on $V(G)$, its f-width is the maximum of $f(A)$ over all cuts $(A, \overline{A})$ of G induced by the edges of T. For a graph G, its f-width, for a cut function f, is the minimum f-width over all branch decompositions of G.

The Maximum-Matching-width (mm-width) $\text{mmw}(G)$ of a graph G was defined by Vatshelle [15] based on the cut function mm defined for $A \subseteq V(G)$ as the size of a maximum matching in $G[A, V \setminus A]$.

The split-matching-width (sm-width) $\text{smw}(G)$ of a graph G was defined by Sæther and Telle [14] based on the cut function sm defined using splits and mm as follows:

$$\text{sm}(A) = \begin{cases} 1 & \text{if } (A, \overline{A}) \text{ is a split of } G \\ \text{mm}(A) & \text{otherwise} \end{cases}$$

A function $f : 2^X \to \mathbb{N}$ is said to be *submodular* if for any subset $A, B \subseteq X$ it holds that $f(A) + f(B) \geq f(A \cup B) + f(A \cap B)$.

3   Solving Hamiltonian Cycle in EPT Time

From Theorem 3 of [6] applied to a *graphic matroid* we have the following corollary, which is the core tool for getting our EPT algorithm.

Corollary 3 ([6]). *Let G be a connected graph on n vertices and S a family of p-sized subsets of $E(G)$. We can for any integer q find a subset $\hat{S}$ of S with $|\hat{S}| \leq 2^n$ so that for any q-sized subset Y of $E(G)$, if there exists a set $X \in S$ disjoint from Y so that $X \cup Y$ is a forest, then there exists a set $\hat{X} \in \hat{S}$ disjoint from Y so that $\hat{X} \cup Y$ is a forest. Furthermore, the set $\hat{S}$ can be computed in $n^{\mathcal{O}(1)}(|S|4^n)$ time.*

This might not seem like something related to finding Hamiltonian Cycles, since finding a largest forest is polynomial time solvable and finding a largest cycle is NP-complete. However, as shown in [6], many NP-complete problems that contain a global connectivity constraint, for instance Steiner Tree and Hamiltonian Cycle, can be solved faster by the use of Corollary 3. In [6], the authors focus more on the Steiner Tree problem and less on Hamiltonian Cycle, so the precise usage of Corollary 3 applied to Hamiltonian Cycle is not very explicit. The following result can, however, be deduced from their paper, but as it is such a crucial part of our paper, we need all the details formally stated.

Lemma 4 ($\star$). *Let G be a graph and $A \subseteq V(G)$ some set of vertices separated from $V(G) \setminus A$ by $C \subseteq V(G)$ of size at least three. Given a family S of subsets of edges of $E(G[A \cup C])$, we can in $n^{\mathcal{O}(1)}(|S|12^{|C|})$ time construct a family $\hat{S} \subseteq S$ of size at most $6^{|C|}$ so that for any set Y of edges in $E(G[C \cup \overline{A}])$, if there exists a set $X \in S$ disjoint from Y so that $Y \cup X$ is a Hamiltonian cycle, then there exists a set $\hat{X} \in \hat{S}$ disjoint from Y so that $\hat{X} \cup Y$ is a Hamiltonian cycle.*

Lemma 4 gives an insight to how it is possible to make EPT algorithms for Hamiltonian Cycle parameterized by treewidth, as done in [6]. The idea is to build a set of partial solutions using dynamic programming in a bottom up manner in a tree decomposition, and at each step use Lemma 4 to reduce the number of partial solutions needed to ensure you will find a Hamiltonian cycle in the end of your algorithm. This works because at each step of the algorithm all partial solutions will be disjoint paths that have all their endpoints inside a small separator (a bag in the tree decomposition). However, when parameterizing by split-matching width, even for cuts of small mm-value and a vertex cover C of small size, the partial solutions (certificates) will not necessarily consist of paths that have endpoints inside C, but possibly in $N(C)$, which could be large. To overcome this problem, we define what we call an *extension*.

An extension of a certificate is a certificate plus some extra edges. The idea is that an extension will encompass how a certificate $C \in \text{cert}(X)$ looks after adding more edges than those in $E(G[X])$. Formally, for a certificate $C \in \text{cert}(X)$ and set of edges E^* disjoint from $E(G[X])$, we say that a set $S \subseteq \{C \cup E' : E' \subseteq E^*\}$ is an extension of C by the set E^*. For a set of certificates P, the set S is an extension of P by E' if it is a union of extensions by E' of the certificates in P. For a set of certificates P we say that an extension S of P by E^* is *preserving* if for any edge set Y not intersecting $E(G[A]) \cup E^*$, if there is a certificate $C \in P$ and $E' \subseteq E^*$ so that $Y \cup C \cup E'$ is a Hamiltonian cycle, then there is an element $C' \in S$ so that $C' \cup Y$ is a Hamiltonian cycle. A preserving extension of a single certificate C is simply a preserving extension of $\{C\}$.

Observation 5. *For $P \subseteq \text{cert}(A)$ and edges $E' \subseteq E(G[A, \overline{A}])$, if S is a preserving extension of P by E', then $\{S \setminus E' : S \in S\}$ preserves P.*

Motivated by Observation 5, we give the following lemma, which will be used to reduce the number of certificates needed to preserve certificate sets over sets of small mm-value. The result of this is captured in Corollary 7.

Lemma 6 ($\star$). *Given a set of certificates $S \subseteq \text{cert}(A)$ and a vertex cover C of $G[A, \overline{A}]$ of size at least three, we can in $n^{\mathcal{O}(1)}(|S|2^{\mathcal{O}(|C|)})$ time create a preserving extension of S by $E^* = E(G[A, C \setminus A])$ of size no larger than $6^{|C|}$.*

Corollary 7. *Given a set $S \subseteq \text{cert}(A)$, and a vertex cover C of $G[A, \overline{A}]$ where $3 \leq |C| = k$, we can in $n^{\mathcal{O}(1)}(|S|2^{\mathcal{O}(k)})$ time find a set $\hat{S} \subseteq S$ so that $\hat{S} \preceq_A S$ and $|\hat{S}| \leq 6^k$.*

Combining Corollary 7 with the result of [14] saying that for a split $(A, \overline{A})$ and a set $S \subseteq \text{cert}(A)$ of certificates, we can in time polynomial in $n^{\mathcal{O}(1)}|S|$-time compute a set $S' \preceq_A S$ of size $n^{\mathcal{O}(1)}$, we get the following.

Corollary 8. *For a set $A \subseteq V(G)$ and set $S \subseteq \text{cert}(A)$, we can in $n^{\mathcal{O}(1)}2^{\mathcal{O}(\text{sm}(A))}$ time compute a set $S' \preceq_A S$ of size $n^{\mathcal{O}(1)}2^{\mathcal{O}(\text{sm}(A))}$.*

Now that we have defined preserving extensions, and already shown how we can use this to reduce the size of a preserving set of certificates, we will show

how we can also use extensions to produce small sets S preserving $\oplus(S_1, S_2)$ for certificates S_1 and S_2. This is the last step needed to create our dynamic programming algorithm for Hamiltonian cycle.

Lemma 9. *For a tri-partition (A, B, W) of $V(G)$, and $S_a \in \text{cert}(A)$ and $S_b \in \text{cert}(B)$, for $k = \max\{\text{mm}(A), \text{mm}(B)\}$ we can in $n^{\mathcal{O}(1)}2^{\mathcal{O}(k)}$ time compute a set $S \subseteq \oplus(S_a, S_b)$ so that $S \preceq_{A \cup B} \oplus(S_a, S_b)$ of size at most $2^{\mathcal{O}(k)}n^{\mathcal{O}(1)}$.*

Proof. The case when both $(A, \overline{A})$ and $(B, \overline{B})$ are splits, we can construct a preserving set S of size polynomial in n in $n^{\mathcal{O}(1)}$ time as shown by [14]. So, we will only give proof for the case when at least one of A and B are not splits.

We first assume that there exists a certificate $S_w \in \text{cert}(W)$ so that the set $\oplus(\oplus(S_a, S_b), \{S_w\})$ contains a witness $H = S_a \cup S_b \cup S_w \cup E'$ where E' is disjoint from the three certificates. Let $X_a \subseteq A$ and $X_b \subseteq B$ be the set of vertices in A and B incident with less than two edges of S_a and S_b, respectively. That is, X_a and X_b are exactly the vertices of A and B, respectively, that are incident with E'.

We now show that if a witness H as described above exists, then $|X_a| \leq 2\,\text{mm}(A)$ and $|X_b| \leq 2\,\text{mm}(B)$. Without loss of generality, we prove that this holds for X_a. As each vertex in X_a must be incident with an edge of E', and each vertex in $\overline{A}$ can be incident to at most two edges of E' since H is a simple cycle, there is a matching in E' of at least half the size of X_a, implying that $|X_a| \leq 2\,\text{mm}(A)$. The same also holds for X_b. Let C be a vertex cover of $G[A, \overline{A}]$. If both $\text{mm}(A), \text{mm}(B) \leq k$, then $C \cup X_a \cup X_b$ is a vertex cover of size $\mathcal{O}(k)$. This means that by Lemma 6 that we can construct a preserving extension $\hat{S}_a$ of S_a by $E(G[A, (C \cup X_b) \setminus A])$ of size $6^{\mathcal{O}(k)}$, which combined with S_b must preserve $\oplus(S_a, S_b)$. That is, $S' = \{S'_a \cup S_b : S'_a \in \hat{S}_a\}$.

For the case when either $\text{mm}(A) > \text{sm}(A)$ or $\text{mm}(B) > \text{sm}(B)$, we need a slightly different argument. Assume without loss of generality that $\text{mm}(B) > \text{sm}(B)$, and thus $(B, \overline{B})$ is a split. As before $|X_a| \leq 2k$, but now $|X_b|$ is possibly very large. What we notice though, is that as each vertex of X_a can be incident to at most two edges of E', the number of edges in E' incident with X_a is at most $2|X_a| \leq 4k$. This means that no more than $4k$ of the paths in S_b will connect directly to S_b by the edges in E'. As all endpoints of all paths in S_b (including isolated vertices, which can be thought of as paths of length zero) have the same neighbourhood in $\overline{B}$ and are interchangeable, we can simply disregard all but $4k$ paths of S_b, and do the same for the remaining $4k$ paths as we did for S_b for the case when there were no splits. This means the set X_b is of size at most $8k$ instead of $4k$, but this constant disappears in the $\mathcal{O}$-notation.   □

We now have the means to prove Theorem 2. The following recursive algorithm will, based on Lemma 9 and Corollary 7 decide Hamiltonian cycle in EPT time given a branch decomposition (T, δ) of sm-width k:

We subdivide an arbitrary edge of T (add a vertex "in the middle" of it) and root T in the new vertex r from the subdivision, and then in a bottom up manner compute for each node $v \in T$ with children $c_1, c_2 \in T$ a set $S_v \preceq_{\delta(v)} \text{cert}(\delta(v))$. We do this by applying Lemma 9 on each pair of certificates of S_{c_1} and S_{c_2} and then bounding the size of S_v by Corollary 7. For the base case where $v \in T$ is a leaf, $\text{cert}(\delta(v))$ can be computed exactly in polynomial time. In the root node, we have computed $S \preceq_{\delta(c_1) \cup \delta(c_2)} \text{cert}(\delta(c_1) \cup \delta(c_2)) = \text{cert}(V(G))$. Deciding the Hamiltonian cycle problem can then be answered easily by checking each certificate in S_r whether or not it is a Hamiltonian cycle, by polynomial amount of work for each certificate. The total amount of work is bounded by the number of nodes in T (which is n), times the work spent at each node, which by Lemma 9 and Corollary 7 is bounded by $n^{\mathcal{O}(1)} 2^{\mathcal{O}(k)}$. This concludes the proof of Theorem 2.

4 Approximating sm-width

In [14], Sæther and Telle gave an algorithm for constructing an algorithm of split-matching-width $\mathcal{O}(\text{sm}(G)^2)$ in FPT time. Their procedure consisted of four main steps;

1. construct a split decomposition $\mathcal{D}$ of G,
2. compute a branch decomposition of sm-width less than $9k$ for each prime graph in $\mathcal{D}$ where $\text{smw}(G) < k$,
3. adjust each branch decomposition slightly so that the lifted-sm-width of each decomposition becomes less than $54k^2$, and then finally
4. combine all the branch decompositions together to form a branch decomposition for G of sm-width less than $54k^2$.

In this paper we will keep the general structure as in [14], but we will replace steps (2) and (3) by a single step where we compute branch decompositions for each prime graph of lifted-sm-width bounded by $18k$ directly. The last part is covered by the following theorem which can be extracted from the proof of Theorem 13 in [14], and the first part was shown by [2] to be computable in polynomial time, so we will focus this section on constructing branch decompositions of low lifted-sm-width for prime graphs.

Theorem 10 ([14]). *Given a graph G and a split decomposition $\mathcal{D}$ over prime graphs $G_1, G_2, \ldots$ with corresponding branch decompositions $(T_1, \delta_1), (T_2, \delta_2), \ldots$ all of lifted-sm-width k, we can in polynomial time construct a branch decomposition for G of sm-width k.*

The difference between the approach of this paper and of [14] is that we in this paper explicitly define the lifted version of the cut function sm, and show that this cut function directly can be approximated to a linear factor of $\text{sm}(G)$.

Lifted sm-width
For a cut function f and prime graph G' of some split decomposition of G, we denote by f^ℓ the value of f *lifted* from G' to G. That is, $f^\ell(X) = f(\text{tot}(X : G'))$. We may also refer to this by simply writing "lifted-f".

Theorem 11 ([14]). *The cut function* mm*-value is submodular.*

Theorem 12 ([12]). *For a symmetric submodular cut-function f and graph G of optimal f-width k, a branch decomposition of f-width at most $3k + 1$ can be found in $n^{\mathcal{O}(1)}(2^{3k+1})$ time.*

From the definition of submodularity, and Theorem 11 saying that mm is submodular, we can deduce that also lifted-mm is submodular. We simply substitute $\mathrm{mm}^\ell(X)$ by $\mathrm{mm}(\mathrm{tot}(X))$ in the submodularity inequality:

$$\mathrm{mm}^\ell(A) + \mathrm{mm}^\ell(B) = \mathrm{mm}(\mathrm{tot}(A)) + \mathrm{mm}(\mathrm{tot}(B))$$
$$\geq \mathrm{mm}(\mathrm{tot}(A \cup B)) + \mathrm{mm}(\mathrm{tot}(A \cap B))$$
$$= \mathrm{mm}^\ell(A \cup B) + \mathrm{mm}^\ell(A \cap B).$$

Corollary 13. *Lifted-*mm*-value is submodular.*

Corollary 14 ($\star$). *For a graph of lifted-*mm*-width k, we can find a branch decomposition of mm^ℓ-width at most $3k + 1$ in $n^{\mathcal{O}(1)}(2^{3k})$ time.*

Lemma 15 ($\star$). *Let G be a graph and $\mathcal{D}$ a split decomposition of G. For any prime graph G' in $\mathcal{D}$ there exists a branch decomposition (T, δ) of G' of sm^ℓ-width $\leq 3\,\mathrm{smw}(G)$.*

The following lemma is an improvement of Lemma 7 of [14], which cascades throughout their paper, improving their analysis to show that the resulting branch decomposition is of sm-width $36\,\mathrm{sm}(G)^2$ instead of $54\,\mathrm{sm}(G)^2$. We may note that without the below lemma, we could use Lemma 7 of [14] to get a 27 approximation instead of a 18 approximation.

Lemma 16 ($\star$). *Let G be a graph of split-matching-width less than k, and G' a non-trivial prime graph in a split decomposition of G. For any vertex v in G' of weight at least $3k$, v is either adjacent to exactly one other vertex of weight at least $3k$ or the* mm*-value of $\mathrm{tot}(v)$ is less than $6k$.*

As we notice from Lemma 16, vertices of weight $3k$ are more restricted than the rest of the vertices. We say that a vertex of weight at least $3k$ is *heavy*, and an edge incident with two heavy vertices is called a *heavy* edge.

For a heavy edge uv, if a branch decomposition of lifted-sm-width less than $3k$ induce a non-trivial cut (A, B), it must be the case that u and v are either both in A or both in B. Otherwise, the lifted-sm-width will be too large. This means that in any branch decomposition of lifted-sm-width less than $3k$, $(\{uv\}, \overline{\{uv\}})$ must be a cut induced by the decomposition.

Corollary 17. *By contracting each heavy edge uv in a prime graph G' of G to single vertices v_{uv}, letting $\mathrm{tot}(v_{uv}) = \mathrm{tot}(\{u, v\})$, we get a graph of mm^ℓ-width at most $6\,\mathrm{sm}(G)$.*

Based on Corollary 17, and Corollary 14 saying that we can 3-approximate the width, we get the a branch decomposition of sm^ℓ-width $18\,sm(G)$ for each prime graph of G.

Lemma 18. *For a graph G and prime graph G' in a branch decomposition of G, a branch decomposition (T, δ) over G' of lifted-sm-width at most 18 times the optimal sm-width k of G can be generated in $n^{\mathcal{O}(1)}(2^{18k})$ time.*

Proof. By first contracting each heavy edge as described in Corollary 17, we have a graph of lifted-mm-width less than $6k$. By Corollary 14, we can construct a branch decomposition of sm^ℓ-width less than $18k$. For each contracted heavy edge uv, we append u and v to the leaf v_{uv} corresponding to the contracted edge. This will only alter the decomposition in the form of adding trivial cuts, i.e., splits. And thus, the lifted sm^ℓ-width of the decomposition remains the same.   □

Lemma 18 completes the part of finding branch decompositions of the prime graphs with lifted-sm-width only a linear factor larger than the original graph. Putting Lemma 18 together with the fact that we can find a split decomposition in polynomial time by [2] and Theorem 10 saying that we can combine lifted-sm-decompositions of prime graphs together to form a branch decomposition of the original graph, we have proved Theorem 1 as promised.

5   Conclusions

We have shown a dynamic programming algorithm solving Hamiltonian Cycle in $n^{\mathcal{O}(1)}2^{\mathcal{O}(k)}$ time when given a branch decomposition of sm-width k. We have also supplied an algorithm for finding a branch decomposition of a graph G of sm-width $\mathcal{O}(sm(G))$ by focusing on lifted-sm-width of prime graphs. This results in an EPT algorithm for Hamiltonian Cycle. In fact, combining the algorithm for finding branch decompositions of low split-matching-width with the three algorithms for solving Chromatic Number, Edge Dominating Set, and Max Cut given in [14], we end up with algorithms of runtime, $n^{\mathcal{O}(1)}k^{\mathcal{O}(k)}$, $n^{\mathcal{O}(1)}2^{\mathcal{O}(k)}$, and $n^{\mathcal{O}(1)}2^{\mathcal{O}(k)}$, respectively, which under the Exponential Time Hypothesis is optimal[8,9][2] (no algorithm where the $\mathcal{O}(k)$'s are exchanged with $o(k)$'s exist).

References

1. Bodlaender, H.L., Cygan, M., Kratsch, S., Nederlof, J.: Deterministic single exponential time algorithms for connectivity problems parameterized by treewidth. In: Fomin, F.V., Freivalds, R., Kwiatkowska, M., Peleg, D. (eds.) ICALP 2013, Part I. LNCS, vol. 7965, pp. 196–207. Springer, Heidelberg (2013)
2. Cunningham, W.H.: Decomposition of directed graphs. SIAM Journal on Algebraic Discrete Methods 3(2), 214–228 (1982)

---

[2] For Edge Dominating Set lower bound, see [13].

3. Cygan, M., Nederlof, J., Pilipczuk, M., Rooij, J.M.M.v., Wojtaszczyk, J.O.: Solving connectivity problems parameterized by treewidth in single exponential time. In: Proceedings FOCS, pp. 150–159. IEEE (2011)
4. Flum, J., Grohe, M.: Parameterized complexity theory, vol. 3. Springer (2006)
5. Fomin, F.V., Golovach, P., Lokshtanov, D., Saurabh, S.: Intractability of clique-width parameterizations. SIAM Journal on Computing 39(5), 1941–1956 (2010)
6. Fomin, F.V., Lokshtanov, D., Saurabh, S.: Efficient computation of representative sets with applications in parameterized and exact algorithms. In: Proceedings SODA, pp. 142–151 (2014)
7. Gajarský, J., Lampis, M., Ordyniak, S.: Parameterized algorithms for modular-width. In: Gutin, G., Szeider, S. (eds.) IPEC 2013. LNCS, vol. 8246, pp. 163–176. Springer, Heidelberg (2013)
8. Lokshtanov, D., Marx, D., Saurabh, S.: Known algorithms on graphs of bounded treewidth are probably optimal. In: Proceedings SODA, pp. 777–789. SIAM (2011)
9. Lokshtanov, D., Marx, D., Saurabh, S., et al.: Lower bounds based on the exponential time hypothesis. Bulletin of EATCS 3(105) (2013)
10. Michael, R.G., David, S.J.: Computers and intractability: a guide to the theory of np-completeness. WH Freeman & Co., San Francisco (1979)
11. Oum, S., Sæther, S.H., Vatshelle, M.: Faster algorithms for vertex partitioning problems parameterized by clique-width. Theoretical Computer Science 535, 16–24 (2014)
12. Oum, S., Seymour, P.: Approximating clique-width and branch-width. Journal of Combinatorial Theory, Series B 96(4), 514–528 (2006)
13. Sæther, S.H.: Solving hamiltonian cycle by an EPT algorithm for a non-sparse parameter, http://www.arxiv.org
14. Sæther, S.H., Telle, J.A.: Between treewidth and clique-width. To appear in Proceedings of WG 2014 (2014) Invited to contribute to special section of Algorithmica
15. Vatshelle, M.: New width parameters of graphs. PhD thesis, The University of Bergen (2012)

Associativity for Binary Parallel Processes: A Quantitative Study[*]

Olivier Bodini[1], Antoine Genitrini[2,3], Frédéric Peschanski[2,3], and Nicolas Rolin[1]

[1] Laboratoire d'Informatique de Paris-Nord, CNRS UMR 7030 - Institut Galilée - Université Paris-Nord, 99, avenue Jean-Baptiste Clément, 93430 Villetaneuse, France
{Olivier.Bodini,Nicolas.Rolin}@lipn.univ-paris13.fr
[2] Sorbonne Universités, UPMC Univ Paris 06, UMR 7606, LIP6, F-75005, Paris, France
[3] CNRS, UMR 7606, LIP6, F-75005, Paris, France
{Antoine.Genitrini,Frederic.Peschanski}@lip6.fr

Abstract. We investigate the common interpretation of parallel processes as computation trees. The basis for our approach is the combinatorics of increasingly labelled structures, and our main objective is to provide quantitative results relying on advanced analytic techniques. Unlike previous works, the combinatorial model we propose captures the following ingredients of the algebraic presentation : a binary parallel operator with associativity law. The switch from general trees to binary encodings in this paper makes everything more complex (but eventually workable). Ultimately, we provide a precise characterization and asymptotic approximations of various measures of parallel processes in the average case, especially the average size of the computation trees and their average number of paths, providing a more meaningful notion of *combinatorial explosion* than in the (rather trivial) worst-case. Beyond the measures, we also provide a precise characterization of the typical combinatorial shape of the computation trees, especially their level-decomposition, an interesting notion of process depth. From a more practical point of view, we develop efficient algorithms for the uniform random sampling of computations. Thanks to our typical shape analysis, it is possible to uniformly sample computation prefixes at a given depth in a very efficient way. Indeed, these algorithms work directly on the syntax trees of the processes and do not require the explicit construction of the state space, hence completely avoiding the combinatorial explosion.

1 Introduction

The combinatorial study of concurrent processes is a relatively recent and quite active area of research. Pure notions of parallelism are studied from different perspectives in the literature. The *shuffle product on regular words* provides an

---

[*] This research was partially supported by the A.N.R. project *MAGNUM*, ANR 2010-BLAN-0204.

S. Ganguly and R. Krishnamurti (Eds.): CALDAM 2015, LNCS 8959, pp. 217–228, 2015.
© Springer International Publishing Switzerland 2015

automata-theoretic interpretation that received much attention (cf. e.g. [1,17]). The *trace monoid* provides another mathematical characterization of pure parallelism, which was also extensively studied (cf. e.g. [15]). On the other hand, we investigate the more common and concrete interpretation of parallel processes as *computation trees*. Despite its straightforward algebraic characterization, the underlying structures based on *increasing labellings* are quite intricate and their study represents a real challenge in terms of analytic combinatorics.

In [4] we provide an interpretation of non-determinism in terms of labelled tree-shaped structures. In the paper [3] (and its extended version [5] currently under submission) we demonstrate a one-to-one correspondence between the subcase of pure parallel processes and the well-known combinatorial class of *general increasing trees* [2]. This leads to many quantitative results, most notably the average number of concurrent runs for syntaxic process, i.e. an average-case analysis of the *combinatorial explosion effect*. We also develop an algorithmic framework for statistical model checking based on the uniform random generation of concurrent runs directly from the syntax, that is without having to construct explicitly the computation trees hence *avoiding* the combinatorial explosion.

However, an important simplification is imposed on the model in these preliminary works. The parallel operator is of an arbitrary arity, which allows us to consider *general trees* in the combinatorial interpretations. In most presentations of process algebras (cf. e.g. [16,9]), on the contrary, a binary parallel operator is considered. It is well-known that general trees can be encoded in a binary form, hence the change is relatively transparent at the syntax level. However, as far as the semantic interpretation is concerned the correspondence between the two is highly non-trivial, as it is emphasized in this paper.

At the technical level, the main contributions of the paper are the following ones. In the theory, we provide a precise characterization and asymptotic approximations of various measures of parallel processes in the average case, especially: (1) the average number of computations (i.e. number of runs), and (2) the average size of computation trees. These provide a rather precise meaning of *combinatorial explosion*. Parallel processes are not just "exponential" in the worst case. Perhaps even more interestingly we characterize the typical combinatorial shape of the computation trees, based on a level-decomposition that proves somewhat unexpectedly *workable*. This provides an interesting interpretation for the notion of *process depth*.

From a more practical point of view, we show that the uniform random generation algorithm of [3,5] can be adapted to the binary model. We also provide, thanks to our level-decomposition of computation trees, an extension of the algorithmic framework to allow efficient uniform samplings of computation prefixes. Hence if one may not explore a complete computation tree given its exponential size, we may exploit the fact that the tree prefixes grow rather "slowly" (although still in an exponential way). Developing these techniques for more expressive concurrent calculi (with e.g. non-deterministic choice as in [4]) would naturally lead to a mix of statistical and bounded exploration techniques for model-checking of large-scale concurrent systems. Indeed, these algorithms work directly on the syntax objects and do not require the explicit construction of the state space, hence avoiding the combinatorial explosion.

2   Context

2.1   Syntax: Process Trees

In this paper, pure parallel processes are specified using the following grammar:

- an atomic action, denoted $a, b, c, \ldots$ is a process,
- the prefixing $a.P$ of an action a and a process P is a process,
- the composition $P_1 \parallel P_2$ of exactly two processes P_1 and P_2, is a process.

The following process is the *running example* that will be used as illustration throughout the paper:

$$(a.b) \parallel [(c \parallel d) \parallel (e.(f \parallel g))].$$

Such process terms can be naturally interpreted as tree structures, as depicted on the left of Fig. 1. Pure parallel processes are composed out of unary action nodes, action leaves and binary parallel nodes. Process terms that are well-formed according to the grammar above will thus be called *process trees* from now on.

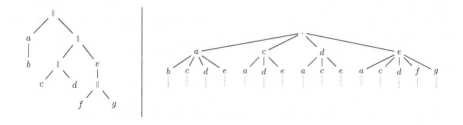

Fig. 1. A process tree of size 7 (left) and the first levels of its computation tree (right)

A grammar for (finite) tree-shaped structures can be almost directly reinterpreted as a *combinatorial class*, only requiring a precise definition for the sizes of the objects belonging to the class. Because for the questions – about pure parallelism – that concern us the identity of the atomic actions will not play any role, our combinatorial specification will abstract from them. This leads to the following specification:

Definition 1. *The combinatorial class of process trees is specified as followed:*

$\mathcal{P} = \mathcal{Z} + \mathcal{Z} \times \mathcal{P} + \mathcal{P} \times \mathcal{P}$, *where $\mathcal{Z}$ marks the nodes containing an action.*

This can be read almost as a grammar: an object in class $\mathcal{P}$ is either a leaf (or external node) marked by $\mathcal{Z}$, a unary internal node also marked by $\mathcal{Z}$ and with a subtree in $\mathcal{P}$, or an unmarked binary nodes with two subtrees in $\mathcal{P}$. The marker $\mathcal{Z}$ explains which nodes must be counted in the resulting size of the object. Here we mark the nodes corresponding to actions in the grammar. Hence, the size of a process tree P is the number of occurrences of actions in the tree, and it is denoted by $|P|$. For example, the size of the running example is 7.

Let us now introduce some notations[1] about the class $\mathcal{P}$ of objects. Let us denote by P_n the number of processes of size n (also called the *counting sequence* of $\mathcal{P}$) and $P(z)$ the ordinary generating function[2] related to the class $\mathcal{P}$: it satisfies $P(z) = \sum_{n>0} P_n z^n$. And the notation for the coefficient extraction of the generating function is $[z^n]P(z) = P_n$. Using Definition 1 and the *symbolic method* (cf.[12]), we deduce a functional equation from the specifications of processes: $P(z) = z + z \cdot P(z) + P(z) \cdot P(z)$. Here we give the first coefficients: $P(z) = z + 2 z^2 + 6 z^3 + 22 z^4 + 90 z^5 + 394 z^6 + \dots$. (e.g. there are 90 trees of size 5 in class $\mathcal{P}$).

Proposition 1. *The combinatorial class $\mathcal{P}$ satisfies:*

$$P(z) = \frac{1 - z - \sqrt{1 - 6z + z^2}}{2} \qquad \text{and} \qquad P_n \sim_{n \to \infty} \sqrt{\frac{3\sqrt{2} - 4}{4 \pi n^3}} \cdot \left(3 - 2\sqrt{2}\right)^{-n}.$$

This is a direct result of applying the symbolic method (cf. [12]), even if our way of counting nodes in not standard. We recall that $\rho_P = 3 - 2\sqrt{2}$ is the dominant singularity of P: it is directly associated to the exponential growth of $(P_n)_n$.

In [3,5] we describe a variant of pure parallel processes in which the parallel operator is of an arbitrary arity. In terms of combinatorics, this is a much simpler setting than the binary case since the process trees can be identified with *general trees* with only one type of nodes: an action followed by a set of sub-processes. However, the binary operator used in the present paper is more faithful to the algebraic presentations of process algebras (cf. e.g. [16,9]). It is well-known that general trees can be encoded in a binary form, hence the change is relatively transparent at the syntax level. The change will be less transparent at the semantic level since we will have to take into account the associative law attached to the parallel operator: $(P_1 \parallel P_2) \parallel P_3 \equiv P_1 \parallel (P_2 \parallel P_3)$. This means, from the semantic view, that process trees that only differ by the left-right succession of parallel nodes must be identified.

---

[1] For all the combinatorial classes that will appear in the paper we will use the same kind of notations like $\mathcal{P}$ for the class, P_n for the counting sequence and $P(z)$ for the generating function.

[2] The exponential generation function $G(z)$ related to the sequence $(G_n)_n$ satisfies $G(z) = \sum_n G_n \frac{z^n}{n!}$.

2.2 Semantics: Computation Trees

Process trees are syntax objects that must be interpreted on a *semantic* domain. Our combinatorial model is to interpret process behaviours as *computation trees* [6]. A run or computation is the result of the merging of the branches of a process tree. For example, using our running example from Fig. 1, we note that the run $\langle a, b, c, d, e, f, g \rangle$ is a computation of the process but $\langle a, b, c, d, f, e, g \rangle$ is not because it is not the result of the merging of the process (action f cannot precede action e).

The whole process behaviour is a tree of all possible computations with all common prefixes shared. The right-hand side of Fig. 1 presents the first levels of the computation tree induced by our running example. It is a well-known fact that the computation trees of pure parallel processes are "exponentially" larger than the syntax trees. We can witness this phenomenon of *combinatorial explosion* on our running example. Indeed, despite its small syntactic size (it has 7 counted nodes), its induced computation tree is of size 2360.

The questions that concern us are firstly of a quantitative nature: we would like to give a precise mathematical – in fact combinatorial – meaning for "combinatorial explosion" that is often used in a somewhat gratuitous way. The most significant measure of the process behaviours is undoubtedly their number of runs *on average*. A finer – and technically more involved – measure is required to properly characterize the amount of prefix-sharing in the computation trees. A more qualitative question is then raised: what is the *typical shape* of the computation trees ? For this we exploit a decomposition of computation trees by levels. A *level* ℓ of a computation tree is the set of nodes that correspond to the ℓ-th occurrence of an action in each of its branches. For example, Fig. 1 depicts the first and second levels of the computation tree of our running example. Ultimately, our theoretical study underlies interesting algorithms for the statistical analysis of computation trees. All these questions shall be now addressed.

3 Typical Binary Processes

In this section, we are interested in typical measures of computations trees in the context of associative binary parallel processes. We first provide the average asymptotic number of runs of processes of size n, when n tends to infinity. We then refine the quantitative study by considering the total size of the computation trees.

3.1 Typical Number of Runs

Theorem 2. *The asymptotic of the average number of runs, denoted by $\bar{G}_n$, induced by binary processes of size n, satisfies when n tends to infinity:*

$$\bar{G}_n \sim_{n \to \infty} 3 \cdot \sqrt{\frac{\ln \frac{3}{2} - \frac{1}{3}}{6\sqrt{2} - 8}} \cdot \left(\frac{3 - 2\sqrt{2}}{3 \left(\ln \frac{3}{2} - \frac{1}{3} \right)} \right)^n \cdot n!.$$

First remark that $(3 - 2\sqrt{2})/(3 \left(\ln \frac{3}{2} - \frac{1}{3}\right)) \approx 0.79287$ thus the average $\bar{G}_n$ is much smaller than $n!$ that corresponds to the number of runs of the worst processes (where all actions are in parallel).

The proof of this theorem follows the general sketch already followed in [3] for the n-ary process trees. However, our calculations relied, there, extensively on *holonomy theory* and we will see that this approach is not possible in the binary case, we must find other calculation ways. In order to compute the number of runs of a process, we exploit an isomorphism between the runs of a process tree and its *increasing labellings*. Let us recall that usually an increasing labelling of a tree is a labelling of each node with an integer from 1 to the size of the tree such that all successors of a node have a strictly greater label compared to the one of this node. Such so-called *increasing trees* are discussed at length in [8, Chapter 1].

In our model of binary processes, only the nodes labelled by an action must be taken into account. But beyond that the isomorphism still holds.

Lemma 1. *Let P be a binary process. There is an isomorphism between the runs (or computations) of P and the increasing labellings of the nodes of P containing an action.*

The proof for this lemma can be adapted from [3] in a straightforward way. Since parallel nodes are not considered for the increasing labelling, the associativity law of parallel does not play any role here.

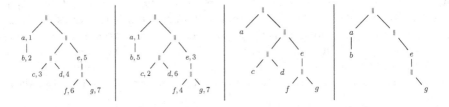

Fig. 2. Two increasingly labelled processes (left) and two admissible cuts (right)

On Fig. 2 the two leftmost increasing trees correspond respectively to the runs $\langle a, b, c, d, e, f, g \rangle$ and $\langle a, c, e, f, b, d, g \rangle$ of the running example.

In analytic combinatorics, the *box product* allows to encode increasing constraints on the labels of nodes. Let $\mathcal{A}$ and $\mathcal{B}$ be two combinatorial labelled classes, then the class $\mathcal{A}^{\square} \star \mathcal{B}$ corresponds to labelled objects such that the smallest label belongs to the first component (in $\mathcal{A}$). For further details, see [12, Chapter II].

Proposition 2. *Let $\mathcal{G}$ be the class of increasingly labelled processes, and $G(z)$ its related exponential generating function. We get:*

$$\mathcal{G} = \mathcal{G} \star \mathcal{G} + \mathcal{Z}^{\square} \star (\mathcal{G} + 1), \quad thus \quad G(z) = -1 - \frac{3}{2} \cdot LambertW\left(-\frac{2}{3} \exp\left(\frac{z-2}{3}\right)\right).$$

The *LambertW*-function satisfies: $\text{LambertW}(z) \cdot \exp(\text{LambertW}(z)) = z$. Many fundamental results about the *LambertW*-function can be found in the paper of

Corless *et al* [7]. Especially its generating function is as follows:

$$\text{LambertW}(z) = \sum_{r \geq 1} w_r \, z^r, \qquad \text{where } w_r = \frac{(-r)^{r-1}}{r!}. \qquad (1)$$

By taking into account that the series $G(z)$ is exponential, we may easily compute the first numbers of increasingly labelled processes (according to the number of actions): $1, 3, 21, 243, 3933, 81819, \ldots$.

Proof. Due to the translation of the boxed product into the formal power series, we know that the exponential generating function $G(z)$ is the solution, analytic at 0 of $T'(z) - 2T(z)T'(z) - T(z) = 1$ such that $T(0) = 0$ and $T'(0) = 1$. By a partial fraction expansion of this differential equation, we can integrate it and thus we get $\frac{2}{3}(1 + G(z))e^{-\frac{2}{3}(1+G(z))} = e^{\frac{z}{3} - \frac{2}{3} \ln \frac{3}{2}}$ Let us define y and x such that $y = \frac{2}{3}(1 + G(z))$ and $x = e^{\frac{z}{3} - \frac{2}{3} - \ln \frac{3}{2}}$, then the equation turns to $ye^{-y} = x$, thus the link with the *LambertW*-function is exhibited. $\qquad \square$

As explained previously all the generating functions that we used in [3] where holonomic. Almost all results have been proved by using "Guess and Prove" strategies that rely on holonomicity. As a reminder, a generating function is holonomic if its coefficients satisfy a homogeneous linear (finite) recurrence (called a *P-recurrence*) with polynomials coefficients. From the first coefficients and the P-recurrence generally obtained through guesses followed by cumbersome calculations, one can compute efficiently the next coefficients.

But switching to the binary parallel operator makes the whole *edifice* collapse since the *LambertW*-function is *not* holonomic, as demonstrated in e.g. [13]. From this we can legitimately suppose that the generating function $G(z)$ of Proposition 2 is not holonomic also. However non-holonomy does not always propagate through function composition so that the following is not a trivial result.

Proposition 3. $G(z)$ *is not holonomic.*

This difficulty can be circumvented by following a more direct approach, which is sketched below. We would like to emphasize, however, the important take away of this section: that a binary encoding of an associative operator is not equivalent to its direct interpretation as a general tree. We have holonomy on the one size, and non-holonomy on the other size.

Proof (of Theorem 2). The dominant singularity of $G(z)$ is reached when the *LambertW*-function reaches $-e^{-1}$. Hence the dominant singularity of G(z) is $\eta = -1 + 3 \ln 3/2$. By basic computations about the *LambertW*-function, we get: $\text{LambertW}(-e^{-1} \cdot (1 - h)) =_{h \to 0} -1 + \sqrt{2\,h} + o(\sqrt{h})$. Together with the Taylor development: $-\frac{2}{3} \exp \left(\frac{z-2}{3} \right) = -e^{-1} \cdot \left(1 - \frac{\eta}{3} \cdot \left(1 - \frac{z}{\eta} \right) \right) + o \left(1 - \frac{z}{\eta} \right)$, we get: $\text{LambertW} \left(-\frac{2}{3} \exp \left(\frac{z-2}{3} \right) \right) = -1 + \sqrt{\frac{2\eta}{3} \left(1 - \frac{z}{\eta} \right)} + o \left(\sqrt{1 - \frac{z}{\eta}} \right)$. The classical transfer theorems due to Flajolet and Odlyzko [11], detailed in [12], give:

$$n! \, [z^n] G(z) \sim_{n \to +\infty} n! \cdot \frac{3}{2} \cdot \sqrt{\frac{\ln \frac{3}{2} - \frac{1}{3}}{2\pi \, n^3}} \cdot \left(\frac{1}{3 \left(\ln \frac{3}{2} - \frac{1}{3} \right)} \right)^n.$$

Finally the stated average value is obtained by normalizing by P_n.   □

3.2 Typical Size of the Computation Trees

Computation trees share the common computation prefixes, which cannot be witnessed by counting its branches (or leaves) as was done in the previous section. The goal of this subsection is to compute the asymptotic average profile of the computation trees: precisely the average number of nodes of each of their levels.

Theorem 3. *Let $\bar{L}_n$ be the average size of the computations trees induced by binary process trees of size n, and $\bar{L}_n^{n-\ell}$ be the average number of nodes at level $n - \ell - 1$ of the computations trees ($\ell \in \{0, \ldots n-1\}$). The asymptotic values of these means, when n tends to infinity, satisfies:*

$$\bar{L}_n \sim_{n \to \infty} e \cdot \bar{G}_n, \qquad \text{and} \qquad \bar{L}_n^{n-\ell} \sim_{n \to \infty} \frac{\bar{G}_n}{\ell!}.$$

Definition 4. *Let P be a process tree. Starting from P, prune iteratively some leaves from the tree structure. If the remaining tree C does not contain leaves labelled by the operator $\|$, then C is called an* admissible cut *of P. The size of an admissible cut is the number of actions it contains.*

The complete process tree T is defined as an admissible cut too, but the empty process (after having removed all nodes of T) is not an admissible cut. On Fig. 2, two admissible cuts obtained from our running example (Fig. 1) are depicted. An increasing labelling of the actions of an admissible cut gives an *increasing admissible cut*.

Lemma 2. *Let P be a process tree. The number of nodes at level $i - 1$, for $i \in \{1, \ldots, |P|\}$, is equal to the number of increasing admissible cuts of size i of the process P.*

Proposition 4. *The following specification enumerates all increasing admissible cuts induced by process trees of the same size.*

$$\mathcal{C} = \mathcal{C} \times \mathcal{C} + 2 \cdot \mathcal{C} \times \mathcal{P} + \mathcal{U}^{\square} \star \mathcal{Z} \times (\mathcal{C} + \mathcal{P} + 1),$$

where $\mathcal{Z}$ marks all nodes and $\mathcal{U}$ the nodes of the increasing admissible cuts. Thus,

$$C(z, u) = -(1 + P(z)) - \frac{3}{2} \cdot LambertW \left(-\frac{2}{3} \cdot (1 + P(z)) \cdot \exp \left(\frac{uz}{3} - \frac{2}{3} (1 + P(z)) \right) \right).$$

An analogous approach as the one presented in the proof of Proposition 2 gives the result.

Proposition 5. *$C(z, u)$ is not holonomic.*

Because of this result, the proof of Theorem 3 is not obvious at all. The result is proved by a detailed analysis of the equation satisfied by $C(z, u)$.

4   Algorithmic Applications

The quantitative study described in the previous sections could misleadingly be seen as only of a purely theoretical interest. A better understanding of the average case – or unbiased – situation comes together with a better understanding of the uniform random distribution of the objects under study. This naturally yields interesting algorithmic applications.

In this section we discuss three such applications: (1) the uniform random generation of runs, (2) the computation of the profile of a computation tree, and (3) the covering of computation prefixes at a given process depth. These algorithms take a fixed process tree, say P, as input. In this section we will mostly consider the process of Fig. 3, which is of size 125 and thus with a very large state space (about 10^{145} distinct runs !).

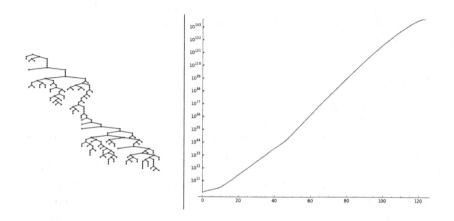

Fig. 3. A process tree of size 125 (left) and the profile of its computation tree (right)

4.1   Uniform Random Generation of Runs

The uniform random generation of runs is the basic algorithmic building block for the statistical analysis of the process behaviours. In [3] we describe a generation algorithm that works with complexity $O(n \log n)$ with n the size of the initial process tree, described as general trees. Indeed, the random sampler generates branches of the computation trees, distributed uniformly, by only considering the syntax trees, hence avoiding the combinatorial explosion. The algorithm relies on the *hook length formula* [14, P. 67] and the implementation uses a dynamic multiset sampler (the implementation is given as an appendix in [2]).

Luckily enough, the hook length formula can be adapted to the case of the binary parallel processes. For a given process P, let us define a *prefixed subprocess*, P_α as a complete subtree of P, rooted at an action node α of P (it cannot be rooted in a parallel node). In our running example, P is not a prefixed subprocess of P, because its root is a parallel node, but $P_e := e.(f \parallel g)$ is one.

Proposition 6. *Hook length formula adapatation: Let P be a binary process. The number of computations of P is:*

$$G_P = \frac{|P|!}{\displaystyle\prod_{P_\alpha \ prefixed \ subprocess} |P_\alpha|}.$$

This proposition is direct once we remark that partially increasing trees of P (only actions are increasingly labelled) are in one-to-one correspondence with runs of P.

Corollary 1. *Let P be a process of size n. Using the hook length formula adaptation, we derive an uniform random generator to sample runs in P. This time complexity is $O(n \log n)$.*

The algorithm is an adaptation of the sampler obtained in [3] based on the multiset sampler. The time complexity is not impacted by the parallel nodes since their number is bounded by n.

4.2   Profile of the Computation Tree

Using the hook length formula adaptation, we get an efficient way to compute the number of nodes of the last level of the computation tree, hence its number of runs. The time complexity of the algorithm is linear in the size of the process tree. However we are not just interested by the number of leaves of the tree, but by its whole *profile* i.e. the number of nodes at each of its levels. This is the qualitative facet of our level decomposition of computation trees, introduced in Section 3. In [5] we develop a naive algorithm to compute the profile of a computation tree. This algorithm has exponential time complexity and can thus only be applied on small process trees. In an experiment we were able to compute the profile of a process tree of size 40 in "a couple of days using a fast parallel computer". We now introduce a much more efficient algorithm (its complexity is linear!). Indeed, the profile of the computation tree corresponding to the process tree of size 125 of Fig. 3 is constructed in less than a second using Sage/Python on a basic PC. Recall that the total size of the computation tree is about 10^{145}. The profile is presented on Fig. 3 with a semilogarithm scale for the number of nodes by level.

In practice, the profile algorithm can be seen as a by-product of a more general algorithm that allows us to sample prefixes of computations uniformly among all the prefixes of the same sizes. It is based on the *recursive method* developed by Nijenhuis and Wilf [18]. It is interesting to note that the algorithm we implemented is very close to (indeed inspired by) the one we develop in [4] for a very different purpose: the uniform random generation of computations for non-deterministic processes. The fundamental idea of the algorithm is to see a process tree as the combinatorial class of all its admissible cuts. To define such a class, we need to introduce two concepts in the process specification. First, the empty subprocess ϵ and then an operator, denoted by $+$, that allows to choose between a prefixed subprocess or this empty subprocess. Thus the subprocess $P_\alpha + \epsilon$ consists of running the subprocess P_α or of stopping the process. Obviously, this $+$ operator has no relation to the nondeterministic choice of [4], although, from an analytic combinatorics view both play the same rôle. In the context of [4], the algorithm allows to generate runs with nondeterministic choice, but it has no relation with the profile problem we address here. Thus, both algorithms are based on some similar polynomial, but their interpretation is completely different. The algorithm's sketch is as follows: For a given process P:

- Define the specification of the combinatorial class of all its admissible cuts: i.e. replace each prefixed subprocess P_α by $(P_\alpha + \epsilon)$.
- Compute the polynomial representation of this combinatorial class: it corresponds to the exponential generating function of the admissible cuts class.
- Using the polynomial representation, one can easily:
  - Compute the computation tree profile.
  - Sample computations prefixes or complete computation according to this distribution by using the recursive method.

By using a directed acyclic graph for the representation of the polynomial, the profile computation is obtained in $O(n)$ time complexity; and the random sampling in $O(n \log n)$. Both complexity results are derived from the proofs of [4].

4.3 Prefix Covering

The previous algorithm allows us to construct the uniform probability distribution on run prefixes of a given length. This opens up an interesting statistical analysis approach that is complementary to the generation of complete runs. It is indeed difficult to talk about *covering* when generating complete runs. In all but the most trivial situations there are simply too many possible runs to even think about covering the full process behaviour. The only guarantee provided is that the sampled computations are truly random and not biased in one way or another. An interesting covering criterion is that of the *process depth*. We now discuss such a covering algorithm that can generate computation *prefixes* of a given length – corresponding to the expected process depth – and uniformly at random. The algorithm finds its justification in the "coupons collector" principle (cf. [10]). When sampling objects whose probability distribution is uniform, then the expected number of samples necessary to collect all the object is equivalent to $n \log n$. This is the smallest possible expectation since the distribution is uniform.

Another covering method would consist in following a *local uniform probability distribution* instead of the global – in fact the *real* – one. When sampling locally uniformly, we only treat each branch of a parallel construct with equal probability, without taking into account the past and the future of the behaviour. But even compared with this naive and deficient approach, it is largely outperformed by our covering algorithm for the global distribution.

For the process of Fig. 3, we conducted some experiments:

Prefix length	1	2	3	4	5	6	7	8
Nb of prefixes	4	14	43	115	265	564	1201	2877
Expected time for covering								
Local uniformity	8	47	223	972	4343	24087	137174	914313
Global uniformity	8	45	187	612	1646	3891	8993	21719
Gain	0%	4.5%	19%	59%	164%	519%	1425%	4110%

The prefixes covering is always much more affordable using our global technique. The larger the prefix length is, the more our approach is unavoidable.

References

1. Bassino, F., David, J., Nicaud, C.: Average case analysis of Moore's state minimization algorithm. Algorithmica 63(1-2), 509–531 (2012)
2. Bergeron, F., Flajolet, P., Salvy, B.: Varieties of increasing trees. In: Raoult, J.-C. (ed.) CAAP 1992. LNCS, vol. 581, pp. 24–48. Springer, Heidelberg (1992)
3. Bodini, O., Genitrini, A., Peschanski, F.: Enumeration and random generation of concurrent computations. In: DMTCS AofA 2012 Proceedings, pp. 83–96 (2012)
4. Bodini, O., Genitrini, A., Peschanski, F.: The Combinatorics of Non-determinism. In: proc. FSTTCS 2013. LIPIcs, vol. 24, pp. 425–436 (2013)
5. Bodini, O., Genitrini, A., Peschanski, F.: Enumeration and random generation of concurrent computations. arXiv/1407.1873 (page under submission, 2014)
6. Clarke, E.M., Emerson, E.A., Sistla, A.P.: Automatic verification of finite-state concurrent systems using temporal logic specifications. ACM Trans. Program. Lang. Syst. 8(2), 244–263 (1986)
7. Corless, R.M., Gonnet, G.H., Hare, D.E.G., Jeffrey, D.J., Knuth, D.E.: On the Lambert W Function. Advances in Computational Mathematics 5, 329–359 (1996)
8. Drmota, M.: Random trees. Springer, Vienna (2009)
9. Duchamp, G., Hivert, F., Novelli, J.-C., Thibon, J.-Y.: Noncommutative Symmetric Functions VII: Free Quasi-Symmetric Functions Revisited. Ann. Comb. 15, 655–673 (2011)
10. Flajolet, P., Gardy, D., Thimonier, L.: Birthday Paradox, Coupon Collectors, Caching Algorithms and Self-Organizing Search. D. A. Math. 39(3), 207–229 (1992)
11. Flajolet, P., Odlyzko, A.M.: Singularity analysis of generating functions. In SIAM J. Discrete Math. 3, 216–240 (1990)
12. Flajolet, P., Sedgewick, R.: Analytic Combinatorics. Cambridge UP. (2009)
13. Gerhold, S.: On Some Non-Holonomic Sequences. Elec. J. Comb. 11(1), 1–8 (2004)
14. Knuth, D.E.: The art of computer programming, 2nd edn. sorting and searching, vol. 3. Addison Wesley Longman Publishing Co., Inc. (1998)
15. Krob, D., Mairesse, J., Michos, I.: Computing the average parallelism in trace monoids. Discrete Mathematics 273(1-3), 131–162 (2003)
16. Milner, R.: A Calculus of Communicating Systems. Springer (1980)
17. Mishna, M., Zabrocki, M.: Analytic aspects of the shuffle product. CoRR (2008)
18. Wilf, H.S., Nijenhuis, A.: Combinatorial algorithms: An update (1989)

A Tight Bound for Congestion of an Embedding

Paul Manuel[1], Indra Rajasingh[2,*], R. Sundara Rajan[2], N. Parthiban[3],
and T.M. Rajalaxmi[4]

[1] Department of Information Science, Kuwait University, Safat, Kuwait, 13060
[2] School of Advanced Sciences, VIT University, Chennai, India, 600 127
[3] School of Computing Science and Engineering, VIT University,
Chennai, India, 600 127
[4] Department of Mathematics, SSN College of Engineering,
Chennai, India, 603 110

Abstract. Graph embedding has been known as a powerful tool for implementation of parallel algorithms or simulation of different interconnection networks. Congestion is one of the main optimization objectives in global routing. In this paper, we introduce a technique to obtain a tight bound for congestion of an embedding. Moreover, we give algorithms to compute exact congestion of embedding the hypercubes into the cylinder and the torus and prove that the bound obtained is sharp.

Keywords: Embedding, congestion, cylinder, torus.

1 Introduction

Graph embedding is an important technique that maps a guest graph into a host graph, usually an interconnection network. Many applications can be modeled as graph embedding. In architecture simulation, graph embedding has been known as a powerful tool for implementation of parallel algorithms or simulation of different interconnection networks. A parallel algorithm can be modeled by a task interaction graph, where nodes and edges represent tasks and direct communications between tasks, respectively [1]. In parallel computing, a large process is often decomposed into a set of small sub-processes that can execute in parallel with communications among these sub-processes. The problem of allocating these sub-processes into a parallel computing system can be again modeled by graph embedding [2].

The quality of an embedding can be measured by certain cost criteria. One of these criteria which is considered very often is the *congestion*. The congestion of an embedding is the maximum number of edges of the guest graph that are embedded on any single edge of the host graph. An embedding with a large congestion faces many problems, such as long communication delay, circuit switching and the existence of different types of uncontrolled noise. Therefore, a minimum

---

* This work is supported by Project No. SR/S4/MS: 846/13, Department of Science and Technology, SERB, Government of India.

S. Ganguly and R. Krishnamurti (Eds.): CALDAM 2015, LNCS 8959, pp. 229–237, 2015.

congestion is a most desirable feature in network embedding [3]. Congestion of an embedding has been well studied for a number of networks [4–8].

A suitable interconnection network is important for the design of a multicomputer or multiprocessor system. This network is usually modeled by a symmetric graph, where the nodes represent the processing elements and the edges represent the communication channels. Desirable properties of an interconnection network include symmetry, embedding capabilities, relatively small degree, small diameter, scalability, robustness, and efficient routing [9].

One of the most efficient interconnection networks is the hypercube due to its structural regularity, potential for parallel computation of various algorithms, and the high degree of fault tolerance [10]. The hypercube has many excellent features and thus becomes the first choice of topological structure of parallel processing and computing systems. The machine based on hypercubes such as the Cosmic Cube from Caltech, the iPSC/2 from Intel and Connection Machines have been implemented commercially [11]. The hypercube embedding problem is the problem of mapping a communication graph into a hypercube multiprocessor. Hypercubes are known to simulate other structures such as grids and binary trees [12, 13].

The rest of the paper is organized as follows. Section 2 gives definitions and other preliminaries. In Section 3, we find a new strategy to compute tight bound for congestion of an embedding. Section 4 deals with the congestion of embedding hypercubes into cylinder and torus. Finally, concluding remarks and future work are given in Section 5.

2   Basic Concepts

In this section we give the basic definitions and preliminaries required for our subsequent work.

Definition 1. [14] *Let G and H be finite graphs. An embedding $\phi = (f, P_f)$ of G into H is defined as follows:*

1. *f is a one-to-one map from $V(G) \to V(H)$*
2. *P_f is a one-to-one map from $E(G)$ to $\{P_f(u, v) : P_f(u, v)$ is a path in H between $f(u)$ and $f(v)$ for $(u, v) \in E(G)\}$.*

For brevity, we denote the pair (f, P_f) as f.

The *expansion* [14] of an embedding f is the ratio of the number of vertices of H to the number of vertices of G. In this paper, we consider embeddings with expansion one.

Definition 2. [14] *The edge congestion of an embedding f of G into H is the maximum number of edges of the graph G that are embedded on any single edge of H. Let $EC_f(e)$ denote the number of edges (u, v) of G such that e is in the path $P_f(u, v)$ between the vertices $f(u)$ and $f(v)$ in H. In other words,*

$$EC_f(e) = |\{(u, v) \in E(G) : e \in P_f(u, v)\}|$$

where $P_f(u, v)$ denotes the path between $f(u)$ and $f(v)$ in H with respect to f.

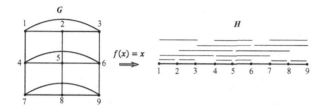

Fig. 1. Wiring diagram of a cylinder G into path H with $EC_f(G, H) = 5$

The *congestion* of an embedding f of G into H is defined as

$$EC_f(G, H) = \max EC_f(e)$$

where the maximum is taken over all edges e of H.
The *congestion* of G into H is defined as

$$EC(G, H) = \min EC_f(G, H)$$

where the minimum is taken over all embeddings f of G into H.

If we think of G as representing the wiring diagram of an electronic circuit, with the vertices representing components and the edges representing wires connecting them, then the edge congestion $EC(G, H)$ is the minimum, over all embeddings $f : V(G) \to V(H)$, of the maximum number of wires that cross any edge of H [15]. See Figure 1.

Definition 3. [16, 17] *For $r \geq 1$, let Q^r denote the r-dimensional hypercube. The vertex set of Q^r is formed by the collection of all r-dimensional binary strings. Two vertices $x, y \in V(Q^r)$ are adjacent if and only if the corresponding binary strings differ exactly in one bit. The vertices of Q^r can also be identified with integers $0, 1, \ldots, 2^r - 1$.*

Definition 4. [18] *An incomplete hypercube on i vertices of Q^r is the subcube induced by $\{0, 1, \ldots, i - 1\}$ and is denoted by L_i, $1 \leq i \leq 2^r$.*

Theorem 1. [19–21] *Let Q^r be an r-dimensional hypercube. For $1 \leq i \leq 2^r$, L_i is an optimal set on i vertices.*

Definition 5. [22] *The 2-dimensional grid is defined as $P_{d_1} \times P_{d_2}$, where $d_i \geq 2$ is an integer for each $i = 1, 2$. The cylinder $C_{d_1} \times P_{d_2}$, where $d_1, d_2 \geq 3$ is a $P_{d_1} \times P_{d_2}$ grid with a wraparound edge in each row.*

It is clear that the vertex set of $P_{d_1} \times P_{d_2}$ is $V = \{x_1 x_2 : 0 \leq x_i \leq d_i - 1, i = 1, 2\}$ and two vertices $x = x_1 x_2$ and $y = y_1 y_2$ are linked by an edge, if $|x_1 - y_1| + |x_2 - y_2| = 1$.

Definition 6. [22, 23] *The torus $C_{d_1} \times C_{d_2}$, where $d_1, d_2 \geq 3$ is a $P_{d_1} \times P_{d_2}$ grid with a wraparound edge in each row and column.*

Remark 1. The cylinder $C_{d_1} \times P_{d_2}$ has $d_1 d_2$ vertices and $2 d_1 d_2 - d_1$ edges. The torus $C_{d_1} \times C_{d_2}$ has $d_1 d_2$ vertices and $2 d_1 d_2$ edges. See Figure 2.

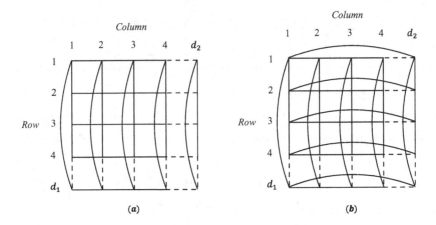

Fig. 2. (a) Cylinder $C_{d_1} \times P_{d_2}$ and (b) Torus $C_{d_1} \times C_{d_2}$

3   Tight Bound

For an embedding f of G into H, the dilation of an embedding is defined as the maximum distance between pairs of vertices of host graph that are images of adjacent vertices of the guest graph. The minimum taken over all the embeddings f is called the *dilation* of embedding G into H. Further the sum of the dilations in H of edges in G is called the wirelength of f. The minimum taken over all the embeddings f is called the *wirelength* of embedding G into H. In 1979, Garey and Johnson proved that embedding problems are NP-complete [24].

The dilation problem, congestion problem and the wirelength problem are different in the sense that an embedding that gives the minimum dilation need not give the minimum congestion or the minimum wirelength and vice-versa. Even though there are numerous results and discussions on the congestion problem, there is no efficient method to compute exact congestion of graph embeddings [4, 5, 15]. In recent years, Manuel et al. obtained a lower bound for dilation of an embedding using minimum wirelength and formulated the result as IPS Lemma [25], and in 2014 the same authors computed an improved bound without using wirelength and formulated the result as Dilation Lemma [26]. The same authors obtained a technique to compute the wirelength of an embedding using the Modified Congestion Lemma [27] in 2014. In this direction, we propose and prove the *Edge Congestion Lemma* to obtain a tight bound for congestion of an embedding. We also prove that the bound is sharp by embedding the hypercube into the cylinder and the torus.

Lemma 1. (Modified Congestion Lemma) [27] *Let f be an embedding of an arbitrary graph G into H. Let S be an edge cut of H such that the removal of edges of S separates H into 2 components H_1 and H_2 and let $G_1 = f^{-1}(H_1)$ and $G_2 = f^{-1}(H_2)$. Furthermore, S satisfies the following conditions:*

(i) *For every edge* $(a, b) \in G_i$, $i = 1, 2$, $P_f(a, b)$ *has no edges in* S.
(ii) *For every edge* (a, b) *in* G *with* $a \in G_1$ *and* $b \in G_2$, $P_f(a, b)$ *has exactly one edge in* S.
(iii) G_1 *and* G_2 *are optimal sets.*

Then $EC_f(S)$ *is minimum and*

$$EC_f(S) = \sum_{v \in V(G_1)} deg_G(v) - 2|E(G_1)| = \sum_{v \in V(G_2)} deg_G(v) - 2|E(G_2)|.$$

Remark 2. When the guest graph G is regular, it is enough to check whether G_1 is an optimal set [12].

The following lemma is the main result of this paper and is formulated as Edge Congestion Lemma which gives a tight bound for the embedding parameter 'congestion'.

Lemma 2. (Edge Congestion Lemma) *Let* G *and* H *be graphs of the same order and let* $f : G \to H$ *be an embedding. Let* S *be an edge cut of* H *satisfying the conditions of the modified congestion lemma. Then*

$$EC(G, H) \geq \frac{EC_f(S)}{|S|}.$$

Proof. By the Modified Congestion Lemma, the inverse images of the sets H_1 and H_2 with respect to the embedding f are maximum subgraphs of G. Therefore $EC_f(S)$ is minimum.

There is at least one edge in S with congestion at least $\frac{EC_f(S)}{|S|}$. Further, for any embedding g of G into H, $EC_g(G, H) \geq EC_f(G, H) \geq \frac{EC_f(S)}{|S|}$. Therefore

$$EC(G, H) = \min_g EC_g(G, H) \geq \frac{EC_f(S)}{|S|}. \qquad \square$$

4   Results and Discussions

Minimum congestion has been obtained for embedding hypercubes into 2 - dimensional grid [12] and n - dimensional grid [28]. In this section we consider cylinder as host graph, obtain congestion on an edge cut and use it to show that the bound obtained from Edge Congestion Lemma is tight. The technique used simplifies the methods employed in [28].

Now we embed hypercube Q^r, $r \geq 2$ into cylinder $C_{2^{r_1}} \times P_{2^{r_2}}$ and torus $C_{2^{r_1}} \times C_{2^{r_2}}$, $r_1 + r_2 = r$ and $r_1 \leq r_2$ with minimum congestion. For proving the main results we need the following definitions and results.

Definition 7. [28] Given a graph G, and a labeling l, let

$$S_k(l) = l^{-1}(\{1, 2, \ldots, k\}),$$

for each k, $1 \leq k \leq |V(G)|$. Thus, $S_k(l)$ is the set of the first k vertices of G to be labeled by l.

Definition 8. [28] For $S \subseteq V(G)$, let

$$\theta_G(S) = |\{(u,v) \in E(G)/u \in S, v \notin S\}|$$

and

$$\theta_G(k) = \min_l \ \theta_G(S_k(l)).$$

For a fixed value of k, the problem of minimizing the value of $\theta_G(S_k(l))$ over all labeling l, can be thought of as a discrete isoperimetric problem.

Remark 3. [28] When G is the r-dimensional hypercube Q^r, $\theta_G(k) = \theta_G(S_k(l^*))$ for each k, where l^* is the labeling of Q^r using lexicographic order [14].

Lemma 3. [28] Let G and H be graphs with same order. Then

$$EC(G,H) \geq \max_{1 \leq k \leq |V(G)|-1} \frac{\theta_G(k)}{\theta_H(k)}.$$

4.1 Hypercube into Cylinder

Embeddings of grids plays an important role in computer architecture. VLSI Layout Problem, Crossing Number Problem, Graph Drawing, and Edge Embedding Problem are all a part of grid embedding. There are very few results in the literature which provide the exact wirelength of embedding graphs into grids [12]. Cylinder is an extension of the grid network. For proving the main result, we need the following result.

Lemma 4. Let G be the r-dimensional hypercube Q^r, $r \geq 2$ and H be the cylinder $C_{2^{r_1}} \times P_{2^{r_2}}$, $r_1 + r_2 = r$ and $r_1 \leq r_2$. Then

$$EC(G,H) = EC(G, P_{2^r}),$$

where P_{2^r} is a path on 2^r vertices.

Proof. Let $k = 2^{r-1} - 2^{r-2} + 2^{r-3} - \cdots + (-1)^{r-r_1+1} \cdot 2^{r_1}$. Note that $2^{r-2} \leq k < 2^{r-1}$ and $k = a \cdot 2^{r_1}$ for some integer a. Consider the $2^{r_1} \times a$ subcylinder F. One has $\theta_H(k) \leq \theta_H(F) = 2^{r_1}$. Also

$$\theta_{Q^r}(k) = \begin{cases} 2^{r-1} + 2^{r-3} + \cdots + 2^{r_1}, & \text{if} \quad r_2 \text{ is odd;} \\ 2^{r-1} + 2^{r-3} + \cdots + 2^{r_1+1}, & \text{if} \quad r_2 \text{ is even.} \end{cases}$$

Therefore

$$\frac{\theta_{Q^r}(k)}{\theta_H(k)} \geq \frac{1}{2^{r_1}} \begin{cases} 2^{r-1} + 2^{r-3} + \cdots + 2^{r_1}, & \text{if} \quad r_2 \text{ is odd;} \\ 2^{r-1} + 2^{r-3} + \cdots + 2^{r_1+1}, & \text{if} \quad r_2 \text{ is even.} \end{cases}$$

$$= \begin{cases} 1 + 2^2 + \cdots + 2^{r_2-1}, & \text{if} \quad r_2 \text{ is odd;} \\ 2 + 2^3 + \cdots + 2^{r_2+1}, & \text{if} \quad r_2 \text{ is even.} \end{cases}$$

$$= \begin{cases} \frac{2^{r_2+1}-1}{3}, & \text{if} \quad r_2 \text{ is odd;} \\ \frac{2^{r_2+1}-2}{3}, & \text{if} \quad r_2 \text{ is even.} \end{cases}$$

$$= EC(Q^{r_2}, P_{2^{r_2}}).$$

Then by Lemma 3, $EC(Q^r, H) \geq EC(Q^{r_2}, P_{2^{r_2}})$. On the other hand, let $l : Q^r \to H$ be given by $lex_i : Q^{r_i} \to P_{2^{r_i}}$, $i = 1, 2$. Thus $EC(Q^r, H) \leq \max_i EC(Q^{r_i})$. Hence the Lemma.

Theorem 2. *Let G be an r-dimensional hypercube Q^r, $r \geq 2$ and H be the cylinder $C_{2^{r_1}} \times P_{2^{r_2}}$, $r_1 + r_2 = r$ and $r_1 \leq r_2$. Then the edge congestion of embedding G into H is given by*

$$
EC(G, H) = \begin{cases} \frac{2^{r_2+1}-2}{3}, & r_2 \text{ is even}; \\ \\ \frac{2^{r_2+1}-1}{3}, & r_2 \text{ is odd}. \end{cases}
$$

Proof. The lexicographic ordering of the vertices of Q^r minimizes $\theta_G(S_k)$ for each k. Therefore

$$
EC(G, P_{2^r}) = \max_{1 \leq k \leq 2^r} \theta_G(S_k(l^*)),
$$

where l^* is the labeling of Q^r using lexicographic order. By Lemma 4, we have $EC(G, H) = EC(G, P_{2^r})$.

We claim that $EC(G, P_{2^r})$ is maximum for some k between 2^{r-2} and 2^{r-1}. Let $k^* = k + 2^{r-2}$, $1 \leq k \leq 2^{r-2}$. Now

$$
\begin{aligned}
\theta_G(S_{k^*}(l^*)) &= 2^{r-1} + \theta_{Q^{r-2}}(S_{k^*-2^{r-2}}(l^*)) \\
&= 2^{r-1} + \theta_{Q^{r-2}}(S_k(l^*)) \\
&\geq 2k + \theta_{Q^{r-2}}(S_k(l^*)) \\
&= \theta_G(S_k(l^*)), \text{ proving our claim.} \qquad \square
\end{aligned}
$$

4.2   Hypercube into Torus

The family of tori is one of the most popular interconnection networks due to its desirable properties such as regular structure, ease of implementation and good scalability. In recent years, the theory of torus embedding has found many applications and has been used in many practical systems such as Cray T3D, Cray T3E, Fujitsu AP3000, Ametak 2010, Intel Touchstone and so on [22].

The proof of the following theorem is similar to that of Theorem 2.

Theorem 3. *Let G be an r-dimensional hypercube Q^r, $r \geq 2$ and H be the torus $C_{2^{r_1}} \times C_{2^{r_2}}$, $r_1 + r_2 = r$ and $r_1 \leq r_2$. An lexicographic embedding [12] $f : G \to H$ given by $f(x) = x$ with minimum congestion.*

5   Concluding Remarks

In this paper, we obtain a new strategy to compute congestion of an embedding. Further, we compute the congestion of embedding the hypercube into the cylinder and the torus. Using the techniques of Section 3, we have the following result.

Theorem 4. *Let G be an r-dimensional folded hypercube FQ^r [25] and H be the grid $P_{2^{r_1}} \times P_{2^{r_2}}$ or cylinder $C_{2^{r_1}} \times P_{2^{r_2}}$, $r_1 + r_2 = r$ and $r_1 \leq r_2$. Then the edge congestion of embedding G into H is given by*

$$EC(G, H) = \frac{(r+1)2^{r-1} + (11 - 6r)2^{r-4}}{2^{r_1}}. \qquad \square$$

Acknowledgement. The authors would like to thank the anonymous referees for their comments and suggestions. These comments and suggestions were very helpful for improving the quality of this paper.

References

1. Opatrny, J., Sotteau, D.: Embeddings of complete binary trees into grids and extended grids with total vertex-congestion 1. Discrete Applied Mathematics 98, 237–254 (2000)
2. Chaudhary, V., Aggarwal, J.K.: Generalized mapping of parallel algorithms onto parallel architectures. In: Proc. Int. Conf. Parallel Process., pp. 137–141 (1990)
3. Dvořák, T.: Dense sets and embedding binary trees into hypercubes. Discrete Applied Mathematics 155(4), 506–514 (2007)
4. Manuel, P.: Minimum average congestion of enhanced and augmented hypercube into complete binary tree. Discrete Applied Mathematics 159(5), 360–366 (2010)
5. Matsubayashi, A., Ueno, S.: Small congestion embedding of graphs into hypercubes. Networks 33(1), 71–77 (1999)
6. Manuel, P., Arockiaraj, M., Rajasingh, I., Rajan, B.: Embedding hypercubes into cylinders, snakes and caterpillars for minimizing wirelength. Discrete Applied Mathematics 159(17), 2109–2116 (2011)
7. Matsubayashi, A.: Separator-Based Graph Embedding into Multidimensional Grids with Small Edge-Congestion, CoRR abs/1402.7293 (2014)
8. Taghavi, T., Yang, X., Choi, B.-K., Wang, M., Sarrafzadeh, M.: Modern circuit placement, Part 3. Springer, US (2011)
9. Ramanathan, P., Shin, K.G.: Reliable broadcast in hypercube multicomputers. IEEE Transactions on Computers 37(12), 1654–1657 (1988)
10. Saad, Y., Schultz, M.H.: Topological properties of hypercubes. IEEE Transactions on Computers 37(7), 867–872 (1988)
11. Choudum, S.A., Sunitha, V.: Augmented cubes. Networks 40(2), 71–84 (2002)
12. Manuel, P., Rajasingh, I., Rajan, B., Mercy, H.: Exact wirelength of hypercube on a grid. Discrete Applied Mathematics 157(7), 1486–1495 (2009)
13. Chen, W.K., Stallmann, M.F.M.: On embedding binary trees into hypercubes. Journal on Parallel and Distributed Computing 24, 132–138 (1995)
14. Bezrukov, S.L., Chavez, J.D., Harper, L.H., Röttger, M., Schroeder, U.-P.: Embedding of hypercubes into grids. In: Brim, L., Gruska, J., Zlatuška, J. (eds.) MFCS 1998. LNCS, vol. 1450, pp. 693–701. Springer, Heidelberg (1998)
15. Bezrukov, S.L., Chavez, J.D., Harper, L.H., Röttger, M., Schroeder, U.P.: The congestion of n-cube layout on a rectangular grid. Discrete Mathematics 213, 13–19 (2000)
16. Xu, J.M.: Topological Structure and Analysis of Interconnection Networks. Kluwer Academic Publishers (2001)

17. Rajasingh, I., Rajan, B., Rajan, R.S.: Embedding of special classes of circulant networks, hypercubes and generalized Petersen graphs. International Journal of Computer Mathematics 89(15), 1970–1978 (2012)
18. Katseff, H.: Incomplete Hypercubes. IEEE Transactions on Computers 37, 604–608 (1988)
19. Harper, L.H.: Global Methods for Combinatorial Isoperimetric Problems. Cambridge University Press (2004)
20. Chen, H.-L., Tzeng, N.-F.: A boolean expression-based approach for maximum incomplete subcube identification in faulty hypercubes. IEEE Transactions on Parallel and Distributed Systems 8, 1171–1183 (1997)
21. Boals, A.J., Gupta, A.K., Sherwani, N.A.: Incomplete hypercubes: Algorithms and embeddings. The Journal of Supercomputing 8, 263–294 (1994)
22. Rajan, R.S., Rajasingh, I., Parthiban, N., Rajalaxmi, T.M.: A linear time algorithm for embedding hypercube into cylinder and torus. Theoretical Computer Science 542, 108–115 (2014)
23. Yang, X., Dong, Q., Tan, Y.Y.: Embedding meshes/tori in faulty crossed cubes. Information Processing Letters 110(14-15), 559–564 (2010)
24. Garey, M.R., Johnson, D.S.: Computers and Intractability, A Guide to the Theory of NP-Completeness. Freeman, San Francisco (1979)
25. Manuel, P., Rajasingh, I., Rajan, R.S.: Embedding variants of hypercubes with dilation 2. Journal of Interconnection Networks 13(1-2), 1–16 (2012)
26. Rajan, R.S., Manuel, P., Rajasingh, I., Parthiban, N., Miller, M.: A lower bound for dilation of an embedding, IEEE Transactions on Computers (submitted)
27. Miller, M., Rajan, R.S., Parthiban, N., Rajasingh, I.: Minimum linear arrangement of incomplete hypercubes, The Computer Journal abstract (2014),
    http://comjnl.oxfordjournals.org/content/early/2014/05/09/
    comjnl.bxu031
28. Bezrukov, S.L., Chavez, J.D., Harper, L.H., Rottger, M., Schroeder, U.-P.: The congestion of n-cube layout on a rectangular grid. Discrete Mathematics 213(1-3), 13–19 (2000)

Auction/Belief Propagation Algorithms for Constrained Assignment Problem

Mindi Yuan, Wei Shen, Jun Li, Yannis Pavlidis, and Shen Li

WalmartLabs, San Bruno, California, USA
University of Illinois, Urbana-Champaign, Illinois, USA
{myuan,wshen,jli1,yannis}@walmartlabs.com,
shenli3@illinois.edu

Abstract. In this paper, we investigate the constrained assignment problem: a set of offers are to be assigned to a set of customers. There are constraints on both the number of available copies for each offer and the number of offers one customer can get. To measure the assignment gain, we have a score for each customer-offer pair, quantifying how beneficial it is for assigning a customer an offer. Additionally, a customer can get at most one copy of the same offer and at most one offer from the same category (an offer is associated with a category in a taxonomy, for example bakery, dairy, and so on). The objective is to optimize the assignment so that the sum of scores (global benefits) are maximized. We developed an auction algorithm for this problem and proved both its correctness and convergence. To show its effectiveness and efficiency, we compared it with heuristic algorithms and one minimum cost flow algorithm (network simplex). To show its scalability, we ran test cases of size up to 200 offers $\times$ 3,000,000 customers on a 16GB machine, where most of the linear/integer programming tools would fail under this setting. Finally, we transformed the auction algorithm into an equivalent belief propagation algorithm, and provided another convergence case for belief propagation on a loopy graph with node constraints.

1 Introduction

A retail company or grocery store periodically issues offers to their customers for items the customers currently buy (reward offers), the customers used to buy but have not bought in a period of time (remind offers), and items the customers do not buy but the company or store believes the customers might enjoy (recommendation offers). At a particular time during the year, like Thanksgiving, the retail company or grocery store could run a large campaign to assign various offers to their customers. Typically, these campaigns will reach the whole customer base. The size of the assignment problem will hence be large. Each kind of offers is limited in terms of its available copies. For example, the company can only provide at most 10,000 offers of kind A. An upper bound is also imposed on the maximum number of offers a customer can get. This number varies at a small scale based on the customer's purchase history. To avoid redundancy and increase diversity, each customer gets at most one offer from the same category.

S. Ganguly and R. Krishnamurti (Eds.): CALDAM 2015, LNCS 8959, pp. 238–249, 2015.

The assignment goals vary according to different objectives. In a reward campaign, the offers for a particular category of items will more likely be issued to the customers who spent a large amount of money in that category. In a recommendation campaign, this score is defined in another way, estimating how likely a customer would use this offer. However, the model of the assignment problem is similar, i.e. the company would always need to solve a constrained assignment problem to maximize the total score among the customers.

Mathematically, the problem can be formulated as the following:

$$\max \sum_i \sum_j s_{ij} x_{ij} \tag{1}$$

$$s.t. \sum_j x_{ij} \le o_i, \ \forall i \tag{2}$$

$$\sum_i x_{ij} \le c_j, \ \forall j \tag{3}$$

$$x_{ij} \in \{0, 1\}, \ \forall i, j \tag{4}$$

where s_{ij} is the score between offer i and customer j, x_{ij} the binary variable indicating if offer i is assigned to customer j ($x_{ij} = 1$) or not ($x_{ij} = 0$), o_i the maximum number of available copies for offer i, and c_j the upper bound for the maximum number of offers customer j can get.

This is an integer programming problem. Generally, integer programming is hard, let alone the size of the problem (almost 0.6 giga variables in our case!). To deal with the challenge, we will develop an efficient auction algorithm and prove both its correctness and convergence.

The rest of the paper is organized as follows. Section 2 discusses the related works. Section 3 derives the auction algorithm. The proof of correctness and convergence for the algorithm is given in section 4. We transform the auction algorithm into an equivalent belief propagation algorithm in section 5. We present the simulation results in section 7. Conclusion is in section 8.

2   Related Works

Zavlanos et al. [13] developed a distributed auction algorithm for the assignment problem. Bertsekas et al. [5] developed auction algorithms for various assignment problems. The assignment problem is essentially a matching problem, which is an easier version of the constrained assignment problem. Bertsekas extended the auction algorithm to the linear transportation problem [4]. The transportation problem has no constraint on the number of same copies of an object (offer) assigned to a particular person (customer). The problem can thus be converted to the matching problem by creating multiple copies of the objects and persons. Bertsekas also developed a generic auction/shortest path algorithm for the linear minimum cost flow problem [5]. However, that algorithm is generally for the minimum cost flow problem, which is not efficient enough when the problem size is big. Refer to [8] for comprehensive performance comparison of different minimum cost flow algorithms. Our algorithm can be viewed as an extension of the auction algorithm.

Another family of algorithms for the assignment problems are the belief propagation algorithms. Bayati *et al.* [3] developed a belief propagation algorithm for the matching problem. They showed their algorithm can be transformed to the auction algorithm. In [2], Bayati *et al.* also provided the first proof of convergence and correctness of an asynchronous BP algorithm for a combinatorial optimization problem. They showed that the BP algorithm converges to the correct solution when the LP relaxation has no fractional solutions. Gamarnik *et al.* [7] developed a belief propagation algorithm for the linear minimum cost flow problem. Again, their algorithm is for the *general* linear minimum cost flow problem. They need to pass a set of functions in each message and the number of messages grows as the iterations increase. Instead, our algorithm only passes a number in the message and the number of messages will not increase in each iteration. In [10], Sanghavi showed the equivalence of LP relaxation and max-product for weighted matching problem in general graphs. Moreover, in [12] and [11], Yuan *et. al.* proposed message passing algorithms for the minimax weight matching and the generalized assignment problem respectively.

Each of the above problems could be potentially solved by generic linear/ integer programming tools. However, if we can cast them into more specific forms, we could develop more specified and hence more efficient algorithms.

3   Basic Auction Algorithm

In this section, we present the auction algorithm. The algorithm views the assignment process as an auction game. Customers take turns to bid on their favorite offers according to the corresponding scores (values). The offers then have prices from the bids they received. As the auction proceeds, the customers adjust their favorite offers according to the current offer values. The algorithm is shown below.

This is an iterative algorithm. In each iteration, a customer computes her current best offers. The current value of an offer to a customer is defined as the difference of that customer's score to the offer and the price of the offer. Then she orders them and starts bidding through from the best one. To calculate the proposed bid, she first estimates the loss of missing this offer. In that case, she can potentially get the $(C[j] + 1)$th best one (the backup). She is thus willing to give a bid, which is the difference of her wished offer and the backup. This is also the maximum bid she can give for that offer. Otherwise, she would rather get the backup, considering the pure profit. When an offer receives that bid, it will add to its maintenance heap $O[i]$ the pair of the customer and her bid. If all the copies of this offer were assigned, it will remove the customer who gave the lowest bid. The price of this offer is the lowest bid it received so far. The price will thus remain 0, if the offer is not all assigned. We also enforce a minimum bid ϵ, prohibiting customers bidding 0 all the time and the algorithm looping forever. We will show this perturbation ϵ would not influence the optimality at termination when it is small enough.

Algorithm 1. Auction

1: **# Initialization**
2: Assignment heap of offer i: $O[i] = []$; Price of offer i: $P[i] = 0$;
3: Basket of customer j: $B[j] = []$; maximum number of offers customer j can get: $C[j]$;
4: Score matrix $S[i][j] = s_{ij}$; Minimum bid: ϵ;
5: **# Auction**
6: **while** algorithm not converged or $maxNumIter$ not reached **do**
7:   **for** each customer j **do**
8:     $curValues = S[:][j] - P[:]$;
9:     **for** each offer i in $curValues.sort()$ **do**
10:       **if** $len(B[j]) < C[j]$ and $i \notin B[j]$ and $curValues[i] > 0$ **then**
11:         $backupValue = max(curValues[C[j] + 1], 0)$;
12:         $bid = curValues[i] - backupValue + \epsilon$;
13:         **if** all copies of offer i assigned **then**
14:           $kickedOut = O[i].pop()$;
15:           $B[kickedOut].remove(i)$;
16:         **end if**
17:         $O[i].push([P[i] + bid, i])$;
18:         $B[j].append(i)$;
19:         $P[i] = min(O[i])$;
20:       **end if**
21:     **end for**
22:   **end for**
23: **end while**

At the beginning of one customer's turn, her basket may not be empty, but she does not need to remove these offers since they are still the best. To see this, note the prices of other offers will not decrease since last iteration and the prices of the offers in her basket will not be greater than her bid. Thus the current values of her occupied offers will still be among the top $C[j]$. For determining the convergence of the algorithm, one compares the bidding results of all the customers at this iteration and those at the previous one. If they all match, the algorithm terminates.

Here are some implementation notes. 0) Use arrays (or lists in Python) to store the score matrix. For example, use the algorithm for storing a compressed sparse matrix with fast row access. This will save both space and accessing time. 1) For tracking the offer assignment, each offer can maintain a minimum heap of pairs $[price, customer]$ sorted by $price$, where the $customer$s are the people currently holding one copy of the offer and the $price$s the bids they had given when they obtained their copies. Once these pairs are stored in a heap, it is convenient to look up the price of the offer, or to remove the customer who bid the lowest price upon heap overflow. 2) To compute the bids at each iteration, customer j sorts her current values ($curValues$), which costs $O(M \log M)$. When the number of offers M is big and $\log M$ is non-trivial, an alternative is to first use a linear algorithm [6] for finding the $(C[j] + 1)$th best offer (the $backupValue$) and then go through the offers again to bid on the top $C[j]$ (compare each one with the *backupValue* and pick up the greater). Thus the total time for bidding will be $O(M)$. However, we should be aware of the big constant before M in this linear solution.

4   Correctness and Convergence

We start by proving the convergence of the algorithm. One observation is that the number of assigned copies will not decrease for each offer. A customer who previously held an offer can only lose this offer when someone is willing to bid higher. In that case, the new person will replace her and continue to hold this offer, i.e. the number of assigned copies remains the same. In the view of the offers, the assigned ones will thus not decrease. On the other hand, each bid can not be 0, hence the price of each offer will eventually increase when all of its copies receive a bid. Therefore, an offer can not continue receiving bids while others not. There must be a moment when this offer, previously very attractive, becomes too expensive to all the customers. At that time, they would turn to other interesting offers and new copies of these offers will be assigned. As a result, the algorithm must terminate when either all the offers are assigned, or all the customers' baskets are full. A shortcut to prove the termination is that: when the current offer values to every customer become negative (the worst case but may not happen), the algorithm must terminate since no one will bid at that time. It takes finite time (through bidding and imposing a minimum bid) to let all the values become negative.

Now we are ready to prove the optimality at termination. When the algorithm terminates, every customer's choice will not change anymore, since

$$\sum_{i \in O_j^*} (s_{ij} - P[i] + \epsilon) \geq \sum_{i \in O_j} (s_{ij} - P[i]), \ \forall j, O_j \tag{5}$$

That is, the set of offers O_j^* customer j occupies at this moment is the best she can obtain under the current prices. For any other *feasible* set of offers O_j, the sum value will not be greater than those in O_j^*. This is true for all the customers. Then

$$\sum_j \sum_{i \in O_j^*} (s_{ij} - P[i] + \epsilon) \geq \sum_j \sum_{i \in O_j} (s_{ij} - P[i]) \tag{6}$$

Note O_j is not only feasible to customer j, i.e. $|O_j| \leq C[j]$, but also satisfies the constraints on the offer side, i.e. the total assigned copies of one offer does not exceed its maximum available number the company or its vendor can fund.

$$\sum_j \sum_{i \in O_j^*} s_{ij} + n\epsilon - \sum_j \sum_{i \in O_j} s_{ij} \geq \sum_j \sum_{i \in O_j^*} P[i] - \sum_j \sum_{i \in O_j} P[i] \tag{7}$$

where $n = \sum_j |O_j^*|$, the total number of offer copies assigned. For all the prices, we can divide them into two categories: zero or non-zero. For those offers whose copies were not all assigned, they have a price of zero according to the definition. For the others whose copies were all assigned, they have a price greater than 0 (more precisely, greater than or equal to ϵ). The sum $\sum_j \sum_{i \in O_j^*} P[i]$ is thus the greatest among all the sets of prices (more precisely, every *multiset/bag* of prices, since at least the copies of the same offer will all have the same price), because it sums up *all* the non-zero prices. Therefore, the right hand of Equation 7 is greater than 0. Consequently:

$$\sum_j \sum_{i \in O_j^*} s_{ij} - \sum_j \sum_{i \in O_j} s_{ij} \geq -n\epsilon \qquad (8)$$

Thus as long as $n\epsilon$ is less than the difference, denoted by w, of the best and the second best solutions, the perturbation will not influence the optimality at termination. That is, if $\epsilon < w/n$, the auction algorithm can converge to the optimum solution at termination. In practice, we can enlarge n to the total number of all the offer copies N (a constant known beforehand). However, we have to estimate w. One trick is to multiply the scores by a common number so that they all become integers. At this moment, the optimum must differ at least 1 from the second best solution. Therefore, as long as $\epsilon < 1/n$ or $\epsilon < 1/N$, $\sum_j \sum_{i \in O_j^*} s_{ij} - \sum_j \sum_{i \in O_j} s_{ij} \geq 0$, optimality guaranteed.

For the complexity of the algorithm, it is not easy to compute a "tight" bound. For estimation, it costs $O(M)$ for a customer to compute all her bids in one iteration, if using the algorithm in [6] as mentioned at the end of section 3. It is therefore an upper bound for calculating one bid. On the other hand, according to the convergence analysis, it needs at most W/ϵ (W is the sum of all the scores) bids to let all the offer values become negative. The complexity upper bound of the entire algorithm is thus $O(MWN/w)$, which is pseudo-polynomial. However, according to the simulations in section 7, the auction algorithm can terminate rather fast in practice.

5   Equivalent Belief Propagation Algorithm

In this section, we transform the auction algorithm into its belief propagation counterpart. The resulting algorithm is shown below.

The function $find(set, k)$ finds the kth largest element in the *set*. If $|set| < k$, it returns 0. a_{O_i} is the number of available copies of offer i, while a_{C_j} is the maximum number of offers customer j can get. Again, $find(set, k)$ has a linear time solution [6], but in practice, one could just sort the vales in *set* and pick up the kth one.

The equivalence of this algorithm and the auction algorithm is straightforward, once we view the message of O_i to C_j at the kth iteration $m_{O_i \rightarrow C_j}^k$ as the current price of offer i at that time and similarly $m_{C_j \rightarrow O_i}^k$ as the bid customer j gives to offer i.

Algorithm 2. Belief Propagation

1: # **Initialization**
2: $m_{O_i \to C_j}^0 = m_{C_j \to O_i}^0 = 0$
3: # **Messages at kth iteration**
4: $m_{O_i \to C_j}^k = find_l(m_{C_l \to O_i}^{k-1}, a_{O_i})$
5: $backupValue = find_l(s_{lj} - m_{O_l \to C_j}^{k-1}, a_{C_j} + 1)$
6: **if** $s_{ij} - m_{O_i \to C_j}^{k-1} > backup$ and number of C_j's non-zero bids $< a_{C_j}$ **then**
7:     $m_{C_j \to O_i}^k = (s_{ij} - m_{O_i \to C_j}^{k-1}) - backupValue + \epsilon$
8: **else**
9:     $m_{C_j \to O_i}^k = 0$
10: **end if**
11: # **Beliefs at kth iteration**
12: $b_{C_j}^k(l) = s_{lj} - m_{O_l \to C_j}^k$
13: # **Assignments**
14: If $b_{C_j}^k(l) > 0$, assign O_l to C_j.

6   Adding More Constraints

In fact, the auction algorithm can be extended with more constraints. In practice, each customer usually has a lower bound on the offers they can get. Moreover, several offers may belong to the same category and we only want to assign at most one offer from each category to a specific customer. As a result, a more constrained problem is defined as the following:

$$\max \sum_i \sum_j s_{ij} x_{ij} \tag{9}$$

$$s.t. \sum_j x_{ij} \le o_i, \ \forall i \tag{10}$$

$$l_j \le \sum_i x_{ij} \le u_j, \ \forall j \tag{11}$$

$$\sum_{i \in C_k} x_{ij} \le 1, \ \forall j, k \tag{12}$$

$$x_{ij} \in \{0, 1\}, \ \forall i, j \tag{13}$$

where l_j and u_j are the lower and upper bounds on the offers customer j can get. C_k is the set of offers belonging to category k. Constraint (12) thus guarantees at most one offer can be assigned from each category k to each customer j.

To incorporate these new constraints in the auction algorithm, we need to do the following two modifications.

For the lower bounds, this modified problem becomes slightly harder. It is not even straightforward to determine if the problem is feasible now. The heuristics proposed in Section 7 will not work in this setting. However, after modifying the bidding process, the auction algorithm will be able to handle the lower bounds.

Once the lower bounds are imposed, the customers may bid on the offers with current negative values. The modified biding rules are: bid on as many positive-value offers as possible from the one with the highest current value to the lowest; if there are not enough positive-value offers, continue bidding until the lower bound satisfied. The backup is the offer right after the last bid one. The bid is still calculated as $bid = currentValue - backupValue$. We should be aware of the following edge cases: 1) If the total number of eligible offers for one customer is less than her lower bound, the problem is infeasible immediately. 2) If the total number of eligible offers equals to a customer's lower bound, that customer will bid infinity on all the offers (since she cannot lose any to construct a feasible solution); 3) If a customer bids on all the offers and the total number of offers is strictly larger than her lower bound, the backup value can be set to 0 (since all the current values must be positive).

For the proof of convergence, if the problem is feasible, the algorithm converges in finite (W/ϵ) iterations due to the same reason for the problem without the lower bounds. Once the previous statement is true, its contrapositive must be true. That is, if the algorithm does not converge in W/ϵ iterations, it must be infeasible. In practice, users can therefore set this upper bound for the number of iterations the algorithm needs to run. However, one might be able to determine the termination of the algorithm earlier. For example, 1) if one customer who bids infinity on an offer is kicked out at some time, the problem must be infeasible; 2) if no one bids in one iteration, the algorithm must terminate already.

For the category constraints, we now need to remember the total categories each customer gets so for. In practice, one can allocate another category array, similar to the *baskets* variable $B[j]$ in Algorithm 1, to record the categories currently in each customer's basket. A customer will bid on a target offer only if its category is not in her category basket. Kindly note that the computation of backup values will change correspondingly. The backup now is not the one right after the last offer each customer bids. Instead, it is the first offer of a different category from all the previous ones after the last target offer, or the second best offer in the same category of the target, whichever is greater.

Again, the algorithm must converge with the category constraints since the prices are always increasing. For the proof of correctness at convergence, it is almost the same as that for the basic problem and hence omitted here.

7   Simulations

In this section, we compare the auction algorithm with the following two algorithms.

- Heuristics. Sort all the scores globally in descending order. Scan the scores from the greatest. Add the customer-offer pair for this score to the solution, if the two constraints are not violated on this customer and this offer. Otherwise, skip. Continue until all offers assigned or all customers' baskets full. There are two variations of this heuristic. We could alternatively do a row-by-row (or column-by-column) sorting and scanning, if it is too memory intensive to sort all the scores together. Name these heuristics $HeuristicG$, $HeuristicR$ and $HeuristicC$ respectively.

- Minimum Cost Flow. Once we view the original problem on a directed graph, it is not hard to develop a minimum cost flow solution for it. The details of the formulation are omitted here due to space limitation.

For the scores, we use the following distributions for testing.

- Uniform distribution $unif(low, up)$. Sample the scores from uniform distribution between low and up.
- Two-extreme distribution $extreme(pivot, popLow, popUp, unpopLow,$ $unpopUp, randFlag)$. It simulates the situation that some of the offers are popular, while others not. The $pivot$ is the cut-off point between popular and unpopular. All the offers with ID's less than $pivot$ are considered popular and otherwise unpopular. For the popular offers, sample scores from $unif(popLow, popUp)$, otherwise from $unif(unpopLow, unpopUp)$. The last parameter $randFlag$ indicates if we need to shuffle the offer ID's ($randFlag = 1$) or not ($randFlag = 0$) before determining if one offer is popular. The popular offers are common among all the customers, if the offers are not shuffled. Otherwise, each customer has their own favorite offers.

7.1 Basic Auction Algorithm

For the experiments, we ran them on a MacBook Pro laptop, with 2.8 GHz Intel Core i7 processor and 16 GB memory. The heuristic algorithms and the auction algorithm were implemented in Python. For the minimum cost flow, we used the network simplex algorithm in the LEMON graph library [9] written in C++. Refer to [8] for the performance of the algorithms in the LEMON package compared to others. In general, network simplex is the most efficient implementation.

Table 1 shows the results for the problem of size 200*30,000, which simulates 200 kinds of offers sent to 30,000 customers. To determine the number of available copies for each offer, we sampled from the distribution constructed by past campaign information. For the maximum number of offers one can get, we set this number to be 20. For the score distributions, shown as the 6 row names in the table, we used the following: $uS = unif(1, 100)$, $uB = unif(1, 10000)$, $eSR = extreme(20, 80, 100, 1, 20, 1)$, $eS = extreme(20, 80, 100, 1, 20, 0)$, $eBR = extreme(20, 8000, 10000, 1, 8000, 1)$ and $eB = extreme(20, 8000, 10000, 1, 8000, 0)$. For each column of the table, it shows the objective values the corresponding algorithm reached under these score distributions. For the column of the auction algorithm, we also show in the brackets the number of iterations needed for convergence. We set $\epsilon = 0.1$ for the auction algorithm. From the table, the auction algorithm was able to reach the optimum (determined by the $minCostFlow$ algorithm) in a limited number of iterations. Generally, $HeuristicR$ performed the worst, but $HeuristicG$ could achieve a solution about only 0.2% away from the optimum. In addition, shuffling the offers did not affect much the global optimum value for the customers.

Table 1. Comparison of different algorithms; Problem size: 200*30,000; Optima shown in bold

	HeuristicG	HeuristicR	HeuristicC	Auction(iter)	MinCostFlow
uS	55819800	45232212	55028023	**55992008**(54)	**55992008**
uB	5552908235	4492064242	5473777875	**5570031496**(121)	**5570031496**
eSR	24259333	22816580	23501866	**24294094**(58)	**24294094**
eS	24257679	22812596	23499044	**24292830**(32)	**24292830**
eBR	4837306376	4259011184	4728390435	**4849251748**(88)	**4849251748**
eB	4837848455	4258960099	4728504919	**4849704958**(53)	**4849704958**

In terms of time cost for each algorithm, both the heuristic and the auction algorithms took about 30 minutes for each of the 6 tests. The $minCostFlow$ took longer time, about 2 hours. Figure 1 shows the per-iteration time cost for a typical run of the auction algorithm. The starting iterations took significantly longer time than the ending iterations. At the beginning of auction, the customers were very active. However, as time elapsed, most of them would get their favorite offers already. A decreasing (generally) number of them would participate in the last rounds of bidding, accelerating the iterations.

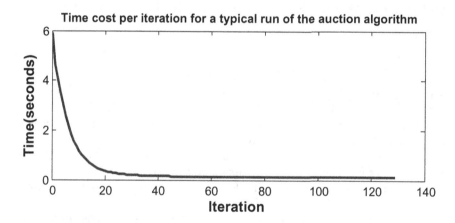

Fig. 1. Problem Size: 200*30,000; Score distribution: unif(1,10000); ϵ : 0.1

Table 2 shows the influence of the minimum bid ϵ. The first row presents the different values of ϵ tested. The second row shows the objective values the algorithm accomplished at convergence. The numbers in the third row are the iterations needed for convergence. If we set this value to 0, the algorithm would be trapped by some non-optimum value after 25 iterations. This is because that a few customers were competing for some common offers but bidding 0 all the time for that offer. With a minimum bid 0.01 or 0.1, the algorithm did converge to the optimum, as determined by the objective value provided by the $minCostFlow$

in Table 1. When $\epsilon = 0.01$, it took more iterations to converge. In this case, the customers were more prudent and they raised the bid very carefully/slowly. For a minimum bid of 1 or 2, the algorithm missed the optimum, but it terminated very fast.

Table 2. Influence of different ϵ

ϵ	0	0.01	0.1	1	2
Value	24110847	24292830	24292830	24283720	24269525
Iter	25	53	32	5	4

Table 3 shows the test cases for problem of size 200*3,000,000. The auction algorithm was consistently better than all the 3 versions of the heuristic algorithm. In this case, the minimum cost flow algorithm ran out of memory.

Table 3. Comparison of different algorithms

	HeuristicG	HeuristicR	HeuristicC	Auction(iter)
uS	55557530697	44978190741	54768390767	55727150799(68)
uB	55527881981940	44953155292963	54739279989179	55697296696459(202)
eSR	24060066476	22617655118	23319463158	24085043572(182)
eS	24060039215	22617680281	23319462482	24085005277(70)
eBR	48373195599982	42604075962732	47282046634892	48491515538543(213)
eB	48373129501238	42604238501775	47281835825158	48491250960360(200)

7.2 Extended Auction Algorithm

In this part, we evaluate the auction algorithm for the more constrained problem as shown in Section 6. In this case, the correctness of the algorithm is of more interests. We already studied the scalability in the previous subsection. We hence do not investigate very large cases here.

Assume 57 offers are to be assigned to 556 customers. In the experiments, the lower bounds are fixed to 1, while the upper bounds are set to 4 and 8 respectively. The number of copies for each offer is set to 10, 20 and 40 respectively. For benchmark purpose, we use the ILP in CVXOPT [1], a software package for convex optimization, to solve the problem and get the optima. The results are shown in table 4. As expected, we observe the auction algorithm and the ILP solver produce the same results. We also notice that the auction algorithm can terminate within 1 minute in all the cases, while the ILP solver takes more than 20 minutes.

Table 4. Correctness of the extended auction algorithm

	10 copies per offer	20 copies per offer	40 copies per offer
1-4 offer(s) per customer	341.58 (1101)	696.73 (96)	1127.94 (449)
1-8 offer(s) per customer	341.58 (1142)	715.10 (28)	1270.95 (47)

8 Conclusion

In this paper, we developed an auction algorithm for the constrained assignment problem and proved its correctness and convergence. Through simulations, we showed both its effectiveness and efficiency. We also transformed the auction algorithm into an equivalent belief propagation algorithm, providing another convergence case of belief propagation algorithm on a loopy graph.

References

1. Andersen, D., Dahl, J., Vandenberghe, L.: CVXOPT: Python software for convex optimization, http://cvxopt.org/index.html
2. Bayati, M., Borgs, C., Chayes, J., Zecchina, R.: Belief propagation for weighted b-matchings on arbitrary graphs and its relation to linear programs with integer solutions. SIAM J. Discrete Math. 25, 989–1011 (2011)
3. Bayati, M., Shah, D., Sharma, M.: Max-product for maximum weight matching: convergence, correctness, and LP duality. IEEE Trans. Info. Theory 54, 1241–1251 (2008)
4. Bertsekas, D.P.: The auction algorithm for the transportation problem. Annals of Operations Research 20(1), 67–96 (1989)
5. Bertsekas, D.P., Tsitsiklis, J.N.: Parallel and Distributed Computation: Numerical Methods. Prentice-Hall (1989)
6. Blum, M., Floyd, B., Pratt, V., Rivest, R., Tarjan, B.: Linear time bounds for median computations. In: STOC, pp. 119–124 (1972)
7. Gamarnik, D., Shah, D., Wei, Y.: Belief propagation for min-cost network flow: Convergence and correctness. Operations Research 60, 410–428 (2012)
8. Kiraly, Z., Kovacs, P.: Efficient implementations of minimum-cost flow algorithms, http://arxiv.org/abs/1207.6381
9. Kiraly, Z., Kovacs, P.: LEMON graph library (COIN OR), http://lemon.cs.elte.hu/trac/lemon
10. Sanghavi, S.: Equivalence of LP relaxation and max-product for weighted matching in general graphs. In: IEEE Info. Theory Workshop, pp. 242–247 (2007)
11. Yuan, M., Jiang, C., Li, S., Shen, W., Pavlidis, Y., Li, J.: Message passing algorithm for the generalized assignment problem. In: Hsu, C.-H., Shi, X., Salapura, V. (eds.) NPC 2014. LNCS, vol. 8707, pp. 423–434. Springer, Heidelberg (2014)
12. Yuan, M., Li, S., Shen, W., Pavlidis, Y.: Belief propagation for minimax weight matching. Tech. rep., University of Illinois (2013)
13. Zavlanos, M.M., Spesivtsev, L., Pappas, G.J.: A distributed auction algorithm for the assignment problem. In: Proceedings of the 47th IEEE Conference on Decision and Control, pp. 1212–1217 (2008)

Bi-directional Search for Skyline Probability

Arun K. Pujari[1], Venkateswara Rao Kagita[1], Anubhuti Garg[2],
and Vineet Padmanabhan[1]

[1] School of Computer and Information Sciences
University of Hyderabad, Hyderabad - 500046, India
{akpcs,vineetcs}@uohyd.ernet.in, venkateswar.rao.kagita@gmail.com
[2] The LNM Institute of Information Technology, Jaipur, India
anubhuti.grg@gmail.com

Abstract. Computing the skyline probability of an object for a database, wherein the probability of *preferences* between pairs of data objects are uncertain, requires computing the probability of union of events from the probabilities of all possible joint probabilities. From the literature it can be seen that for a database of size n it requires computation of 2^n joint probabilities of all possible combinations. All known algorithms for probabilistic skyline computation over uncertain preferences attempt to find inexact value of skyline probability by resorting to sampling or to approximation schemes. In this paper we use a concept called *zero-contributing set* of a power set lattice to denote portion of the lattice (a sub-lattice) such that the signed aggregate of joint probabilities corresponding to this set is zero. When such sets can be implicitly identified, the corresponding terms can be removed, saving substantial computational efforts. We propose an efficient heuristic that employs a bi-directional search traversing level wise the power set lattice from top and from bottom and prunes the exponential search space based on zero-contribution.

1 Introduction

Skyline queries aim at retrieving the most preferred objects from a database with respect to a given *dominance* relation. Skyline queries return objects which are not *dominated* by any other object. In a multidimensional space where the dimension of domains are ordered, a point p dominates a point q if p is strictly better than q on at least one dimension and p is equal or better than q on the remaining dimensions. Preference relation can be a valid dominance relation and it is possible that in a database the preferences of data items are uncertain. This paper addresses the problem of efficient computation of skyline points in a database with uncertain preferences. The main idea is to compute the skyline probability of an object in the database when probability of preferences between pairs of data objects are given. The problem of computing skyline probability for database with uncertain preferences was first proposed in [13], where it is shown that it is necessary to compute all joint probabilities and to use inclusion-exclusion principle to compute the skyline probability. This process

S. Ganguly and R. Krishnamurti (Eds.): CALDAM 2015, LNCS 8959, pp. 250–261, 2015.

requires computation of 2^n joint probabilities for a database of size n. The algorithm proposed in [13] traverses the power set lattice in a bottom-up manner. A zero-contributing set is a portion of the lattice (a sub-lattice) such that the signed aggregate of joint probabilities corresponding to this set is zero. When such sets can be implicitly identified, the corresponding terms can be removed, saving substantial computational efforts. Based on this concept, we propose an efficient heuristic that employs a bi-directional search traversing level wise the power set lattice from top and from bottom and prunes the exponential search space based on zero-contribution. The efficiency of the proposed heuristic relies on the observation that in any real-life database the domain size of attributes are relatively small even when the size of the database is large. In addition to trimming the exponential search space, substantial saving in computational effort is achieved by separating the actual calculation of joint probabilities from the task of pruning the lattice. We first prune the lattice to determine essential set of terms and then calculate the joint probabilities only for the identified terms.

The rest of the paper is organized as follows. In Section 2, the probabilistic preference model is introduced and earlier work on the problem is reviewed. Section 3 outlines the characterization of zero-contributing set. Section 4 explains about intersection of zero-contributing sub-lattices. Section 5 describes bi-directional search algorithm for skyline probability over uncertain preferences. Experimental analysis of the proposed algorithm are reported in Section 6. Section 7 outlines related work.

2 Problem Definition

2.1 Notations and Definitions

Let $\Theta = \{Q^1, Q^2, \ldots, Q^n\}$ be a set of n d-dimensional objects and O be another d-dimensional object not in Θ. $Q^i_j(O_j)$ denotes the value of Q^i, (O, respectively) on dimension j. Define $distinct(S, j)$ for any $S \in 2^\Theta$, and $1 \le j \le d$, as set of distinct attribute values on dimension j of all objects in S and can be written formally as follows.

$$distinct(S, j) = \{a \mid a = Q^i_j \ for \ some \ Q^i \in S\}$$

When we consider all dimensions together, we have,

$$distinct(S) = \cup_j distinct(S, j)$$

Distinct values of $S \subseteq \Theta$ on dimension j with reference to another object $O \notin \Theta$ is defined as follows in [13].

$$distinct(O; S, j) = \{a \mid a = Q^i_j \ for \ some \ Q^i \in S \ and \ a \ne O_j\}$$

and $distinct(O; S) = \cup_j distinct(O; S, j)$. If $S' \subseteq S \subseteq \Theta$ then $distinct(S') \subseteq distinct(S)$. If $S \subseteq \Theta$ and $O \notin \Theta$, $distinct(O; S) \subseteq distinct(S)$. For $S', S \subseteq \Theta$

and $O \notin \Theta$, if $distinct(S') = distinct(S)$ then $distinct(O; S') = distinct(O; S)$, but converse does not hold.

Given two distinct attribute values a and b, we introduce the notation $a \preccurlyeq b$ to denote that a is preferred to b. We say object Q^i dominates object O ($Q^i \preccurlyeq O$) iff Q^i is preferred to O on any dimension and is different from O i.e., $Q^i_j \preccurlyeq O_j$, for all $1 \leq j \leq d$ and $Q^i \neq O$. The object O is said to be a skyline object with respect to set Θ of objects, if no object in Θ dominates O. Given two distinct attribute values a and b, the probabilistic model as outlined in [13] to describe the uncertain preferences between them can be given as follows :

$$Pr(a \preceq b) + Pr(b \preceq a) \leq 1$$

Let e_i denote the event $Q^i \preccurlyeq O$. The probability of e_i is a joint probability of attribute value preferences:

$$Pr(e_i) = Pr\left(\bigcap_{j=1}^{d} (Q^i_j \preceq O_j) \right)$$

If we assume that attribute value preferences of different dimensions are mutually independent then we get

$$Pr(e_i) = \prod_{j=1}^{d} Pr(Q^i_j \preceq O_j)$$

The skyline probability of an object O, denoted as $sky(O)$, is defined in [13] as the probability that object O is not dominated by objects in Θ.

$$sky(O) = Pr\left(\bigcap_{i=1}^{n} \overline{e_i} \right) = 1 - Pr\left(\bigcup_{i=1}^{n} e_i \right) \tag{1}$$

Let 2^Θ denote the power set of Θ and for a given $J \in 2^\Theta$, $E_J = \bigcap_{Q^i \in J} e_i$. Using inclusion-exclusion principle, we get

$$sky(O) = 1 + \sum_{k=1}^{n} (-1)^k \sum_{J \in 2^\Theta, |J| = k} Pr(E_J) \tag{2}$$

where

$$Pr(E_J) = Pr\left(\bigcap_{j=1}^{d} \bigcap_{v \, \in \, distinct(O;J,j)} Pr(v \preceq O_j) \right) \tag{3}$$

Assuming that preferences are independent, equation (3) is further simplified in [13] as follows.

$$Pr(E_J) = \prod_{j=1}^{d} \prod_{v \in \, distinct(O;J,j)} Pr(v \preceq O_j) \tag{4}$$

When two sets have same set of distinct attribute values, the corresponding joint probabilities are equal. In other words, if $distinct(O; S, j) = distinct(O; S', j)$ for all $1 \leq j \leq d$, then $Pr(S) = Pr(S')$.

2.2   ZYLZ Algorithm to Compute Skyline Probability

In [13], an algorithm is proposed to compute skyline probability with predefined uncertain preferences. We refer this method as ZYLZ algorithm[1] which traverses the power set lattice in a bottom-up level wise manner. At each level it computes the joint probabilities making use of the calculation at the lower level. In order to compute the joint probabilities at level k, it makes use of $distinct(O; J)$ for all $|J|$=k. ZYLZ algorithm can be described step-wise in Algorithm 1. The complexity of the algorithm is $O(d.2^n)$. The exponential factor appears as all subsets in 2^Θ are considered in Equation 2.

Algorithm 1. ZYLZ algorithm

Input: O, Θ, and uncertain preference probabilities $Pr(a \preccurlyeq b)$ for all
pairs a, b.
Output: $sky(O)$
for $k = 1 \ldots n$ **do**
  for *each* $J \in 2^\Theta$, $|J| = k$ **do**
    compute $distinct(O, J, j)$;
    compute $Pr(E_J)$ using $Pr(E_{J'})$ for $|J'| = k - 1$ computed earlier;
  end
end
Compute $sky(O)$ as in Equation 2

Motivating Example: Consider the core 5 objects of 2-dimensions(Table1). We have preference probability values of attributes A as $Pr(a1 \preccurlyeq a2) = Pr(a2 \preccurlyeq a1) = Pr(a1 \preccurlyeq a3) = Pr(a3 \preccurlyeq a1) = 0.5$, Similarly the preference probabilities for values of attributes of B are $Pr(b1 \preccurlyeq b2) = Pr(b2 \preccurlyeq b1) = Pr(b1 \preccurlyeq b3) = Pr(b3 \preccurlyeq b1) = 0.5$.

Table 1. Objects with two attributes

Objects	A	B
O	a1	b1
Q^1	a3	b3
Q^2	a1	b3
Q^3	a2	b2
Q^4	a3	b1

---

[1] Based on initials of authors.

The problem is to find skyline probability of object O, $sky(O)$, with respect to $\{Q^1, Q^2, Q^3, Q^4\}$. Using Equation 2, we have

$sky(O) = 1 - [Pr(e_1) + Pr(e_2) + Pr(e_3) + Pr(e_4)] + [Pr(e_1 \cap e_2) + Pr(e_1 \cap e_3) + Pr(e_1 \cap e_4) + Pr(e_2 \cap e_3) + Pr(e_2 \cap e_4) + Pr(e_3 \cap e_4)] - [Pr(e_1 \cap e_2 \cap e_3) + Pr(e_1 \cap e_2 \cap e_4) + Pr(e_1 \cap e_3 \cap e_4) + Pr(e_2 \cap e_3 \cap e_4)] + [Pr(e_1 \cap e_2 \cap e_3 \cap e_4)].$

Using Equation (4) we get $Pr(e_1) = Pr(a1 \preccurlyeq a3) \times Pr(b1 \preccurlyeq b3) = 0.25$, similarly

$Pr(e_2) = Pr(b1 \preccurlyeq b3) = 0.5; Pr(e_3) = Pr(a1 \preccurlyeq a2) \times Pr(b1 \preccurlyeq b2) = 0.25;$
$Pr(e_4) = 0.5; Pr(e_1 \cap e_2) = Pr(a1 \preccurlyeq a3) \times Pr(b1 \preccurlyeq b3) = 0.25;$
$Pr(e_1 \cap e_3) = Pr(a1 \preccurlyeq a2) \times Pr(a1 \preccurlyeq a3) \times Pr(b1 \preccurlyeq b2) \times Pr(b1 \preccurlyeq b3) = 0.0625.$

Thus $sky(O) = 1 - [1.5] + [1.0625] - [0.4375] + [0.0625] = 0.1875$.

We observe that $Pr(e1 \cap e3) = Pr(e1 \cap e2 \cap e3) = Pr(e1 \cap e3 \cap e4) = Pr(e1 \cap e2 \cap e3 \cap e4) = 0.0625$. The sub-aggregate with respect to these terms is zero. It is unnecessary to compute joint probabilities of these terms, if we know in advance that these terms taken together do not contribute to the overall aggregate. This observation prompted us to design a new method.

3   Characterization of Zero-Contributing Set

Consider the power set lattice with elements as subsets of Θ and set inclusion as the partial order. We denote a lattice by its minimal elements followed by its maximal elements. Most often, in the present study, the sub-lattice has a unique minimal element and a unique maximal element. In this sense, the power set lattice is denoted as $L = [\varnothing, \Theta]$. Note that every element of L corresponds to a term in Equation 2. We represent α, β, S, x and y as subsets of Θ.

Definition 1. *Zero-contributing set: A set of elements in L is said to be a zero-contributing subset of L with respect to O and Θ if the terms corresponding to these elements in Equation (2) have no contribution when considered together irrespective of the attribute values.*

In the above example, the sub-lattice $[\{1,3\}, \{1,2,3,4\}]$ is a zero-contributing sub-lattice. We are interested in zero-contributing sub-lattices of L. A sub-lattice $[\alpha, \beta]$ of L is said to be a *non-degenerate* sub-lattice if $\alpha \neq \beta$ and otherwise, it is a degenerate sub-lattice. Clearly, a degenerate sub-lattice cannot be a zero-contributing sub-lattice. Hence the smallest zero-contributing lattice has at-least two elements.

Definition 2. *A zero-contributing sub-lattice $[\alpha, \beta]$ is said to be atomic zero-contributing sub-lattice if $|\alpha| + 1 = |\beta|$ and $[\alpha, \beta]$ contains exactly two elements, namely α and β.*

We state some important results in the following lemmas.

Lemma 1. *If $[\alpha, \beta]$ is a sub-lattice of L and $distinct(O; \alpha) = distinct(O; \beta)$ then $distinct(O; x) = distinct(O; \alpha)$ for all $x \in [\alpha, \beta]$.*

Proof- Since $\alpha \subseteq x \subseteq \beta$, $distinct(O; \alpha) \subseteq distinct(O; x) \subseteq distinct(O; \beta)$. Since $distinct(O; \alpha) = distinct(O; \beta)$, proof follows.

A characterization of zero-contributing sub-lattice of L is given in the following lemma.

Lemma 2. *If $[\alpha, \beta]$ is a non-degenerate sub-lattice of L and $distinct(O; \alpha) = distinct(O; \beta)$ then $[\alpha, \beta]$ is a zero-contributing set of L with respect to O and Θ.*

Proof- The part of Equation 2 corresponding to $[\alpha, \beta]$ is

$$Pr(E_\alpha) + \sum_{k=1}^{n'} (-1)^k \sum_{\alpha \subseteq J \subseteq \beta, |J| = k + |\alpha|} Pr(E_J)$$

where $n' = |\beta| - |\alpha|$.

By Lemma 1, $distinct(O; J) = distinct(O; \alpha)$, for all $\alpha \subseteq J \subseteq \beta$. $Pr(E_J) = Pr(E_\alpha)$, for all $\alpha \subseteq J \subseteq \beta$. Thus, above expression can be simplified (by Binomial theorem) as

$$Pr(E_\alpha) \left(1 + \sum_{k=1}^{n'} (-1)^k \binom{n'}{k} \right) = 0.$$

This completes the proof.

We use the precondition of the above lemma as the primary characterization of zero-contributing set. It may be noted that it is a sufficient condition but not necessary. The condition is necessary and sufficient for atomic zero-contributing lattices. It is simple to construct a counter example to show that for non-atomic zero-contributing lattice the condition given in the above lemma is only sufficient but not necessary. Consider the sub-lattice $[\alpha, \beta]$ such that $\beta = \alpha \cup \{x, y\}$ and $distinct(O; \alpha) \neq distinct(O; \beta)$ but $distinct(O; \alpha) = distinct(O; \alpha \cup \{x\})$ and $distinct(O; \alpha \cup \{y\}) = distinct(O; \beta)$. It is easy to check that $[\alpha, \beta]$ is a zero-contributing set and $[\alpha, \beta]$ does not satisfy the condition of above lemma. $[\alpha, \beta]$ can be viewed as the union of atomic zero-contributing lattices. In that sense, our characterization is complete. We give a stronger characterization below.

Lemma 3. *Let $[\alpha, \beta]$ be a non-degenerate sub-lattice of L. If $distinct(O; \alpha) = distinct(O; \beta)$ then $[\alpha, \beta]$ is a zero-contributing set of L with respect to O and Θ. Conversely, if $[\alpha, \beta]$ is zero-contributing set of L then either $distinct(O; \alpha) = distinct(O; \beta)$ or $[\alpha, \beta]$ is union of atomic zero-contributing sub-lattices $[\alpha^i, \beta^i]$ satisfying $distinct(O; \alpha^i) = distinct(O; \beta^i)$.*

We use characterization of Lemma 2 in our proposed algorithm.

4   Intersection of Zero-Contributing Sub-lattices

Having characterized zero-contributing lattices, it becomes clear that all terms of $[\alpha, \beta]$ can be deleted from the equation without affecting the aggregate.

An important issue of invariance of zero-contribution on set-intersection arises while deleting multiple zero-contributing sub-lattices.

If $[\alpha,\ \beta]$ and $[\alpha',\ \beta']$ are two *non-degenerate* sub-lattices, the intersection $[\alpha,\ \beta]\cap[\alpha',\ \beta']$ is a sub-lattice and is given by $[\alpha\cup\alpha',\beta\cap\beta']$. If $\alpha\cup\alpha'$ is not a subset of $[\beta\cap\beta']$ then we say that $[\alpha,\ \beta]$ and $[\alpha',\ \beta']$ do not intersect. We use this criterion to check whether two lattices intersect or not.

Lemma 4. *Let $[\alpha,\ \beta]$ and $[\alpha',\ \beta']$ be two intersecting non-degenerate sub-lattices. If $distinct(O;\alpha) = distinct(O;\beta)$ and $distinct(O;\alpha') = distinct(O;\beta')$ then*

i $[\alpha,\ \beta]$ and $[\alpha',\ \beta']$ are zero-contributing sub-lattices.
ii $distinct(O;\alpha) = distinct(O;\beta) = distinct(O;\alpha') = distinct(O;\beta')$ and
iii If $[\alpha,\ \beta]\cap[\alpha',\ \beta']$ is non-degenerate then it is a zero-contributing sub-lattice.

Proof of Lemma 4 can be routinely derived from other results proved above.

Let $[\alpha,\ \beta]$ and $[\alpha',\ \beta']$ be zero-contributing intersecting sub-lattices. If the intersection is non-degenerate, it suffices to delete terms in $[\alpha,\ \beta]$ and $[\alpha',\ \beta']\setminus[\alpha,\ \beta]$. These terms are non-overlapping and overall aggregate is not affected. However, it is not easy to check non-degeneracy of intersection of any number of sub-lattices. On the other hand, it is safe to restrict ourselves to non-intersecting zero-contributing family of sub-lattices.

5   Bi-directional Algorithm

Our algorithm maintains two lists, namely *min-list* and *sup-list*. Initially, *sup-list* contains only Θ and *min-list* contains $\varnothing$. The algorithm is a bi-directional search such that *min-list* is generated level-wise in a bottom-up fashion and *sup-list* moves top-down. At any stage of the algorithm, if an α from *min-list* and a β from *sup-list* are found satisfying the precondition of Lemma 2 and if $[\alpha,\ \beta]$ does not intersect with any of sub-lattices deleted earlier then α and β are deleted from the respective lists. All elements of sub-lattice $[\alpha,\beta]$ are implicitly deleted from equation (2). The deleted set of sub-lattices are maintained in another list to ensure that new candidate sub-lattice does not intersect with any of the deleted sub-lattices. This list is used while generating candidates list. The algorithm has three major steps. At every level, it generates a tentative set of α's in candidate generation step. We follow the standard technique of candidate generation of Apriori algorithm given in page number 339 of [12]. The sup-list maintains a set of candidate β's. Algorithm looks for a pair of α and β to identify zero-contributing sub-lattice, in which case α and β are deleted from respective lists. New α's are generated by candidate generation. New β's are generated by a separate procedure as follows.

When the sub-lattice $[\alpha,\ \beta]$ is deleted, the new maximal elements are generated as β^x for each $x \in \alpha$, where $\beta^x = \beta\setminus x$. At any given time, for any $\alpha' \notin [\alpha,\beta]$ with β as its least upper bound (lub), one of β^x is next lub for α', after deletion of $[\alpha,\beta]$. Algorithm 2 outlines the steps of the proposed algorithm.

Algorithm 2. Bi-directional Search

Input: Θ, O
Output: set of terms as L
$NL \leftarrow \emptyset$;
$delList \leftarrow \emptyset$;
$suplist \leftarrow \Theta$;
$L_1 \leftarrow [\{Q^1\}, \{Q^2\}, , \ldots, \{Q^n\}], (Q^i \in \Theta)$;
for $k = 1$ *to* n **do**
    if $L_k \neq \emptyset \wedge suplist \neq \emptyset$ **then**
        for *each* $\alpha \in L_k$ **do**
            if $\exists \beta \in suplist$ *such that*
            $\alpha \subsetneq \beta \wedge distinct(O; \alpha) = distinct(O; \beta) \wedge$
            $[\alpha, \beta] \cap [\alpha', \beta'] = \emptyset, \forall [\alpha', \beta'] \in delList$ **then**
                delete α from current minlist: $L_k \leftarrow L_k \setminus \{\alpha\}$;
                delete β from current suplist: $suplist \leftarrow suplist \setminus \{\beta\}$;
                store $[\alpha, \beta]$ in delList: $delList \leftarrow delList \cup \{[\alpha, \beta]\}$;
                generate new elements for suplist:
                $suplist \leftarrow suplist \cup NewElement(\alpha, \beta, L_k, suplist)$;
            end
        end
        $L_{k+1} \leftarrow GenerateCandidates$;
        $L_{k+1} \leftarrow Prune(L_{k+1}, delList)$;
    end
    else
        return $L \leftarrow \bigcup_{i=1}^{k} L_i$;
    end
end

Algorithm 3. NewElement

Input: $\alpha, \beta, L_k, suplist$
Output: New elements list NL
$NL = \emptyset$;
for *each* $a \in \alpha$ **do**
    $NL \leftarrow NL \cup \beta \setminus \{a\}$;
end
for *each* $c \in NL$ **do**
    if $c \in L_k$ *or* $Superset(c) \in suplist$ **then** $NL \leftarrow NL \backslash c$;
end
return NL;

Algorithm 4. Prune

Input: L_k, $delList$
Output: L_k
for *each* $c \in L_k$ **do**
   for *each* $[\alpha, \beta] \in delList$ **do**
      if $\alpha \subseteq c \wedge c \subseteq \beta$ **then** $L_k \leftarrow L_k \setminus c$;
   end
end
return L_k;

Computation of Joint Probabilities. Algorithm 2 returns a pruned lattice L of subsets of Θ and for each subset x we have the associated $distinct(O; x)$. Using equation (3), $Pr(E_x)$ is computed as the product of individual preference probabilities. While computing for all subsets in L, it is possible to make use of level-wise calculation to save computational efforts.

Table 2. Best case example

Objects	A	B	C	D
O	a1	b1	c1	d1
Q^1	a2	b2	c2	d2
Q^2	a1	b2	c2	d2
Q^3	a1	b1	c2	d2
Q^4	a1	b1	c1	d2

Table 3. Worst case example

Objects	A	B	C	D
O	a1	b1	c1	d1
Q^1	a2	b2	c2	d2
Q^2	a3	b2	c2	d2
Q^3	a4	b1	c2	d2
Q^4	a5	b1	c1	d2

5.1 Analysis

The proposed algorithm is essentially a heuristic relying on efficiently identifying zero-contributing sets. We analyze the best case and the worst case analysis of the algorithm.

Best Case: The best case instance occurs when maximum possible candidates are removed at earliest possible occasion. Consider set of objects given in Table 2. $[\{1\}, \{1, 2, 3, 4\}], [\{2\}, \{2, 3, 4\}]$ and $[\{3\}, \{3, 4\}]$ are different zero-contributing sub-lattices. Residual term is only $\{4\}$. Hence number of residual terms in best case is only one.

Worst Case: The worst case instance occurs when no candidate is eliminated and the entire lattice is required for the computation. If we consider the example given in Table 3, no sub-lattice is a zero-contributing set. Total number of residual terms in this case is 15 ($2^4 - 1$). Similarly for n objects total number of residual terms possible in worst case is $2^n - 1$. In an average case half of the terms will be eliminated. Thus, the efficiency of the algorithm can be measured as the proportion of candidates removed from the lattice. We show below empirically that large number of terms are eliminated.

The complexity of the algorithm depends on number of levels traversed by bottom-up part of the search. Estimating the possible level for a given dataset may require computationally intensive analysis. We study this aspect empirically.

6   Experimental Results

In this section, we demonstrate the experiments carried out to evaluate the effectiveness and efficiency of our proposed algorithms using both real and synthetic datasets. We selected moderate sized dataset having categorical attributes without missing values from UCI ML repository[2]. Experimental results related to Lenses dataset is presented here. Our empirical study also examines synthetic data of size 25 with dimensions varying in the range [2, 8], which are randomly generated with normal distribution for each dimension and assuming mutual independence of dimensions. This data is generated with less possible domain size[3] and with no duplicates. It is obvious that for less value of d domain size is relatively high. The object O is selected randomly and $sky(O)$ is computed. This process is repeated 5 times by selecting 5 random $O's$ and the experimental results presented here are taken from those 5 random observations. Our experimental results shows that there is a substantial reduction in number of terms to be computed, termination level and therefore time. The size of Lenses dataset is 24. If we take one object as O and compute skyline probability with respect

Table 4. Efficiency of our algorithm over number of terms and termination level

Dataset	n	d	% of reduction in no. of terms	Termination level
Lenses	24	5	91.52 - 99.56	15-20
Synthetic	25	5	92.93 - 99.99	10-20

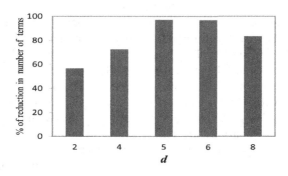

Fig. 1. % of average reduction in number of terms by varying d

---

[2] https://archive.ics.uci.edu/ml/datasets.html
[3] Number of values in each attribute.

to other objects, number of terms needed for the computation by using ZYLZ algorithm is $2^{23} = 8388608$. Best observed results on number of terms needed for computation by using our proposed algorithm is 36339, which is 0.04% of 8388608. Hence our algorithm is able to reduce 99.56% of terms needed for computation. Results summarization on different datasets is shown in Table 4. We conduct second set of experiments to see the behaviour of our algorithm by varying dimensions of dataset. Experimental results are reported in Figure 1. It can be seen that domain size and d are influencing the performance of our algorithm. For high domain size(at $d = 2$, 4) and for very high value of d, our algorithm could not reduce more number of terms. But still percentage of reduction is quite good.

7 Related Work

Devising efficient methods to retrieve skyline objects in a database with pre-specified dominance relation has become an important area of database research and there have been several proposals of skyline query processing [3–7, 9, 11]. In addition to conventional skyline operator, numerous skyline query variants have also been explored in the literature including skyline query for databases with uncertainties [1, 2, 8, 10, 14, 15]. In 2007, Pei et al [10], proposed a method of skyline computation where the database consists of uncertain attributes and developed bounding-pruning-refining techniques to deal with it efficiently. Efficient skyline computation against sliding windows on uncertain data streams was proposed by Zhang et al. [15]. Lian and Chen [8] studied efficient reverse skyline query processing on uncertain data in both monochromatic and bi-chromatic fashion. Atallah et al. [1] investigated asymptotic time complexity of computing all skyline probabilities for data with discrete uncertainty. Recently, Bartolini et al., [2] introduced a novel concept of *P-domination* for tuples with probabilistic relations. In this setting, skyline computation is concerned with computing a probability, called *skyline probability*, that a tuple is a skyline point. One may consider preference as a dominance relation and instead of data uncertainties, it is pertinent to have uncertainty of preferences. Thus skyline computation can also be relevant when the attribute preferences are uncertain [13].

Recently, Zhang et al [13] proposed a detailed investigation of computation of skyline probability with uncertain preferences. The algorithm given in [13], the only known method to compute skyline probability of a given object, relies on *inclusion-exclusion principle* and computes joint probabilities of all elements of the power set. It essentially involves exponential number of terms and hence, skyline object cannot be determined even for a database of moderate size. Zhang et al [13] also proposed an approximation scheme to make it a polynomial time algorithm. In addition a preprocessing step called absorption is proposed to reduce the time complexity. Integrating preprocessing step with our algorithm is part of our future work.

8   Conclusions and Future Work

In this paper we propose a new algorithm to compute skyline probability of an object in a database with uncertain preferences. Our algorithm implicitly enumerates the terms which eventually vanish while computing joint probabilities. These terms, when ignored, result in substantial saving of computational effort. Thus our algorithm uses much fewer terms than earlier algorithm and hence takes less time. Using absorption as a preprocessing step to our algorithm is part of our future work.

References

1. Atallah, M.J., Qi, Y., Yuan, H.: Asymptotically efficient algorithms for skyline probabilities of uncertain data. ACM Trans. Database Syst. 36(2), 12 (2011)
2. Bartolini, I., Ciaccia, P., Patella, M.: The skyline of a probabilistic relation. IEEE Trans. Knowl. Data Eng. 25(7), 1656–1669 (2013)
3. Börzsönyi, S., Kossmann, D., Stocker, K.: The skyline operator. In: ICDE, pp. 421–430 (2001)
4. Chomicki, J., Godfrey, P., Gryz, J., Liang, D.: Skyline with presorting. In: ICDE, pp. 717–719 (2003)
5. Chomicki, J., Godfrey, P., Gryz, J., Liang, D.: Skyline with presorting: Theory and optimizations. In: Intelligent Information Systems, pp. 595–604 (2005)
6. Godfrey, P., Shipley, R., Gryz, J.: Maximal vector computation in large data sets. VLDB, 229–240 (2005)
7. Kossmann, D., Ramsak, F., Rost, S.: Shooting stars in the sky: An online algorithm for skyline queries. VLDB, 275–286 (2002)
8. Lian, X., Chen, L.: Reverse skyline search in uncertain databases. ACM Trans. Database Syst. 35(1) (2010)
9. Papadias, D., Tao, Y., Fu, G., Seeger, B.: Progressive skyline computation in database systems. ACM Trans. Database Syst. 30(1), 41–82 (2005)
10. Pei, J., Jiang, B., Lin, X., Yuan, Y.: Probabilistic skylines on uncertain data. VLDB, 15–26 (2007)
11. Tan, K.-L., Eng, P.-K., Ooi, B.C.: Efficient progressive skyline computation. VLDB, 301–310 (2001)
12. Tan, P.-N., Steinbach, M., Kumar, V.: Introduction to Data Mining. Addison-Wesley (2005)
13. Zhang, Q., Ye, P., Lin, X., Zhang, Y.: Skyline probability over uncertain preferences. In: EDBT/ICDT, pp. 395–405 (2013)
14. Zhang, W., Lin, X., Zhang, Y., Cheema, M.A., Zhang, Q.: Stochastic skylines. ACM Trans. Database Syst. 37(2), 14 (2012)
15. Zhang, W., Lin, X., Zhang, Y., Wang, W., Yu, J.X.: Probabilistic skyline operator over sliding windows. In: ICDE, pp. 1060–1071 (2009)

Cumulative Vehicle Routing Problem:
A Column Generation Approach

Daya Ram Gaur[1] and Rishi Ranjan Singh[2]

[1] Department of Mathematics and Computer Science,
University of Lethbridge,
4401 University Drive, Lethbridge, Alberta, Canada
gaur@cs.uleth.ca

[2] Department of Computer Science and Engineering,
Indian Institute of Technology Ropar,
Nangal Road, Rupnagar, Punjab - 140001, India
rishirs@iitrpr.ac.in

Abstract. Cumulative vehicle routing problems are a simplified model of fuel consumption in vehicle routing problems. Here we study computationally, an approach for constructing approximate solutions to cumulative vehicle routing problems based on rounding solutions to a linear program. The linear program is based on the set cover formulation, and is solved using column generation. The pricing sub-problem is solved using dynamic programming. Simulation results show that the simple scalable strategy computes solutions with cost close to the lower bound given by the linear programming relaxation.

Keywords: Routing, Transportation, Linear programming, Traveling salesman, Column generation.

1 Introduction

The variant of VRP that we study in this paper is known as the *cumulative vehicle routing problem* (Cu-VRP), first defined by Kara et al. [16]. In cumulative VRPs, the objective is to minimize the "cumulative" cost. The cumulative cost per unit distance is proportional to the total weight of the vehicle, weight of the empty vehicle plus the weight of the load. Cumulative VRPs are also referred in the literature as linear fuel consumption model for capacitated VRPs by Xiao et al. [26] and as energy minimizing model for vehicle routing problem by Kara el al. [17]. The cumulative VRP is a simplified model for fuel consumption [23,16,26] in vehicle routing problems.

Fuel cost is an important fraction of the transportation cost and depending on the medium, it can be as high as 60% (32% for cargo in the seas, 46% for the railroads and 60% for the road transportation) [22]. Therefore, by minimizing the fuel consumption, we can reduce the entire transportation cost. Cumulative VRPs generalize two well known and well studied computationally hard problems.

S. Ganguly and R. Krishnamurti (Eds.): CALDAM 2015, LNCS 8959, pp. 262–274, 2015.

Next, we define the problem. The definitions are from Kara et al. [16], and Gaur et al. [11]. We are given a complete graph $G(V, E)$ with weights on the edges. The edge weights satisfy the triangle inequality. A special node numbered 0 is the depot and all the client nodes are numbered from $\{1, 2, \cdots, n\}$. Stationed at the depot is a vehicle with capacity Q. There is a positive and integral demand d_i at each customer node i. We can think of d_i as the weight of the goods to be transported from the client node to the depot. The objective is to schedule the vehicle starting at depot such that the vehicle collects all the items from the customers and deposits them at the depot and the total cumulative cost is minimized. We allow the vehicle to offload the cargo at the depot an arbitrary number of times. The objects to be transported are indivisible. The cumulative cost per unit distance is defined next. Let a be the cost of moving empty vehicle per unit distance and b be the cost of moving unit weight per unit distance. The cost of moving a vehicle unit distance with weight w is $a + bw$, where w is the weight of the cargo. The solution to Cu-VRP is a collection of k directed cycles $T = \{T_1, T_2, T_3, \ldots, T_k\}$. Each customer node is in exactly one of the cycles. The total weight of the objects picked in a cycle T_j is at most Q. Each cycle starts and ends at the depot node and visits at least one customer. Let l_i denote the distance traveled by the vehicle after picking the object at customer i and offloading it at the depot, in some cycle T_j. Let $|T_j|$ be the length of directed cycle T_j. The total cumulative cost $C(T)$ of the travel schedule given by solution T is

$$C(T) = a \sum_{j=1}^{k} |T_i| + b \cdot \sum_{i=1}^{n} w_i l_i.$$

Note that the vehicle travels a total distance of $\sum_{j=1}^{k} |T_i|$. Each object i with weight w_i travels a distance of l_i in T. Our objective is to find a solution T^* such that there is no other solution with a strictly smaller cumulative cost.

Kara et al. [16,17] studied the deterministic version of the problem under the names 'energy minimizing vehicle routing problem' and 'cumulative VRP'. They give polynomial size integer programming formulations for the collection and delivery cases of the problem. Furthermore, they noted that the cumulative VRPs generalize several previously studied problems in literature such as the minimum latency problem (MLP) [6,5] and the k-traveling repairmen problem (k-TRP) [12]. A special case of deterministic cumulative VRPs occurs when there, only a single offload is allowed, the capacity of the vehicle is infinite, and the travel schedule is single TSP tour of graph G. Constant factor approximation algorithms have been given for this case by Blum et. al. [6]. Recently, Gaur et al. [11] studied the cumulative VRP problem and gave constant factor approximation algorithms for four different variations.

A need for approximation algorithms and good heuristics is evident for cumulative VRPs. Randomized rounding [19] is a tool used to design provably good approximation algorithms for NP-hard problems. In this approach we compute an approximate answer given the optimal solution to a suitable linear programming relaxation. The fractional values are treated as probabilities with which the variables take integral values. Several rounding experiments are conducted to obtain a feasible integral solution (for minimization problems). An upper bound on the expected value of the integral solution is implied by the union bound. For successful applications of randomized rounding to numerous problems see the books by Vazirani [24] and the recent one by Williamson and Shmoys [25].

Our Contributions: In this paper we study the cumulative vehicle routing problem (Cu-VRP). In the next section we give the mathematical formulation for the Cu-VRP problem. Next we decompose the problem into the master problem and the sub-problem, so that the column generation can be applied. In the third section we give three algorithms. First we describe a dynamic programming algorithm to solve the pricing sub-problem. This dynamic program computes an approximate solution to the sub-problem. The solution is used by the column generation to introduce a new column into the basis for the master problem. Using simulation we establish that this approximate solution to the linear program (LP) is quite close to the optimal solution to the linear programming relaxation for Cu-VRP. In fact, in most of the cases the bounds are the same. We use randomized rounding to obtain integral solutions. Using simulations on well known instances of CVRP data sets, we show that good approximate solutions can be computed to Cu-VRP instances. Given that large scale LP relaxations are routinely solved using column generation these days, our approach has the added benefit of being truly scalable. We can scale the dynamic progam as well. A branch cut and price based exact approach will deliver an exact answer but is typically not scalable. However, it should be noted that our approach delivers only approximate solutions. We give a summary and discussion of the simulation results in the final section.

2   Formulation

Mathematical formulation below is due to Kara et al. [16,17]. We modify the objective function to account for the cumulative cost as defined in [11].

Recall that n is the number of customers and Q is the capacity of the vehicle. a is the cost of moving empty vehicle per unit distance and b is the cost of moving unit weight cargo per unit distance. The demand at customer node i is d_i. We assume integral demands. Let c_{ij} be the distance between customer i and customer j. We use two decision variables x_{ij} and y_{ij}. The decision variable $x_{ij} = 1$, if the vehicle visits customer j just after visiting customer i, otherwise $x_{ij} = 0$. y_{ij} is the weight of the cargo carried by the vehicle from customer i to customer j. The mixed integer linear programming formulation for the cumulative vehicle routing problem is as follows:

$$min : \sum_{i=0}^{n} \sum_{j=0}^{n} ((a \cdot x_{ij} + b \cdot y_{ij}) c_{ij}) \tag{1}$$

$$s.t. : \sum_{i=0}^{n} x_{ij} = 1 \qquad (j = 1, 2, \cdots, n) \tag{2}$$

$$\sum_{i=0}^{n} x_{ip} - \sum_{j=0}^{n} x_{pj} = 0 \qquad (p = 1, 2, \cdots, n) \tag{3}$$

$$\sum_{j=0}^{n} y_{pj} - \sum_{i=0}^{n} y_{ip} = d_p \qquad (p = 1, 2, \cdots, n) \tag{4}$$

$$y_{ij} \leq Q \cdot x_{ij} \qquad (i, j = 1, 2, \cdots, n) \tag{5}$$

$$x_{ij} \in \{0, 1\} \qquad (i, j = 1, 2, \cdots, n) \tag{6}$$

$$y_{ij} \geq 0 \qquad (i, j = 1, 2, \cdots, n) \tag{7}$$

Constraint (2) ensures that each customer node will have exactly one incoming edge. Constraint (3) ensures that in-degree is equal to out-degree for all the customer nodes. Equation (4) is a flow constraint which ensures that the difference between the sum of the weight carried by vehicle on the edges going outside from a customer node and the sum of weight carried on the edges going inside to the customer node is equal to the the supply at that customer node. This constraint ensures that the supply at each customer is picked. Equation (5) is the capacity constraint and ensures that the weight carried by the vehicle on an edge is always less than the capacity. This constraint is active only on the edges that belong to some cycle in the solution. The last two constraints state that x_{ij} is a binary variable and that y_{ij} is positive. Next we describe the master problem based on the set cover formulation.

Let j be a subset of the customer nodes. A subset j is feasible if the sum of the demands at the clients in j is at most Q. z_{ij} is a binary variable that represents whether or not customer node i is in subset j. θ_j is the minimum cumulative cost for serving the demand of the customers in subset j from the depot. Stated otherwise it is the cost of the route, visiting the customers in an optimal order. α_j is a binary variable that represents whether or not subset j is in the optimal solution. Let R be the set of all the subsets of the customer nodes. The master problem can now be formulated as :

$$min : \sum_{j \in R} \theta_j \cdot \alpha_j \tag{8}$$

$$s.t. : \sum_{j \in R} z_{ij} \cdot \alpha_j \geq 1 \qquad (i = 1, 2, \cdots, n) \tag{9}$$

$$\alpha_j \in \{0, 1\}. \tag{10}$$

Note that even the computation of θ_j is an NP-hard problem. This method of formulation for VRP is due to Balinski and Quandt [4]. Such a formulation was first used successfully by Cullen et al. [10] to develop heuristics for CVRP problem. Bramel and Simchi-Levi [7] studied such a formulation for VRP with time windows (VRPTW). They show that for any distribution of service times, time window, customer loads etc the integrality gap becomes arbitrarily small as the number of customers increases. Their results explain the efficient behaviour of branch and bound type algorithms on the set cover formulation for VRPTW. Next we state the mixed integer linear programming formulation (MILP) for the sub-problem.

Let $\pi = \{\pi_1, \pi_2, \cdots, \pi_n\}$ be the set of dual prices given the current set of columns in the LP relaxation of the master problem. The following pricing sub-problem uses π to generate a new route with minimum reduced cost.

$$min : \sum_{i=0}^{n} \sum_{j=0}^{n} ((a \cdot x_{ij} + b \cdot y_{ij})c_{ij}) - \sum_{i=1}^{n} (\pi_i \cdot \sum_{j=0}^{n} x_{ji}) \tag{11}$$

$$s.t. : \sum_{j=1}^{n} x_{0j} = 1 \tag{12}$$

$$\sum_{j=1}^{n} x_{j0} = 1 \tag{13}$$

$$\sum_{i=0}^{n} x_{ip} - \sum_{j=0}^{n} x_{pj} = 0 \qquad (p = 1, 2, \cdots, n) \tag{14}$$

$$\sum_{j=0}^{n} y_{pj} - \sum_{i=0}^{n} y_{ip} = d_p \cdot \sum_{k=0}^{n} x_{kp} \qquad (p = 1, 2, \cdots, n) \tag{15}$$

$$y_{ij} \leq Q \cdot x_{ij} \qquad\qquad (i, j = 1, 2, \cdots, n) \tag{16}$$

$$x_{ij} \in \{0, 1\} \qquad\qquad (i, j = 1, 2, \cdots, n) \tag{17}$$

$$y_{ij} \geq 0 \qquad\qquad (i, j = 1, 2, \cdots, n) \tag{18}$$

As before, y_{ij} is the weight of the cargo being moved from node i to node j. Most of the constraints in the MILP sub-problem are same as for the main problem. Constraints (12) and (13) ensures that the solution is a cycle. There is one outgoing edge from the depot, and one incoming edge into the depot. Constraint (14) is the same as the constraint (3) in the MILP main problem. Constraint (15) is the flow constraint. Constraint (16) states that total weight of the pickups at the nodes in the cycle is at most Q. π_i are the dual-values in the objective function. The objective function represents the reduced cost of a cycle. The optimal solution to the sub-problem is a route with minimum reduced cost.

In the next section we describe the column generation algorithm used to solve the LP relaxation of the master problem. The sub-problem, if solved using dynamic programming might not return the most negative reduced cost column. Therefore the linear programming solution to the master problem need not be optimal. Using a simple randomized rounding scheme we convert the fractional LP solution to an integral solution for the master problem.

3   Algorithm

First we describe the column generation approach in brief. Next we give an algorithm which computes an approximate solution to the sub-problem using dynamic programming scheme. Finally, we will see how to obtain a good feasible integral solution to the master problem using randomized rounding and pruning.

A column in (9) represents a route. The ith element of the column j tells whether the ith customer is in route j. The order in which clients are visited in the route is not specified. We assume that the clients are visited in the order that minimizes the cumulative cost of the cycle. The optimal order is computed when the cost θ_j is computed. We start the column generation algorithm with an initial feasible set of columns, say the identity matrix. Identity matrix corresponds to routes, from the depot to client i and back to the depot. We then solve the LP relaxation of the master problem with respect to the current basis and get the optimal dual prices given the current basis. We pass the dual prices to the sub-problem as input and generate a new column with minimum reduced cost, if one exists. The sub-problem may be solved exactly by using an MILP solver such as CPLEX or solved approximately using the dynamic programming algorithm described below. If the objective function value of the sub-problem is less than 0 then a column (cycle) is added to the current basis in the master problem. If no column has negative reduced cost then the column generation procedure stops.

We solve the sub-problem by dynamic programming. Our approach is similar to the one of Lysgaard and Wohlk [18] for the SPPRC problem. The paths in SPPRC are not elementary. Our sub-problem has the requirement that the paths are elementary. Similar approaches were used to solve the sub-problems (ESPPRC) for column generation for VRP with time windows by Rousseau et al. [21], Feillet et al. [13], and Chabrier [8]. We use dynamic programming to compute an elementary route with minimal reduced cost.

Recall that a route starts at the depot and ends at the depot. Let the depot node be r. We assume that the routes are elementary, i.e. no node is re-visited on the route. The demand at node i is d_i, for $i \in [1..n]$. Let $C(i, q, x)$ be the cost of the minimal cost route that collects q units of goods, visits a total of x clients and the last node visited before returning to the depot is client i. Let the route that achieves this minimum cost be $R(i, q, x)$. Note that there might be more than one route which attains the minimal cost. $C(i, q, x)$ can be initialized, for $i \in [1..n]$, and $q \in [d_1, d_2, \ldots, d_n]$, and $x = 1$, as follows:

$$C(i, q, x) = a.c(r, i) + (a + b.d_i).c(i, r) - y_i.$$

where $c(r, i)$ is the shortest distance between the depot r and node i. This follows from the definition of the cost, for the route $[r, i, r]$. The cost of transporting the vehicle is $a.2.c(r, i)$ and the cost of transporting d_i units of goods from client i to the depot r is $b.d_i.c(r, i)$. The dual value associated with client i is y_i.

$C(i, q, x)$ can be updated as follows.

$$C(i, q, x) = \min_{j \neq i, i \notin R(j, q-d_i, x-1)} \left\{ \begin{array}{l} C(j, q - d_i, x - 1) - (a + b(q - d_i))c(j, r) + \\ (a + b(q - d_i))c(j, i) + (a + b.q)c(i, r) - y_i \end{array} \right\}$$

$R(i, q, x)$ has to be updated accordingly. Next we note that the principle of optimality may not hold for the recurrence relation above. Let R be the route that determines the minimum cost $C(i, q, x)$. As before j is the node before client i in route R. Let us consider the route that uses the prefix in R till node j, then it returns to r. We label this route $R(j)$. We have to establish that $R(j)$ is a route with minimal cost that ends at node j, moves a total of $q - d_i$ units of goods, and visits a total of $x - 1$ nodes. Suppose there is some other route $R'(j)$ that visits j then r, and has a smaller cost. This route $R'(j)$ also moves a total of $q - d_i$ units of goods, and visits $x - 1$ vertices. The suffix cost, from j to r, is same on both the routes, and the prefix cost is smaller on $R'(j)$. If i does not belong to $R'(j)$, replacing the prefix till node j in R, with the prefix from $R'(j)$, we obtain a route with cost smaller than R, a contradiction. Therefore in this case no $R'(j)$ with cost strictly lower than $R(j)$ exists. However it is plausible that $R'(j)$ visits i, and ends in j.

The capacity constraint is enforced implicitly, as the computation is restricted for $q \leq Q$. Even though the recurrence relation does not consider all the paths, there is ample evidence, in the section on simulation, that the paths returned by the dynamic programming algorithm are close to the optimal. The LP solution to the master problem computed using dynamic programming are quite close to the optimal LP solutions for CVRP instances considered. Dynamic programming algorithms that return optimal paths from a subset of constrained paths are of independent interest, see for example Balas et al. [3], and the results on approximate constrained shortest paths due to Hassin [14]. The details about how to obtain a good feasible integral solution to the master problem using randomized rounding and pruning are available in the full version on the paper.

4   Simulation Results

The objective function in the Cumulative VRPs generalize the objective functions of the capacitated VRPs and the minimum latency problem (CMLP). In the objective function, multiplier a express the part of CVRP's objective and multiplier b expresses the part of CMLP's objective. It was noted in [11], that when a/b is equal to Q, two terms in the objective function balance each other out. Using this property they [11] gave constant factor approximation algorithms for four variants. Therefore, we perform simulation on three different pairs of values for (a, b); $(a = 1, b = 0)$, the CVRP case, $(a = 0, b = 1)$, the CMLP with at most n offloads allowed, and $(a = Q, b = 1)$, the special case mentioned above, where Q is the capacity of the vehicle.

4.1   Instances

For simulation, we used instances available at http://neo.lcc.uma.es/vrp/vrp-instances/capacitated-vrp-instances. Most of the instances and their optimal CVRP solutions are also available at http://www.branchandcut.org. The instances are as below.

- Augerat et al. [2]
  - *A-set:* We use all the 27 instances from this data set. The capacity constraint on the vehicle was 100 units for all instances.
  - *B-set:* We use all the 23 instances from this data set. The capacity constraint on the vehicle was 100 units for all instances.
  - *P-set:* We use all the 24 instances from this data set. The capacity constraint on the vehicle was different for all instances.
- Christofides and Eilon [9] (*E-set:*) This data set contains 16 instances with different capacity constraints for different instances.
- *Rinaldi and Yarrow* [20] (RY-instance) This data set contains a single instance with the capacity constraint of 15 units on the vehicle. The demand of customer nodes is unit each.

Instances in the data sets A-set, B-set, P-set and E-set are labelled $x - np - kq$, where $x \in \{A, B, P, E\}$ denotes the set to which the instance belongs, p denote the number of total nodes including the depot and q denotes the number of offloads at the depot for the CVRP case. For each instance the coordinates of the nodes are provided in the data file. We compute the Euclidean distance in the two dimensional plane between each pair of nodes to obtain the distance matrix. For each instance, also given are the demands, the vehicle capacity, the number of customer nodes and the distance matrix. We relax the constraint on the number of offloads. We allow an arbitrary number of offloads at the depot. Due to space limitations we report the data on only the E instances. Simulation data for A, B, P and RY instances is available in the full version on the paper. The E instances appear to be hardest in the computational study, hence we choose to report them here. We use a fixed time-out of 3 hours for all the instances.

We implemented the algorithms in MATLAB V7.12.0.635 (R2011a). The MILP formulation for the sub-problem was also solved using CPLEX Studio 125 in MATLAB. All the simulations were performed on a 32 bit Ubuntu machine with 3.00GHz Intel Core 2 Duo E8400 processor and 3.4 GB RAM. Twenty randomized rounding were performed for each instance to obtain the average cost of of the integral solution obtained using rounding. The discussion over the simulation results is next.

We will discuss the results in three parts; CVRP's objective function (Table 1), CMLP's objective function (Table 2), Cu-VRP special case of $\frac{a}{b} = Q$ (Table 3). A plot of the average running times for solving the sub-problems is also provided. The missing data points represent time out of three hours.

Table 1 is the table for the results of Cu-VRPs when $a = 1$ and $b = 0$. This has the affect of modelling the CVRP objective function. In the first column is the name of the instance. Included in the second column is the cost of the optimal CVRP solution, available from *http://www.branchandcut.org*. Note that the optimal CVRP solution is computed using a fleet of fixed size, whereas in our simulations we relax the constraint on the fleet size (an arbitrary number of offloads are allowed). Note that the optimal CVRP cost (CVRP OPT) is listed only in Table 1. Table 2 is the table for the results of Cu-VRPs when $a = 0$ and $b = 1$. This reduces the objective function to the objective function for the minimum latency problem, with arbitrary number of offloads allowed. Table 3 is the result for the special case used in the proof in [11] in which we set $a = Q$ and $b = 1$. In some tables the second column is the capacity Q of the vehicle. In tables where the capacity of the vehicle is not listed, the capacity is 100 units.

Next ten columns are divided into two blocks. First five columns are the computational results when the MILP sub-problem is solved using CPLEX. Next five columns are results when the sub-problem is solved using dynamic programming (DP). MILP LB value is computed by solving each sub-problem optimally using CPLEX. DP LB is the value of the fractional solution to the LP relaxation of the master problem, when dynamic program is used to solve each sub-problem. The DP LB values can be larger than MILP LB values. Empty entries in columns indicate timeout events. The next four columns are: total number of generated columns (NOC); total time spent to compute the fractional solution (TT); total time spent in the computation of the dual variables (DT); total time spent in the sub-problems (SPT). The next column (Avg R-value) is the average cost of the integral solutions over 20 randomized rounding of solution. The next column (Avg factor) in Table 1, is the ratio of Avg R-value to the CVRP optimal cost.

CVRP: (Table 1) When dynamic programming is used to generate columns with negative reduced cost, the LP solution to the master problem (DP LB values) are close to the optimal LP solution (MILP LB values). There are several instances on which column generation using CPLEX times out after three hours. The rounding process is quite fast. The average performance ratio (Avg-factor values) is close to 1. Note that for these instances the optimal CVRP solution is computed with constraint on the fleet size, and we relax the fleet size constraints in our simulations. It is also to be noted that the best known approximation algorithm for CVRPs has performance ratio $1 + \alpha$, where α is the approximation ratio for traveling salesman problem (TSP). For Euclidean space $\alpha = 1 + \epsilon$ for any constant ϵ, as a PTAS exists. If one uses Christofides heuristic for approximating the TSP, then the theoretical bound in Altinkemer and Gavish [1] is 2.5. Performance ratio observed in our simultations is below the theoretical worst-case bounds from [1].

CMLP: Table 2 summarize the results, when the objective function is similar to the minimum latency problem with arbitrary number of offloads. Here the optimal solution are not available on branchandcut.org. Integral solutions obtained after rounding have almost always the same cost as the fractional LP solutions. Once again dynamic programming when used to solve the pricing sub-problem is very effective. We relax the constraint on the fleet size in our formulation. The problem even without the fleet size constraint is NP-Complete, as it subsumes Bin Packing.

Consider unit demands, and vehicle capacity of a unit. The cost of the optimal solution, under this restriction, for CMLP objective function is simply the sum of the shortest path from depot to each client. If the underlying packing problem is not hard, then the instances are easy (polynomially solvable). Perhaps that is why we see exact solutions, almost always, when dynamic programming is used to solve the pricing problem.

Cu-VRP: Table 3 summarize the results when we set $a = Q$ and $b = 1$. Once again dynamic programming for approximate pricing columns performs as good as CPLEX to price the columns optimally. The average approximation ratio is close to 1. Only for one instance the approximation ratio is 2.504 (E-n31-k7). For Euclidean space, Gaur et al. [11] gave $3.414 + \epsilon$, factor approximation algorithm for Cu-VRPs using $1 + \epsilon$ approximation factor TSP tour. Their algorithm uses the iterated tour partitioning heuristic and does not use randomized rounding. The performance ratio in the simulations agree well with the only known worst-case theoretical bound on the approximability of the problem.

Table 1. CVRP's $(a = 1, b = 0)$ and instances from E-set

Instances	Q	CVRP OPT	MILP					DP					Avg R-value	Avg Fact
			LB	NOC	TT	DT	SPT	LB	NOC	TT	DT	SPT		
E-n7	3		100.00	3	0.9	0.3	0.6	100.00	3	0.3	0.3	0.0	116.20	
E-n13-k4	6000	247	247.00	23	10.3	0.5	9.8	247.00	23	60.5	0.4	60.0	247.00	1.00
E-n22-k4	6000	375	373.71	37	48.4	0.7	47.7	373.71	40	634.0	0.7	633.3	412.75	1.10
E-n23-k3	4500	569	558.95	62	340.0	1.3	338.6	558.95	73	1034.8	1.6	1033.2	749.80	1.32
E-n30-k(3, 4)	4500	534	484.10	112	6954.1	3.0	6951.1	484.10	103	3539.4	2.4	3536.9	692.15	1.30
E-n31-k7	140	379	309.00	43	65.7	0.8	64.8	309.00	49	46.8	0.9	45.8	643.80	1.70
E-n33-k4	8000	835												
E-n51-k5	160	521	517.06	284	8154.2	34.5	8119.5	517.08	309	1840.7	47.7	1792.9	685.60	1.32
E-n76-k7	220	682												
E-n76-k8	180	735												
E-n76-k10	140	830						812.45	397	7013.9	65.9	6947.9	1140.60	1.37
E-n76-k(14,15)	100	1021						1002.75	283	3408.8	15.9	3392.8	1412.55	1.38
E-n101-k8	200	815												
E-n101-k14	112	1071												

Table 2. CMLP's $(a = 0, b = 1)$ and instances from E-set

Instances	Q	MILP					DP					Avg R-value
		LB	NOC	TT	DT	SPT	LB	NOC	TT	DT	SPT	
E-n7	3	72.00	2	0.5	0.3	0.2	72.00	2	0.1	0.0	0.0	76.00
E-n13-k4	6000	429400.00	5	2.1	0.3	1.7	429400.00	5	15.1	0.1	15.1	429400.00
E-n22-k4	6000	628700.00	1	0.9	0.2	0.6	628700.00	1	30.9	0.0	30.9	628700.00
E-n23-k3	4500	407318.00	1	1.0	0.2	0.7	407318.00	1	28.3	0.2	28.0	407318.00
E-n30-k(3, 4)	4500	577025.00	3	2.7	0.3	2.4	577025.00	3	136.5	0.3	136.1	577025.00
E-n31-k7	140	10071.00	42	90.1	0.7	89.3	10071.00	44	41.8	0.5	41.2	14536.00
E-n33-k4	8000	2296050.00	8	12.6	0.3	12.2	2296050.00	11	960.9	0.4	960.5	2296050.00
E-n51-k5	160	18017.00	3	7.6	0.3	7.3	18017.00	3	23.5	0.3	23.2	18017.00
E-n76-k7	220	32010.00	3	43.0	0.3	42.7	32010.00	7	224.8	0.1	224.6	32010.00
E-n76-k8	180	32010.00	3	40.8	0.3	40.5	32010.00	7	183.1	0.3	182.7	32010.00
E-n76-k10	140	32010.00	4	40.3	0.3	40.0	32010.00	7	139.9	0.4	139.5	32010.00
E-n76-k(14,15)	100	32010.00	3	23.7	0.3	23.4	32010.00	7	96.6	0.3	96.2	32010.00
E-n101-k8	200						36614.00	20	1257.0	0.5	1256.5	37043.45
E-n101-k14	112						36614.00	20	690.2	0.2	690.0	36969.55

Table 3. Cu-VRP's $(a = Q, b = 1)$ and instances from E-set

Instances	Q	MILP					DP					Avg R-value
		LB	NOC	TT	DT	SPT	LB	NOC	TT	DT	SPT	
E-n7	3	388.00	2	0.5	0.2	0.2	388.00	2	0.3	0.3	0.0	456.00
E-n13-k4	6000	2067300.00	19	8.0	0.4	7.5	2067300.00	19	50.5	0.4	50.0	2067300.00
E-n22-k4	6000	3123000.00	38	50.4	0.7	49.7	3129800.00	31	495.1	0.6	494.5	3336200.00
E-n23-k3	4500	3243577.00	51	1084.3	1.0	1083.3	3243577.00	26	377.3	0.5	376.7	3243577.00
E-n30-k(3, 4)	4500						2992922.06	50	1735.6	1.0	1734.5	3368842.50
E-n31-k7	140	3243577.00	41	91.9	0.8	91.1	58766.00	44	42.1	0.8	41.3	147134.85
E-n33-k4	8000											
E-n51-k5	160						116705.95	267	1576.7	26.3	1550.3	155593.95
E-n76-k7	220											
E-n76-k8	180						182427.33	459	10617.0	124.0	10492.9	261691.30
E-n76-k10	140						160979.54	336	5917.6	32.2	5885.3	234395.35
E-n76-k(14,15)	100						141236.88	261	3142.7	10.8	3131.8	187975.15
E-n101-k8	200											
E-n101-k14	112											

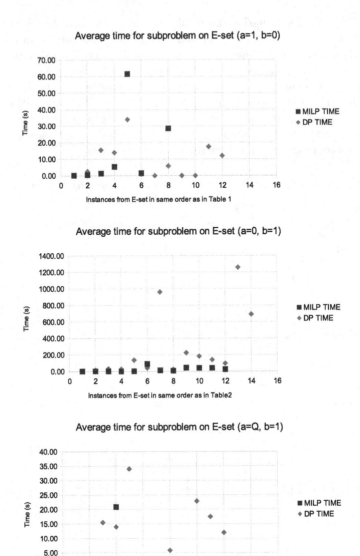

The time plots for the CVRP case and the special case of Cu-VRP, are very similar. The average time taken by the dynamic program is less than the average time taken by the MILP. For the case when ($a = 0, b = 1$) the time taken by the dynamic program is more than the time taken by the MILP solver. In this case the weighted sum of the shortest path from the each node to the depot is the cost of the optimal solution, as there is no constraint on the number of tours. Packing problem we assume is easy in some sense. We surmise that the MILP in this case is easy (poly time), whereas the dynamic program takes pseudo-polynomial time.

To the best of our knowledge, this is the first empirical study of approximation algorithm for cumulative VRP using column generation and randomized rounding. Such an empirical study was posed as a question by the authors in [15]. In our simulation columns with negative reduced cost are added one at a time. It is possible to speed up the column generation algorithm by adding more than one column in each iteration. This modification increases the number of columns in the basis, and rounding takes longer.

5 Conclusion

In this paper we empirically evaluate the performance of column generation based approximation algorithm for the cumulative vehicle routing problem. To the best of our knowledge this is a first such study. We solve a set cover type formulation for the cumulative VRP problem using column generation. Dynamic programming is used to solve the pricing problem. Simulations are performed on well known instances. Simulations study three variants of the problem; the objective function for the CVRP, objective function for the minimum latency problem with arbitrary number of offloads, and a special case important to the theoretical analysis of performance ratio. Large scale LP relaxations are routinely solved using column generation these days. Hence, the approach presented here is truly scalable, as opposed to a branch cut and price based exact approach. However, it should be noted that our approach delivers approximate solutions whereas a branch cut and price based approach will deliver an exact answer. The simulation results are better than the worst-case bounds on the approximation algorithms developed using the iterated tour partitioning technique in [11]. Finding theoretical worst case bounds on the performance ratio of the column generation based randomized algorithm remains open.

Acknowledgements. Partial support for this work was provided to DG by Natural Sciences and Engineering Research Council of Canada (NSERC) under the Discovery Grant (DG) Program. Work of RS was supported by IIT Ropar under the Institute scholar scheme.

References

1. Altinkemer, K., Gavish, B.: Technical Note: Heuristics for Delivery Problems with Constant Error Guarantees. Transportation Science 24(4), 294–297 (1990)
2. Augerat, P., Belenguer, J.M., Benavent, E., Corberan, A., Naddef, D., Rinaldi, G.: Computational results with a branch and cut code for the capacitated vehicle routing problem, Tech. Rep. 1 RR949-M, ARTEMIS-IMAG, Grenoble France (1995)
3. Balas, E.: New classes of efficiently solvable generalized traveling salesman problems. Annals of Operations Research (56), 529–558 (1999)
4. Balinski, M.L., Quandt, R.E.: On an integer program for a delivery problem. Operations Research 12, 300–304 (1964)
5. Bianco, L., Mingozzi, A., Ricciardelli, S.: The traveling salesman problem with cumulative costs. Networks 23(2), 81–91 (1993)

6. Blum, A., Chalasani, P., Coppersmith, D., Pulleyblank, W.R., Raghavan, P., Sudan, M.: The minimum latency problem. In: STOC, pp. 163–171 (1994)
7. Bramel, J., Simchi-Levi, D.: Probabilistic Analysis and Practical Algorithms for the Vehicle Routing Problem with Time Windows. Operations Research 44, 501–509 (1996)
8. Chabrier, A.: Vehicle Routing Problem with elementary shortest path based column generation. Computers & Operations Research 33(10), 2972–2990 (2006)
9. Christofides, N., Eilon, S.: An algorithm for the vehicle-dispatching problem. In: OR, pp. 309–318 (1969)
10. Cullen, F.H., Jarvis, J.J., Ratliff, H.D.: Set partitioning based heuristics for interactive routing. Networks 11(2), 125–143 (1981)
11. Gaur, D.R., Mudgal, A., Singh, R.R.: Routing vehicles to minimize fuel consumption. Operations Research Letters 41(6), 576–580 (2013)
12. Fakcharoenphol, J., Harrelson, C., Rao, S.: The k-traveling repairman problem. In: Proceedings of the fourteenth Annual ACM-SIAM Symposium on Discrete Algorithms, pp. 655–664. Society for Industrial and Applied Mathematics (January 2003)
13. Feillet, D., Dejax, P., Gendreau, M., Gueguen, C.: An exact algorithm for the elementary shortest path problem with resource constraints: Application to some vehicle routing problems. Networks 44(3), 216–229 (2004)
14. Hassin, R.: Approximation schemes for the restricted shortest path problem. Mathematics of Operations Research 17(1), 36–42 (1992)
15. Irnich, S., Desaulniers, G.: Shortest path problems with resource constraints. In: Desaulniers, G., Desrosiers, J., Solomon, M.M. (eds.) Column generation, GERAD 25th Anniversary Series, ch. 2, pp. 33–65. Springer (2005)
16. Kara, İ., Kara, B.Y., Yetiş, M.K.: Cumulative Vehicle Routing Problems. In: Caric, T., Gold, H. (eds.) Vehicle Routing Problem, pp. 85–98. I-Tech Education and Publishing KG, Vienna (2008)
17. Kara, İ., Kara, B.Y., Yetis, M.K.: Energy Minimizing Vehicle Routing Problem. In: Dress, A., Xu, Y., Zhu, B. (eds.) COCOA. LNCS, vol. 4616, pp. 62–71. Springer, Heidelberg (2007)
18. Lysgaard, J., Wohlk, S.: A branch-and-cut-and-price algorithm for the cumulative capacitated vehicle routing problem. European Journal of Operational Research (August 2013)
19. Raghavan, P., Thompson, C.D.: Randomized Rounding: A Technique for Provably Good Algorithms and Algorithmic Proofs. Combinatorica 7(4), 365–374 (1987)
20. Rinaldi, G., Yarrow, L.: Optimizing a 48-city traveling salesman problem: A case study in combinatorial problem solving, New York University, Graduate School of Business Administration (1985)
21. Rousseau, L.M., Gendreau, M., Pesant, G., Focacci, F.: Solving VRPTWs with constraint programming based column generation. Annals of Operations Research 130(1-4), 199–216 (2004)
22. Sahin, B., Yilmaz, H., Ust, Y., Guneri, A.F., Gulsun, B.: An approach for analyzing transportation costs and a case study. European Journal of Operational Research 193, 1–11 (2009)
23. Toth, P., Vigo, D.: The Vehicle Routing Problem. SIAM (2001)
24. Vazirani, V.V.: Approximation Algorithms. Springer, New York (2001)
25. Williamson, D.P., Shmoys, D.B.: The Design of Approximation Algorithms, 1st edn. Cambridge University Press, New York (2011)
26. Xiao, Y., Zhao, Q., Kaku, I., Xu, Y.: Development of a fuel consumption optimization model for the capacitated vehicle routing problem. Computers & Operations Research 39(7), 1419–1431 (2012)

Energy Efficient Sweep Coverage with Mobile and Static Sensors

Barun Gorain and Partha Sarathi Mandal

Department of Mathematics
Indian Institute of Technology Guwahati, India
{b.gorain,psm}@iitg.ernet.in

Abstract. Sweep coverage provides solution for the applications in wireless sensor networks, where periodic monitoring is sufficient instead of continuous monitoring. For a given set of points in the plane, the objective of the sweep coverage problem is to minimize number of sensors required in order to guarantee sweep coverage for a given set of points of interest. Instead of using only mobile sensors for sweep coverage, use of both static and mobile sensors can be more effective in terms of energy utilization. In this paper we introduce the EEGSweep coverage problem, where objective is to minimize energy consumption by a set of sensors (mobile and/or static) with guaranteed sweep coverage for a given set of points. We prove that the EEGSweep coverage problem is NP-hard and cannot be approximated within a factor of 2. We propose an 8-approximation algorithm to solve the problem. A 2-approximation algorithm is also proposed for a special case of this problem.

Keywords: Sweep Coverage, Mobile Sensor, Static Sensor, k-TSP, k-MST, Approximation Algorithm, Wireless Networks.

1 Introduction

Coverage is one of the most important issues for environmental monitoring in wireless sensor networks (WSNs). In general, coverage is defined as the measurement of the quality of surveillance of sensing function. Information from a target field is collected by deploying sensor nodes in different locations of the field. After deployment, sensor nodes form a network through which the collected data are propagated to a sink. The quality of the collected information depends on how well the target field is covered by the set of sensor nodes. Depending on the subject to be covered, coverage problems are broadly categorized in three types. First one is point coverage [6,13,16], where a set of discrete points is continuously monitored, second one is area coverage [2,17,20], where all points within a bounded region are continuously monitored, and third one is barrier coverage [7,18], where specified path or boundary of a region is continuously monitored by sensor nodes. For example, in forest monitoring, [1] every location of the forest must be covered by at least one sensor node in order to immediately detect any unusual activities like forest fire, activities of poachers, etc. Similarly, covering

S. Ganguly and R. Krishnamurti (Eds.): CALDAM 2015, LNCS 8959, pp. 275–285, 2015.

the boundary of the forest allows controlling and elimination of the poaching activities, and illegal entry through the boundary. A continuous monitoring with static sensor nodes is required for the aforementioned types of coverage problems.

But there are typical applications where only periodic patrol inspections are sufficient for a certain set of points of interest instead of continuous monitoring like traditional coverage. This type of coverage scenario are termed as *sweep coverage*. Sweep coverage concept in WSNs is introduced in the literature by Li et al. in [15], where periodic patrol inspections are required for a given set of points of interest (PoIs) by a set of mobile sensors. As per the given definition, a point is said to be *t-sweep covered* if and only if at least one mobile sensor visits the point within every t time period, where t is called *sweep period* of the point. Objective of the sweep coverage problem is to find minimum number of mobile sensors to guarantee sweep coverage for the set of PoIs. In [11], the authors have shown that instead of using only mobile sensors, use of both static and mobile sensor can be more effective in terms of total number of sensor used. In this paper we introduce an energy efficient sweep coverage problem named as EEGSweep coverage. The objective of this problem is to guarantee sweep coverage by a combined set of mobile and static sensors such that total energy consumption by the deployed sensors is minimized.

Our Contribution: In this paper we introduce an energy efficient sweep coverage problem, in short EEGSweep coverage problem where the objective is to guarantee sweep coverage for a given set of PoIs with combination of static and mobile sensors such that total energy consumption by the set of sensors is minimized. We prove that the EEGSweep coverage problem is NP-hard and cannot be approximated within a factor of 2 unless P=NP. We propose an 8-approximation algorithm to solve the problem. We propose another 2-approximation algorithm for a special case of the same problem, where all mobile sensors visit same subset of PoIs. This approximation factor is the best achievable approximation factor for this special case of EEGSweep coverage problem.

The rest of the paper is organized as follows. In Section 2, we discuss about related works. We define the EEGSweep coverage problem in Section 3 and the algorithm for solving the EEGSweep Coverage Problem is proposed in Section 4. A solution for a special case of EEGSweep coverage problem is proposed in Section 5. Finally, we conclude in Section 6.

2   Related Work

Several approaches can be found in the literature to overcome coverage problems in WSNs. Since most of the coverage problems presented in [7,14,15] are NP-complete, several heuristics [4,17,20] and approximation algorithms [6,7] have been proposed to solve the problems. The point coverage problems are studied in [6,13,16]. In [6], the authors considered the geometric version of the point coverage problem called unit disk cover problem. They discussed about the computational complexity of the problem which is NP-hard. A constant factor approximation algorithm is provided to solve the problem. Area coverage problems are studied in [3,4,17,21]. In [4], the authors proposed a fully distributed coverage preserving scheme for wireless sensor networks with a set of static sensor nodes. The proposed algorithm extend network lifetime by sleep wakeup scheduling the sensor nodes. The algorithm guarantees network coverage without any off-duty conflict. In paper [21], Wang et al. proposed three movement assisted algorithms for area coverage, which are vector based algorithm (VEC), Voronoi based algorithm (VOR) and Minimax. In this paper authors used Voronoi diagram to identify coverage holes. These movement strategies provide efficient improvement of the coverage with mobile sensors. in [17], Ma et al. proposed a distributed heuristic where in each iteration sensor nodes move in such a way that the overall topology becomes closer to an equilateral triangulation of the plane which is the optimal layout for area coverage problem.

The concept of sweep coverage initially comes from the context of robotics [3]. Recently there have been several papers [5,8,15,22] found in the area of sweep coverage problem. In [8], Du et al. proposed two different heuristics for different movement constraints on the mobile sensors. In the first heuristic *MinExpand*, mobile sensors move in the same path in every time period and in the second heuristic *OSweep*, the mobile sensor nodes move in different paths in different time periods. The theoretical aspects of sweep coverage problem is studied in [15] by Li et al. The authors proved that finding minimum number of mobile sensors to sweep coverage for a set of discrete points is NP complete. The authors showed that this problem is equivalent to solve the Traveling salesman problem (TSP) and proved that the sweep coverage problem cannot be approximated less than a factor of 2; and a $(2 + \epsilon)$-approximation and a 3-approximation algorithm are proposed. The authors remarked on the impossibility to design distributed local algorithm, which can guarantee the required sweep coverage, *i.e.*, a mobile sensor cannot locally determine whether all PoIs are sweep covered without global information. To extend lifetime of sweep coverage, Yang et al. in [23] utilized base station as a power source for refueling or replacing battery of the mobile sensors periodically. The authors proposed two heuristics for sweep coverage with one base station and multiple base stations, respectively. In [10], Gorain et al. proposed a 2-approximation algorithm for sweep coverage of a set of PoIs when all the mobile sensors visit all the PoIs. The authors discussed a distributed version of the proposed algorithm where a set of static sensors are considered as PoIs. The static sensors computes the number of mobile sensors and their movement scheduling with initial deployment locations by exchange of messages. Authors introduced the area sweep coverage problem in this paper and proposed a 2-approximation algorithm for rectangular bounded areas. In [12], sweep coverage for a set of line segments, called line sweep coverage is introduced and a 2-approximation algorithm is proposed in order to solve the problem. As an application of line sweep coverage problem, a data gathering problem is formulated and solved with an approximation factor 3. In [11], the authors remark on the flaw of 3-approximation algorithm proposed in [15] for sweep coverage with mobile sensors and proposed a new sweep coverage problem, called GSweep coverage, where instead of using only mobile sensors, both mobile and static sensors are used for sweep coverage. A 3-approximation algorithm is proposed to guarantee sweep coverage for the given set of PoIs.

3 Preliminaries and Problem Definition

Definition 1 (t-GSweep coverage [11]). *Let* $\mathcal{U} = \{u_1, u_2, \cdots, u_n\}$ *be a set of PoIs,* $S = \{s_1, s_2, \cdots, s_p\}$ *be a set of static sensors and* $M = \{m_1, m_2, \cdots, m_q\}$ *be a set of mobile sensors. Let* v *be the uniform speed of the mobile sensors. For a given time period* $t > 0$*, a PoI* u_i *is said to be t-GSweep covered if and only if either of the following two cases happens:*

1. *A static sensor is deployed at* u_i *which continuously monitors* u_i*.*
2. *At least one mobile sensor visit* u_i *in every* t *time period.*

The time period t is called the *sweep period* of the set of points $\mathcal{U}$.

   We define the following problem, called energy efficient GSweep coverage problem , for short EEGSweep coverage problem.

Definition 2 (EEGSweep Coverage Problem). *Let* $\mathcal{U} = \{u_1, u_2, \cdots, u_n\}$ *be a set of PoIs,* $S = \{s_1, s_2, \cdots, s_p\}$ *be a set of static sensors and* $M = \{m_1, m_2, \cdots, m_q\}$ *be a set of mobile sensors. Let* v *be the uniform speed of the mobile sensors and* λ*,* μ *be the energy consumptions in every unit time for a static sensor and mobile sensor, respectively. Find* $X \leq p$ *and* $Y \leq q$*, such that using* X *static sensors and* Y *mobile sensors, t-GSweep coverage for every PoI in* $\mathcal{U}$ *can be guaranteed and* $\lambda X + \mu Y$ *is minimized.*

   The following theorem gives the complexity result for the EEGSweep coverage problem.

Theorem 1. *The EEGSweep coverage problem is NP-hard and cannot be approximated within a factor of 2 unless P=NP.*

Proof. For $\lambda = \mu$, the EEGSweep coverage problem reduces to the sweep coverage problem presented in [15], where the objective is to minimize total number of sensors. According to Theorem 1 in [15], the sweep coverage problem is NP-hard and cannot be approximated within a factor of 2 unless $P = NP$. Therefore, EEGSweep coverage problem is also NP-hard and cannot be approximated within a factor of 2 unless P=NP.   □

   To get the solution to EEGSweep coverage problem, we use solutions of k-TSP [9] and k-MST [9,19] problems.

Definition 3 (k-MST problem [9]). *Let* $G = (V, E, w)$ *be a weighed graph, where the edge weights are positive real number and* k *be a given positive integer. Find a minimum weighted tree of* G *that spans any* k *vertices of* G*.*

Definition 4 (k-TSP problem [9]). *Let* $G = (V, E, w)$ *be a weighed graph, where the edge weights are positive real number and* k *be a given positive integer. Find a minimum weighted tour of* G *visiting any* k *vertices of* G*.*

Both of the above problems are NP-hard and the best known solutions for these problems are proposed by Garg in [9] where 2-approximation algorithms for both of the problems are provided.

4   Solution to the EEGSweep Coverage Problem

In this section, we propose an approximation algorithm for the EEGSweep coverage problem with approximation factor 8. Our algorithm use the 2-approximation algorithm for k-MST [9] as a subroutine.

Consider a given set of PoIs $\mathcal{U} = \{u_1, u_2, \cdots, u_n\}$ on 2D plane. Let $G = (V, E, w)$ be the complete weighted graph with each PoI as a vertex and the line segment joining two PoIs on the plane as edge. The weight w of an edge $e \in E$ is equal to the Euclidean distance between the vertices and it is denoted by $w(e)$. For any subgraph H of G, we denote $w(H)$ as the sum of the edge weights of H.

The inputs of our following proposed Algorithm 1 (EEGSWEEPCOVERAGE) are the graph G, speed v, sweep period t, energy consumption per unit time for static and mobile senors λ and μ, respectively. The output of the algorithm is the number of mobile and static sensors with the deployment locations for the static sensors and the movement schedule for the mobile sensors. The formal description of the algorithm is as follows.

At the beginning of the algorithm we create another graph $G' = (V, E, w')$ from $G = (V, E, w)$ by changing weight of every edge $e \in E$ as follows. For all $e \in E$, if $w(e) \leq vt$ then assign $w'(e) = \frac{w(e)}{vt}$, otherwise assign $w'(e) = 1$. Next, for each k, $1 \leq k \leq n$, we do the following steps.

We compute a tree $MST_k^{G'}$ which spans k vertices of G' using 2-approximation algorithm [9] for k-MST. Let MST_k^G be the corresponding tree of $MST_k^{G'}$ in G. Now, we delete all the edges of weight more than vt from MST_k^G. This may split the tree MST_k^G into several connected components. Find tours on each of the components by doubling the edges and short cutting. Then partition the tours into several parts of length vt. Compute the value of $F_k = \lambda X_k + \mu Y_k$, where Y_k is the total number of partitions over all tours and $X_k = n - k$. Let $k = k_0$ for which the value of F_k, $1 \leq k \leq n$, be minimum. We select the number of static sensors $X = X_{k_0}$ and number of mobile sensors $Y = Y_{k_0}$ as the output of our algorithm. The deployment locations of static sensors and the movement schedules of the mobile sensors are calculated as follows.

There are $n - k_0$ vertices, which are not covered by the $MST_{k_0}^G$, so $n - k_0$ static sensors are deployed at the $n - k_0$ vertices, one for each. For deployment of mobile sensors, assume that $MST_{k_0}^G$ splits into h components $C_1, C_2, \cdots, C_h$ and $T_1, T_2, \cdots, T_h$ are corresponding tours after doubling the edges and short cutting. Then, for each j, $1 \leq j \leq h$, partition the tour T_j into $\left\lceil \frac{w(T_j)}{vt} \right\rceil$ parts of length at most vt and deploy one mobile sensor at each of the partitioning points on T_j. All mobile sensors deployed on T_j start their movement along T_j at the same time in the same direction.

Algorithm 1: EEGSWEEPCOVERAGE

1: Construct graph $G' = (V, E, w')$ from $G = (V, E, w)$
2: **for** $k = 1$ **to** n **do**
3:    Find a tree $MST_k^{G'}$ which spans k vertices of G' using the
      2-approximation algorithm for k-MST [9] .
4:    Remove all the edges with weight $> vt$ from MST_k^G. Assume that
      after removal of the edges MST_k^G splits into k_c components
      $C_1, C_2, \cdots, C_{k_c}$.
5:    Find tours $T_1, T_2, \cdots, T_{k_c}$ by doubling the edges of each components
      and short cutting.
6:    Set $Y_k = \sum_{i=1}^{k_c} \left\lceil \frac{w(T_i)}{vt} \right\rceil$ and $X_k = n - k$ then define $f_k = \lambda X_k + \mu Y_k$.
7: **end for**
8: $f = \min\{f_1, f_2, \cdots, f_k\}$ and k_0 is the index such that $f = f_{k_0}$.
9: For $k = k_0$ partition the tours $T_1, T_2, \cdots T_h$ into $\sum_{i=1}^{h} \left\lceil \frac{w(T_i)}{vt} \right\rceil$ parts of
    length vt (found in Step 5).
10: Deploy one mobile sensor at each of the partitioning points of every tour
    and deploy one static sensor at each of the remaining $n - k_0$ vertices.
11: The mobile sensors start their movement at the same time along the
    corresponding tour in the same direction.

4.1   Analysis

Theorem 2. *Algorithm 1 ensures t-GSweep coverage for each of the PoIs.*

Proof. To prove the statement of the theorem we consider any vertex u_i and
show that it is t-GSweep covered. According to the Algorithm 1, either of the
following two cases happen for u_i.

Case 1. One static sensor is deployed at u_i. In this case the statement is trivially
true.

Case 2. u_i is visited by some mobile sensors. Let u_i is visited by a mobile sensor
at time t_0. According to the algorithm, the mobile sensors are deployed within
distance at most vt along the tour in which u_i belongs. So u_i will again be visited
by the following mobile sensor within time $t + t_0$.   □

To bound approximation factor of the proposed algorithm, let mst_{opt}^k be the
the minimum weighted tree on G' that spans k vertices of G'. Let X_{opt} and Y_{opt}
be the number of static and mobile sensors in the optimal solution of EEG sweep
coverage problem for the graph G. Let Y_{opt} mobile sensors visit k_{opt} vertices of
G and remaining $n - k_{opt}$ vertices are sweep covered by $X_{opt} = n - k_{opt}$ static
sensors. Following lemma establish a relation between $w'(mst_{opt}^{k_{opt}})$ and Y_{opt}.

Lemma 1. $\left\lceil w'(mst_{opt}^{k_{opt}}) \right\rceil \leq 2Y_{opt}.$

Proof. Let us consider the movement of the mobile sensors in the optimal solution on G. Let $P_1, P_2, \cdots, P_{Y_{opt}}$ be the movement paths of the Y_{opt} mobile sensors in a time interval $[t_0, t + t_0]$. Since a mobile sensor can move at most vt distance during the time interval, no P_i contains any edge with weight more that vt. Therefore, in G', $w'(P_i) = \frac{w(P_i)}{vt} \leq 1$ and thus $\sum_{i=1}^{Y_{opt}} w'(P_i) \leq Y_{opt}$.

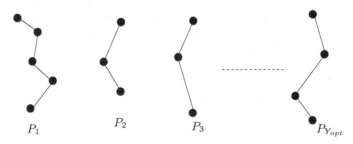

Fig. 1. Movement of the mobile sensors in the optimal solution where $w(P_i) \leq vt$ for $i = 1, 2, \cdots, P_{Y_{opt}}$

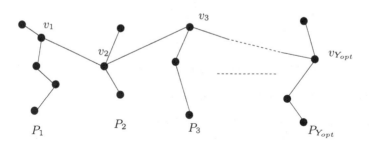

Fig. 2. The tree H' spans k_{opt} vertices of G'

Consider a vertex $v_i \in P_i$ for $i = 1$ to Y_{opt}. Let H' be the subgraph of G' which has vertex set V and edge set of H' contains all edges of the paths $P_1, P_2, \cdots, P_{Y_{opt}}$ in G' in addition with the edges in $\{(v_j, v_{j+1}) | j = 1 \text{ to } Y_{opt} - 1\}$. An example of movement paths of the mobile sensors in the optimal solution is shown in Fig. 1 and formation of the graph H' is shown in Fig. 2. Note that H' is a tree which spans k_{opt} vertices of G' and $w'(H) = \sum_{i=1}^{Y_{opt}} w'(P_i) + \sum_{i=1}^{Y_{opt}-1} w'(v_i, v_{i+1}) \leq Y_{opt} + (Y_{opt} - 1) \leq 2Y_{opt} - 1$. Since $mst_{opt}^{k_{opt}}$ is the minimum weighted tree on G' that spans k_{opt} vertices of G', therefore $w'(mst_{opt}^{k_{opt}}) \leq w'(H) \leq 2Y_{opt} - 1$. Hence $\left\lceil w'(mst_{opt}^{k_{opt}}) \right\rceil \leq w'(mst_{opt}^{k_{opt}}) + 1 \leq 2Y_{opt}$ □

Theorem 3. *The approximation factor of the Algorithm 1 is 8.*

Proof. The Algorithm 1 finds tree $MST_{k_{opt}}^{G'}$ for the iteration $k = k_{opt}$ using 2-approximation algorithm of k-MST [9] on G'. Then $w'(MST_{k_{opt}}^{G'}) \leq 2w'(mst_{opt}^{k_{opt}})$. After computing the tree $MST_{k_{opt}}^{G'}$, all the edges of weight more than vt are deleted from the tree $MST_{k_{opt}}^G$. Let m be the number of deleted edges. First we consider the case when at least one edge is deleted, i.e., $m \neq 0$. So, the tree $MST_{k_{opt}}^G$ splits into $m + 1$ connected components $C_1, C_2, \cdots, C_{m+1}$ in G. The corresponding tours $T_1, T_2, \cdots, T_{m+1}$ are computed after doubling the edges of each component and short cutting. According to the step 6 of the Algorithm 1, total number of mobile sensor needed is $Y_{k_{opt}}$, which is

$Y_{k_{opt}} = \sum_{i=1}^{m+1} \left\lceil \frac{w(T_i)}{vt} \right\rceil \leq \sum_{i=1}^{m+1} \frac{2w(C_i)}{vt} + (m+1)$.

Also, $w'(MST_{k_{opt}}^{G'}) = \sum_{i=1}^{m+1} \frac{w(C_i)}{vt} + m$. Therefore,

$$Y_{k_{opt}} \leq \sum_{i=1}^{m+1} \frac{2w(C_i)}{vt} + (m+1)$$

$$\leq \sum_{i=1}^{m+1} \frac{2w(C_i)}{vt} + 2m$$

$$\leq 2w'(MST_{k_{opt}}^{G'})$$

$$\leq 4w'(mst_{opt}^{k_{opt}})$$

$$\leq 8Y_{opt} \qquad \text{by Lemma 1}$$

Now, we consider the case when no edge from $MST_{k_{opt}}^G$ is deleted i.e., $m = 0$.

Then $Y_{k_{opt}} \leq 2 \left\lceil \frac{w\left(MST_{k_{opt}}^G\right)}{vt} \right\rceil \leq 4 \left\lceil w'(mst_{opt}^{k_{opt}}) \right\rceil \leq 8Y_{opt}$.

Therefore from the above two cases we have
$\lambda X_{k_{opt}} + \mu Y_{k_{opt}} \leq \lambda X_{opt} + 8\mu Y_{opt} \leq 8(\lambda X_{opt} + \mu Y_{opt})$. Hence, Algorithm 1 is an 8-approximation algorithm. □

5   A Special Case of EEGSweep Coverage Problem

A special case of the EEGSweep coverage problem named as SEEGSweep coverage problem is proposed in this section. In this variant all mobile sensors visit the same subset of PoIs to guarantee t-GSweep coverage while for the remaining PoIs, static sensors are used to guarantee t-GSweep coverage. We define the SEEGSweep coverage problem as follows.

Definition 5 (SEEGSweep Coverage Problem). *Let $\mathcal{U} = \{u_1, u_2, \cdots, u_n\}$ be a set of PoIs, $S = \{s_1, s_2, \cdots, s_p\}$ be a set of static sensors and $M = \{m_1, m_2, \cdots, m_q\}$ be a set of mobile sensors. Let v be the uniform speed of the mobile sensors and λ and μ be the energy consumption in every unit of time for a static sensor and mobile sensor, respectively. Find X and Y, such that*

1. *X static sensors guarantee t-GSweep coverage of X PoIs,*
2. *remaining n − X PoIs are t-GSweep covered by Y mobile sensor such that each mobile sensor visits all n − X PoIs and*
3. *$\lambda X + \mu Y$ is minimum.*

Note that the SEEGSweep coverage problem is also NP-hard and cannot be approximated within a factor of 2, unless P=NP, which follows from the proof of Theorem 1 in [15] for $\lambda = \mu$. Our proposed Algorithm 2 (SEEGSWEEPCOVERAGE) to solve SEEGSweep coverage problem uses the 2-approximation algorithm for k-TSP [9] as a subroutine. The inputs of the Algorithm 2 are the graph G, speed v, sweep period t, and energy consumption per unit time for static and mobile senors λ, μ, respectively. The output of the algorithm is the number of mobile and static sensors with the deployment locations for the static sensors and the movement schedule for the mobile sensors.

Algorithm 2: SEEGSWEEPCOVERAGE

1: **for** $k = 1$ to n **do**
2:     Find a tour TSP_k^G of G visiting k vertices using the k-TSP 2-approximation algorithm [9].
3:     Set $Y_k = \left\lceil \frac{w(TSP_k^G)}{vt} \right\rceil$ and $X_k = n - k$ then define $f_k = \lambda X_k + \mu Y_k$.
4: **end for**
5: $f = \min\{f_1, f_2, \cdots, f_k\}$ and k_0 is the index such that $f = f_{k_0}$
6: For $k = k_0$, partition the tours $TSP_{k_0}^G$ into $\left\lceil \frac{w(TSP_{k_0}^G)}{vt} \right\rceil$ parts of length vt
7: Deploy one mobile sensor at each of the partitioning points of the tour and deploy one static sensor at each of the remaining $n - k_0$ vertices.
8: The mobile sensors start their movement at the same time along the tour in the same direction.

According to the Algorithm 2, all the mobile sensor move along a single cycle which covers a subset of PoIs of G and at the remaining PoIs, static sensors are deployed. Therefore, by the similar argument given in Theorem 2, each PoI in $\mathcal{U}$ are t-GSweep covered.

Theorem 4. *The Algorithm 2 is a 2-approximation algorithm.*

Proof. Let X_{opt} and Y_{opt} be the number of static and mobile sensors in the optimal solution. Let Y_{opt} mobile sensors visit k_{opt} vertices of G and remaining $n - k_{opt}$ vertices are sweep covered by $X_{opt} = n - k_{opt}$ static sensors. Let $tsp_{opt}^{k_{opt}}$ be the minimum weighted tour in G covering k_{opt} vertices. The tour TSP_k^G of G visiting k vertices is computed using the k-TSP 2-approximation algorithm [9]. Then $w(TSP_{k_{opt}}^G) \leq 2w(tsp_{opt}^{k_{opt}})$. Since Y_{opt} mobile sensors visit k_{opt} vertices of G and all mobile sensors visit all the k_{opt} vertices, therefore,

$Y_{opt} \cdot vt \geq w(tsp_{opt}^{k_{opt}})$, which is equivalent to $\left\lceil \frac{w(tsp_{opt}^{k_{opt}})}{vt} \right\rceil \leq Y_{opt}$. According to

the Algorithm 2, when $k = k_{opt}$, $Y_{k_{opt}} = \left\lceil \frac{w(TSP_{k_{opt}}^G)}{vt} \right\rceil \leq 2 \left\lceil \frac{w(tsp_{opt}^{k_{opt}})}{vt} \right\rceil \leq 2Y_{opt}$.
Therefore, $\lambda X_{k_{opt}} + \mu Y_{k_{opt}} \leq \lambda X_{opt} + 2\mu Y_{opt} \leq 2(\lambda X_{opt} + \mu Y_{opt})$. Hence the Algorithm 2 is a 2-approximation algorithm.   □

6   Conclusion

In this paper we have proposed an energy efficient sweep coverage problem. The objective of the problem is to guarantee sweep coverage for a given set of points by a set of sensors (static and/or mobile) with minimum energy consumption. We have proposed an 8-approximation algorithm to solve the problem. Another solution for a special case of this problem is proposed, which achieves the best possible approximation factor 2. The sweep coverage in presence of obstacles might be a interesting problem to investigate in future.

References

1. Al-Turjman, F.M., Hassanein, H.S., Ibnkahla, M.A.: Connectivity optimization for wireless sensor networks applied to forest monitoring. In: Proceedings of the 2009 IEEE international conference on Communications, ICC 2009, pp. 285–290. IEEE Press, Piscataway (2009)
2. Bartolini, N., Calamoneri, T., Fusco, E.G., Massini, A., Silvestri, S.: Push & pull: autonomous deployment of mobile sensors for a complete coverage. Wireless Networks 16(3), 607–625 (2010)
3. Batalin, M.A., Sukhatme, G.S.: Multi-robot dynamic coverage of a planar bounded environment. In Technical Report, CRES-03-011 (2002)
4. Boukerche, A., Fei, X., de Araujo, R.B.: An optimal coverage-preserving scheme for wireless sensor networks based on local information exchange. Computer Communications 30(14-15), 2708–2720 (2007)
5. Chu, H.-C., Wang, W.-K., Lai, Y.-H.: Sweep coverage mechanism for wireless sensor networks with approximate patrol times. In: Symposia and Workshops on Ubiquitous, Autonomic and Trusted Computing, pp. 82–87 (2010)
6. Das, G.K., Fraser, R., Lòpez-Ortiz, A., Nickerson, B.G.: On the discrete unit disk cover problem. In: Katoh, N., Kumar, A. (eds.) WALCOM 2011. LNCS, vol. 6552, pp. 146–157. Springer, Heidelberg (2011)
7. Dash, D., Bishnu, A., Gupta, A., Nandy, S.C.: Approximation algorithms for deployment of sensors for line segment coverage in wireless sensor networks. In: COMSNETS, pp. 1–10 (2012)
8. Du, J., Li, Y., Liu, H., Sha, K.: On sweep coverage with minimum mobile sensors. In: Proceedings of the 2010 IEEE 16th International Conference on Parallel and Distributed Systems, ICPADS 2010, pp. 283–290. IEEE Computer Society, Washington, DC (2010)
9. Garg, N.: Saving an epsilon: a 2-approximation for the k-mst problem in graphs. In: Gabow, H.N., Fagin, R. (eds.) Proceedings of the 37th Annual ACM Symposium on Theory of Computing, Baltimore, MD, USA, May 22-24, pp. 396–402. ACM (2005)
10. Gorain, B., Mandal, P.S.: Approximation algorithms for sweep coverage in wireless sensor networks. J. Parallel Distrib. Comput. 74(8), 2699–2707 (2014)

11. Gorain, B., Mandal, P.S.: Brief announcement: Sweep coverage with mobile and static sensors. In: Proceedings of 16th International Symposium on Stabilization, Safety, and Security of Distributed Systems (SSS 2014), Paderborn, Germany, September 28 - October 1, pp. 346–348 (2014)

12. Gorain, B., Mandal, P.S.: Line sweep coverage in wireless sensor networks. In: Sixth International Conference on Communication Systems and Networks, COMSNETS 2014, Bangalore, India, January 6-10, pp. 1–6. IEEE (2014)

13. Gu, Y., Ji, Y., Li, J., Zhao, B.: Fundamental results on target coverage problem in wireless sensor networks. In: GLOBECOM 2009, pp. 1–6 (2009)

14. Li, J., Wang, R., Huang, H., Sun, L.: Voronoi-based coverage optimization for directional sensor networks. Wireless Sensor Network 1(5), 417–424 (2009)

15. Li, M., Cheng, W.-F., Liu, K., Liu, Y., Li, X.-Y., Liao, X.: Sweep coverage with mobile sensors. IEEE Trans. Mob. Comput. 10(11), 1534–1545 (2011)

16. Lu, M., Wu, J., Cardei, M., Li, M.: Energy-efficient connected coverage of discrete targets in wireless sensor networks. In: Lu, X., Zhao, W. (eds.) ICCNMC 2005. LNCS, vol. 3619, pp. 43–52. Springer, Heidelberg (2005)

17. Ma, M., Yang, Y.: Adaptive triangular deployment algorithm for unattended mobile sensor networks. IEEE Trans. Computers 56(7) (2007)

18. Meguerdichian, S., Koushanfar, F., Potkonjak, M., Srivastava, M.B.: Coverage problems in wireless ad-hoc sensor networks. In: INFOCOM 2001, pp. 1380–1387 (2001)

19. Ravi, R., Sundaram, R., Marathe, M.V., Rosenkrantz, D.J., Ravi, S.S.: Spanning trees short or small. In: Proceedings of the Fifth Annual ACM-SIAM Symposium on Discrete Algorithms. Arlington, Virginia, January 23-25, pp. 546–555 (1994)

20. Wang, G., Cao, G., LaPorta, T.: A bidding protocol for deploying mobile sensors. In: Proceedings. 11th IEEE International Conference on Network Protocols (ICNP 2003), pp. 315 – 324 (November 2003)

21. Wang, G., Cao, G., LaPorta, T.F.: Movement-assisted sensor deployment. IEEE Trans. Mob. Comput. 5(6), 640–652 (2006)

22. Xi, M., Wu, K., Qi, Y., Zhao, J., Liu, Y., Li, M.: Run to potential: Sweep coverage in wireless sensor networks. In: ICPP, pp. 50–57 (2009)

23. Yang, M., Kim, D., Li, D., Chen, W., Du, H., Tokuta, A.O.: Sweep-coverage with energy-restricted mobile wireless sensor nodes. In: Ren, K., Liu, X., Liang, W., Xu, M., Jia, X., Xing, K. (eds.) WASA 2013. LNCS, vol. 7992, pp. 486–497. Springer, Heidelberg (2013)

Generation of Random Digital Curves
Using Combinatorial Techniques

Apurba Sarkar[1], Arindam Biswas[2], Mousumi Dutt[3,*],
and Arnab Bhattacharya[4]

[1] Department of Computer Science and Technology,
Indian Institute of Engineering Science and Technology, Howrah, India
[2] Department of Information Technology,
Indian Institute of Engineering Science and Technology, Howrah, India
sarkar@cs.becs.ac.in, barindam@gmail.com
[3] Department of Information Technology,
Indian Institute of Information Technology, Kalyani, India
duttmousumi@gmail.com
[4] Department of Computer Science and Engineering,
Indian Institute of Technology, Kanpur, India
arnabb@cse.iitk.ac.in

Abstract. A fast linear-time algorithm to generate non-intersecting closed random orthogonal (4-connected) digital curves of finite length imposed on a background grid is proposed in this paper. A novel timestamp-based combinatorial technique is used so that the curve grows freely without intersecting itself. The combintaorial constraints are further modified to generate 8-connected digital curves. The time complexity of the algorithm is linear in length of the curve, as decisions are made locally based on current-neighbourhood points of the digital curve. The algorithm has been implemented and tested exhaustively and an analysis of the results is also presented.

Keywords: Random Digital Curves, testing of algorithms, Combinatorial Technique.

1 Introduction

Apart from the theoretical interest, generation of random digital curve has applications in testing and verification in many areas of image processing, computational geometry, and graph algorithms [2,9,6]. It involves the testing of correctness and evaluation of execution time of different algorithms. When enough practically relevant data is not available, a large number of random inputs may be used for testing an algorithm. Various applications of random curves involve creating scenes in animation, interactive arts, etc. Random curves may also be useful in graphical applications [4] which intend to generate textures of the nature: like clouds and landforms.

---

* Corresponding author.

S. Ganguly and R. Krishnamurti (Eds.): CALDAM 2015, LNCS 8959, pp. 286–297, 2015.

Several works on generating random polygon and random curves have been reported in the literature [10,5,1,12,11,4]. A heuristic approach for generation of simple polygons is presented in [10], which also presents several tests to verify the uniformity of random simple polygon generator. Zhu et al. [12] presented an algorithm for generation of x-monotone polygons on a given set of vertices. Auer et al. [1] analyzed heuristics to generate simple and star-shaped polygons on a given set of points. Two algorithms are presented in [11], one generates a random polygon starting from a cell using an inflate-cut technique, the other employs a constraint programming technique with a control on the characteristics of the generated polygon. In [4], an algorithm is presented which generates random n-gons in $O(n^2 \log n)$ time. In most of these works, a polygon is generated from a random set of vertices. A more recent work on generating random digital closed curve which computes each vertex on the fly has been presented in [3].

A random curve is called orthogonal (isothetic or 4-connected) when it consists of alternate axis parallel edges whereas in case of an 8-connected random digital curve eight directions (including the diagonal directions) are permissible. In this paper, a linear-time algorithm is proposed based on combinatorial technique devoid of any backtracking. It generates random simple orthogonal closed curve imposed on the background grid where the granularity of the generated closed orthogonal curve can be controlled by varying the grid size. With modified combinatorial technique the algorithm also generates a random simple 8-connected closed digital curve. In both cases, a move from a given grid point is chosen randomly from a set of 'safe' directions determined on the basis of the local arrangements such that the curve does not traverse into a 'dead end' to avoid intersection with itself. Two instances of 4- and 8-connected random curves generated by the proposed algorithm are shown in Fig. 1.

The paper is organized as follows. Required definitions are presented in Sec. 2. The basic principle behind formulation of the combinatorial rules for traversal is discussed in Sec. 3. The methods for generation of random digital curves are presented in Sec. 4. Section 5 presents the experimental results with analysis and the conclusion is presented in Sec. 6.

2   Definitions

The *digital plane*, $\mathbb{Z}^2$, is the set of all points having integer coordinates. A point in the digital plane is called a *digital point*, or called a *pixel* in the case of a digital image. Henceforth, the terms "point" and "digital point" will be used interchangeably.

A *digital grid* (henceforth referred simply as a *grid*) $\mathcal{G}$ consists of a set $\mathcal{H}$ of horizontal (digital) grid lines and a set $\mathcal{V}$ of vertical (digital) grid lines, where, $\mathcal{H} = \{\ldots, l_H(j-2g), l_H(j-g), l_H(j), l_H(j+g), l_H(j+2g), \ldots\} \subset \mathbb{Z}^2$ and $\mathcal{V} = \{\ldots, l_V(i-2g), l_V(i-g), l_V(i), l_V(i+g), l_V(i+2g), \ldots\} \subset \mathbb{Z}^2$, for a *grid size* $g \in \mathbb{Z}^+$. Here, $l_H(j) := \{(i', j) : i' \in \mathbb{Z}\}$ denotes the horizontal grid line and $l_V(i) := \{(i, j') : j' \in \mathbb{Z}\}$ denotes the vertical grid line intersecting at the point $(i, j) \in \mathbb{Z}^2$, called the *grid point*, where i and j are multiples of g.

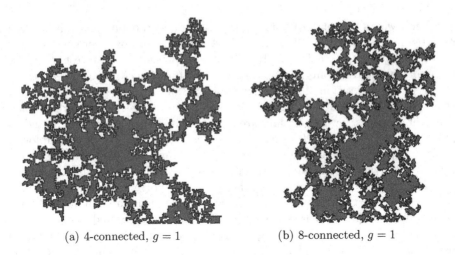

(a) 4-connected, $g = 1$ (b) 8-connected, $g = 1$

Fig. 1. Two instances of 4- and 8-connected digital random curves. The inside of the curves are shown in gray for better visibility.

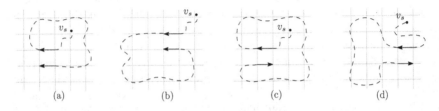

(a)   (b)   (c)   (d)

Fig. 2. (a) and (b): curves are in same direction, start of $\mathcal{C}$ is unreachable, (c) and (d): in opposite direction, $\mathcal{C}$ can reach its start

The objective is to generate random simple closed digital curves (4- and 8-connected) imposed on the background grid $\mathcal{G}$ where the vertices are grid points.

3   Basic Principle

The proposed orthogonal closed random digital curve, $\mathcal{C}$, is generated by choosing a random direction from a set of "safe" directions, S, from a grid point, $p(i, j)$, where S is computed on the basis of the presence (and orientation) of $\mathcal{C}$ in $N_8(p)$[1]. Eventually the traversal concludes at the start point, producing a closed non-intersecting curve with alternate axis-parallel sides. The set of safe directions, S, is determined using a combinatorial analysis in such a way that $\mathcal{C}$ never moves

---

[1] Two points p and q are said to be k-connected ($k = 4$ or 8) in a set S if and only if there exists a sequence $p = \langle p_0, p_1, \ldots, p_n = q \rangle \subseteq S$ such that $p_i \in N_k(p_{i-1})$ for $1 \leqslant i \leqslant n$. The 4-neighborhood of a point (x, y) is given by $N_4(x, y) = \{(x', y') : |x - x'| + |y - y'| = 1\}$ and its 8-neighborhood by $N_8(x, y) = \{(x', y') : \max(|x - x'|, |y - y'|) = 1\}$.

into a trap where it cannot proceed without intersecting itself. The algorithm is designed with the invariant that under no condition, for any two grid points on C, separated by unit grid length, either of incoming or outgoing direction of one will be same as either of incoming or outgoing direction of the other. Portion of C between these two grid points describes a closed region. If the directions are same, then the C propagates into a region where the start point does not lie, thus moves into a dead end, hence it is avoided. For example, as shown in Fig. 2(a) and (b), when the two directions are same, in either of the cases, the curve can never enter the region where lies its start point, v_s. On the other hand, movements in the opposite directions are permissible as the curve can grow to meet v_s eventually (Fig. 2(c) and (d)). In order to capture this fact, a timestamp t_i is assigned in increasing order to each vertex v_i as C grows. If v_i is generated earlier than v_j then t_i is less than t_j.

Initially a random grid point, v_s, is chosen as the start point of the digital curve, with a timestamp of $t_s = 0$. At a given (current) point, v_c, S is computed based on the occupancy and orientation of C at the earliest visited amongst the designated neighbours of v_c. Then, the next direction of propagation of C from v_c, denoted by d_c, is randomly chosen from S. As the curve advances to the new grid point along d_c, the corresponding timestamp is increased by one. The procedure is repeated until the curve returns to v_s. In order to generate a random curve of finite length, a bounding rectangle, B, is chosen, and boundary conditions are imposed such that if C takes a clockwise (anticlockwise) turn when it meets the boundary for the first time (which is perpendicular to a side of B), then it will take a clockwise (anticlockwise) turn in all subsequent cases whenever it meets the boundary again. Forthcoming sections describe the methods for generating 4-connected and 8-connected digital curves.

4   Generation of Simple Random Digital Curves

The rules for generation of random curves are discussed in this section. Sec. 4.1 explains the rules for generating 4-connected curves and the rules for generating 8-connected curves are explained in Sec. 4.2. Corresponding time complexity is discussed at the end of this section.

4.1   Generation of 4-Connected Curves

Let the eight neighbours of a point are denoted by the set $N = \{n_0, n_1, \ldots, n_7\}$ as shown in Fig. 3(a) where n_d is the neighbour in the direction d. Let v_c be the current point, and v_l and v_r be the front-left and front-right points (Fig. 3(b)). An array **b** of size 8 is considered to signify the 8 directions initialized to 1, which means all directions are initially permissible. In case of 4-connected curves, **b**[1], **b**[3], **b**[5], and **b**[7] are marked as invalid by default (set to '0' indicating not permissible), as there cannot be any move in these directions from v_c. Also, if a neighbour n_d is already visited then **b**[d] is invalidated.

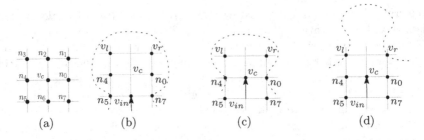

Fig. 3. (a) 8 neighbours, (b) trap involving n_5 and n_7, (c) n_0 and n_4, and (d) v_l and v_r

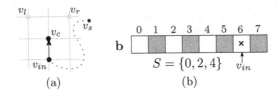

Fig. 4. Case I: (a) None of v_l and v_r visited (b) Direction array **b**

For generation of 4-connected random digital curves, the neighbors v_l and v_r of v_c are considered to find S. The reason is to avoid potential traps involving neighbours of the current point, v_c. W.l.o.g. let C reaches v_c from v_{in} as shown in Fig. 3(c), it indicates that the existence of a possible loop between v_{in} and either n_4 or n_0, whichever is visited earlier, is already checked. Now, for a safe move from v_c, we need to check only for a possible loop between v_c and the earliest visited of v_l and v_r (Fig. 3(d)). There exist three different cases as stated below.

Case I: *Neither v_l nor v_r is visited.* The curve can move in any direction except in the direction of the visited neighbors.
For example, in Fig. 4(a), none of v_l and v_r is visited. Directions **1, 3, 5**, and **7** are invalid by default for a 4-connected curve, shown in Fig. 4(b) as gray cells, and $\mathbf{b}[6]$ is marked as invalid, '×', because n_6 is already visited. Thus, the curve can move in one of the directions randomly chosen from $S = \{0, 2, 4\}$, shown as white cells in Fig. 4(b).

Case II: *At least one in $\{v_l, v_r\}$ is visited.* Let v_m *from* $\{v_l, v_r\}$ *has the lower timestamp, and v_{m+1}, the grid point next to v_m on C, lies in N.* For each visited neighbor n_d of v_c, $\mathbf{b}[d]$ is marked as invalid. Then, a traversal is made in **b** from v_m in the direction of v_m to v_{m+1} (the immediate next cell in **b**) upto the farthest visited vertex in $\{v_l, v_r, v_{in}\}$ invalidating all the directions encountered during traversal (if not already invalidated) and the unmarked directions constitute the set of permissible directions, S, from v_c. Fig. 5(a) depicts one such situation when only v_r is visited and its next grid point v_{ro} is in N. Fig. 5(b) shows the initial marking; shaded cell denotes the default invalid directions and $\mathbf{b}[6]$ and $\mathbf{b}[2]$ are marked as invalid as $v_{in}(= n_6)$ $v_{ro}(= n_2)$ are already visited. The traversal is made in the direction from

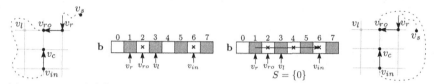

(a) v_r is visited   (b) Initial marking of **b**   (c) Final marking of **b**   (d) v_r is visited

Fig. 5. Illustration of Case II for 4-connected curve

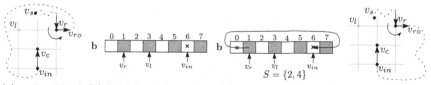

(a) v_r is visited   (b) Initial marking of **b**   (c) Final marking of **b**   (d) v_r is visited

Fig. 6. Illustration of Case III for 4-connected curve

v_r to v_{ro}, towards right, in **b** till v_{in} thereby invalidating **b**[4], others being already invalid, thus $S = \{0\}$. Note that **b** should be treated as circular during the traversal. It may be noted here that the curve shown in the example Fig. 5(a) from v_{ro} the curve reaches v_{in} in an anticlockwise manner, thus the loop excludes v_s. If the traversal was clockwise, the loop would include v_s, thus in both the cases, 0 is the only possible safe direction.

Case III: v_l, v_r, *or both are visited and let* v_m *amongst* $\{v_l, v_r\}$ *has the lower timestamp,* v_{m+1}, *the grid point next to* v_m, *does not lie in* N. First, the visited neighbours of v_c are marked as invalid in **b**. Then, a traversal is made from v_m in forward (backward) direction if there is a clockwise (anticlockwise) turn at v_m till the farthest visited vertex in $\{v_l, v_r, v_{in}\}$ thereby invalidating all the directions on the way of traversal, yielding S.

In Fig. 6(a) $v_m = v_r$ and $v_{ro} \notin N$ and there is an anticlockwise turn at v_m. Hence, a backward traversal is made from v_r to v_{in} which is the farthest visited vertex in v_l, v_r, v_{in}, as v_l is not visited at all. Thus, $S = \{2, 4\}$.

4.2   Generation of 8-Connected Curves

The 8-connected random digital curves are generated by a similar method used for generating 4-connected random curves only that the least recently visited neighbor of a different set of neighbors is considered for determining S. There are two distinct cases depending on the incoming direction to v_c, denoted by d_{ci}, i) d_{ci} is axis parallel ($d_{ci} \in \{0, 2, 4, 6\}$): the least recently visited of the three neighbours of v_c, on left v_l, on right v_r, and at front v_f (shown in Fig. 7 (a)) or ii) when d_{ci} is diagonal ($d_{ci} \in \{1, 3, 5, 7\}$): the least recently visited of $\{v_{fl}, v_{fr}\}$ (shown in Fig. 7 (b)) is considered to find S.

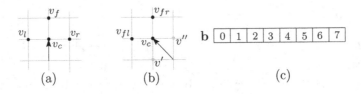

Fig. 7. Two cases of 8-connected curves, a) d_{ci} is axis parallel, b) d_{ci} is diagonal

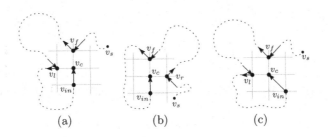

Fig. 8. Illustration of why v_l, v_f and v_f are considered

A. d_{ci} Is Axis Parallel: The three orthogonal neighbors, $\{v_l, v_f, v_r\}$, are considered because a possible loop between them has to be checked. For example, assume that v_r is visited earlier, it means the curve has traversed from v_r to v_c forming a path between these two points. Now, depending on the direction of traversal C from v_r, it can be determined in which side of the loop the start point, v_s, lies, accordingly the directions are invalidated. It may be noted that, v_c and v_r being orthogonal neighbors, there is no scope of moving from one side of the loop to the other side, which is not true for the diagonal neighbors, when d_{ci} is axis parallel. Thus, only the orthogonal neighbors are considered. Also, only the least recently visited neighbor is considered because the loop described by the earliest visited invalidates all the unsafe directions.

Case I: *None of $\{v_l, v_r, v_f\}$ are visited.* The curve can proceed in any direction except in the directions of visited neighbours.

Case II: *At least one of $\{v_l, v_r, v_f\}$ is visited and let v_m has the lowest timestamp amongst them and its next vertex, say v_{m+1}, is in N.* A traversal is made in **b** from v_m in the direction of v_m to v_{m+1} to the farthest (in **b**) visited vertex in $\{v_l, v_r, v_f, v_{in}\} \setminus \{v_m\}$ and the corresponding directions are invalidated.

In Fig. 9(a) all three v_l, v_f and v_r are visited. Here, $v_m = v_r = n_0$, and its next vertex, $n_7 \in N$, hence the traversal is made from n_0 in the direction of n_7 till $v_f(n_2)$ which is the farthest visited vertex in the set $\{v_l, v_{in}, v_f\}$ as shown in Fig. 9(c) leaving 1 as the only permissible direction.

Case III: *At least one of $\{v_l, v_r, v_f\}$ is visited and let v_m has the lowest timestamp and $v_{m+1} \notin N$.* Then, a traversal is made from v_m in **b** in forward (backward) direction if there is a clockwise (anticlockwise) turn at v_m till

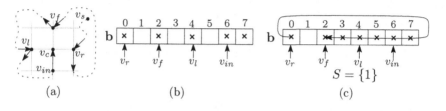

Fig. 9. (a) All three v_l, v_f, v_r are visited, (b) Visited neighbors marked, and (c) Permissible set after traversal

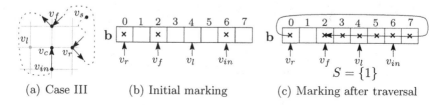

Fig. 10. v_l, v_f are visited

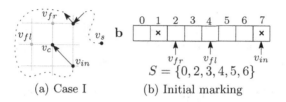

Fig. 11. None of v_{fl}, v_{fr} are visited

the farthest visited vertex in $\{v_l, v_r, v_f, v_{in}\} \setminus \{v_m\}$ thereby invalidating all the directions on the way of traversal.

As shown in Fig. 10(a) $v_r (= v_m)$, has the lowest timestamp with a counter clock-wise move. So a backward traversal is made from $v_m = v_r = n_0$ to $v_f = n_2$ which is the farthest visited vertex in $\{v_l, v_{in}, v_f\}$ which gives $S = \{1\}$ (Fig. 10(c)).

B. d_{ci} Is Diagonal: When $\mathcal{C}$ proceeds from v_{in} to v_c, the two orthogonal neighbors of v_c, namely, v' and v'' (shown in Fig. 7(b)), if visited, are already checked for a possible loop and a safe move has been made. Hence, at v_c only the two neighbors, the front left (v_{fl}) and front right (v_{fr}), as shown in Fig. 7(b), are checked for a possible loop and the safe directions are decided. Also, like the earlier case, if both v_{fl} and v_{fr} are visited, then the earliest visited of them is considered for generating the safe directions.

Case I: None of $\{v_{fl}, v_{fr}\}$ are visited, so the curve can move in any direction except in the directions of visited neighbors, i.e., v_1 and v_7 (Fig. 11).

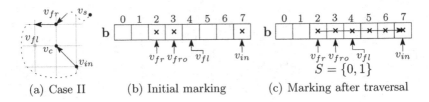

Fig. 12. v_{fr} is visited

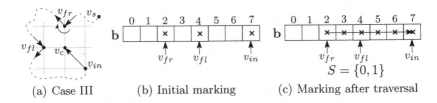

Fig. 13. v_{fr}, v_{fl} both are visited

Case II: The vertex with the lowest timestamp in $\{v_{fl}, v_{fr}\}$ has its next vertex in N, the traversal from this vertex in the direction of its next vertex determines P (illustrated in Fig. 12).

Case III: The vertex with the lowest timestamp in $\{v_{fl}, v_{fr}\}$ does not have its next vertex in N, when it has a clockwise (anticlockwise) turn the backward (forward) traversal from this vertex yields P (illustrated in Fig. 13).

Note that if $v_s \in N_8(v_c)$, the direction towards v_s from v_c is included in S, such that if that direction is chosen randomly then the curve will conclude. As a finite size canvas is considered for generating the random curve considering the boundary conditions and as each move is randomly selected from the set of safe directions which are determined using combinatorial techniques such that the start vertex v_s is always reachable, the curve $\mathcal{C}$ will always converge to v_s.

The portion of $\mathcal{C}$ between the current point, v_c, and the least recently visited front diagonal neighbor, v_m, (which is v_r both in Fig. 5 and Fig. 6) describes a closed region, $\mathcal{R}$. Let us consider two grid points, $(v_j, v_k) \in \mathcal{C}$ and $v_j \in N_8(v_k)$. As per the invariant (stated in Sec. 3), if d_{ji} and/or $d_{jo} \notin \{d_{ki}, d_{ko}\}$, where d_{ji} and d_{jo} denote the incoming and outgoing directions at v_j, then $\mathcal{C}$ propagates to the region where v_s belongs thus ensures that $\mathcal{C}$ never moves into dead end. As $\mathcal{C}$ always propagates into the region containing v_s, $\mathcal{C}$ eventually concludes. $N_8(v_c)$ forms the closet possible circle (essentially a square here) with v_c at center. When $\mathcal{C}$ touches this circle at v_m in an anticlockwise (clockwise) manner then the neighbors of v_c are invalidated till the last visited neighbor in that direction, in the same anticlockwise (clockwise) manner thus avoids v_i and v_k having same direction. All cases for 4- and 8-connected curves are formulated keeping the above in view. Thus, the algorithm always produces a random closed curve and ensures that the propagation of the curve never enters a dead end.

Time Complexity: A constant number of grid points are checked to determine the next direction from a grid point lying on $\mathcal{C}$, thus the algorithm is linear in the number of grid points on the length of the generated curve, given by $O(|\mathcal{C}|/g)$.

5   Experimental Results and Analysis

The proposed algorithm is implemented in C in Ubuntu 12.04, 64-bit, kernel version 3.5.0-43-generic, the processor being Intel i5-3570, 3.4 GHz FSB. Six instances of 4-connected and 8-connected random curves are shown in two columns in Fig. 14, first row shows two random curves for $g = 1$ whereas the second and third rows show two random curves each for $g = 2$ and $g = 4$ respectively. The figures depict randomness of the curves and also the number of vertices for each curve. It may be noted that by 8-connected, we mean that the vertices of curves (which are also grid points) are 8-connected and the vertices are connected by straight lines. It is evident from the results that by varying the grid sizes the resolution of the curves can be controlled.

Randomness of the generated curves are tested using one-sample runs test [7] in which we select 100 equidistant points on $\mathcal{C}$ in each iteration and 0 is assigned for a point lying on a horizotal grid line and 1 is assigned when it lies on a vertical line. Thus, a sequence of 0 and 1 of length 100 is generated. From the sequence for each random curve, we determine n_1, the number of 0s, n_2, the number of 1s, and r, the number of runs. Mean and standard error of the r-statistic are given by $\mu_r = \frac{2n_1 n_2}{(n_1+n_2)} + 1$ and $\sigma_r = \sqrt{\frac{2n_1 n_2(2n_1 n_2 - n_1 - n_2)}{(n_1+n_2)^2(n_1+n_2-1)}}$. The corresponding z-value is computed as $z = \frac{r - \mu_r}{\sigma_r}$. The sampling distribution is approximated by normal distribution, and as per the table of standard normal probability distribution [8] the significance level α is determined. Table 1 presents the data related to 20 random curves and their level of significance for the hypothesis that the curve is random. It is evident from Table 1 that the hypothesis is true for the curves with α varying from 66% to 95% with 15 of them having α more than 80%.

Table 1. Runs test statistic for 20 random curves: $n_1 = \#$ 0s, $n_2 = \#$ 1s, $r = \#$ runs, μ_r and σ_r are mean and standard error of the r statistic respectively, and α is the significance level corresponding to the z value

#$\mathcal{C}$	n_1	n_2	r	μ_r	σ_r	z	α in %
$\mathcal{C}_1$	46	54	51	50.68	4.9425	0.0647	94.70
$\mathcal{C}_2$	53	47	50	50.82	4.9565	−0.1654	86.20
$\mathcal{C}_3$	45	55	51	50.50	4.9244	0.1015	92.01
$\mathcal{C}_4$	51	49	51	50.98	4.9726	0.0040	100.00
$\mathcal{C}_5$	62	38	48	48.12	4.6852	−0.0256	98.00
$\mathcal{C}_6$	48	52	51	50.92	4.9666	0.0167	99.01
$\mathcal{C}_7$	49	51	49	50.98	4.9726	−0.3981	69.40
$\mathcal{C}_8$	62	38	49	48.12	4.6852	0.1878	85.14
$\mathcal{C}_9$	49	51	51	50.98	4.9726	0.0040	100.00
$\mathcal{C}_{10}$	53	47	53	50.82	4.9565	0.4398	66.02
$\mathcal{C}_{11}$	54	46	50	50.68	4.9425	−0.1375	89.00
$\mathcal{C}_{12}$	48	52	52	50.92	4.9666	0.2174	83.11
$\mathcal{C}_{13}$	60	40	49	49.00	4.7736	0.0000	100.00
$\mathcal{C}_{14}$	61	39	50	48.58	4.7314	0.3001	76.41
$\mathcal{C}_{15}$	52	48	50	50.92	4.9666	−0.1852	85.50
$\mathcal{C}_{16}$	50	50	51	51.00	4.9746	0.0000	100.00
$\mathcal{C}_{17}$	44	56	50	50.28	4.9023	−0.0571	96.01
$\mathcal{C}_{18}$	53	47	52	50.82	4.9565	0.2380	81.02
$\mathcal{C}_{19}$	44	56	49	50.28	4.9023	−0.2611	79.50
$\mathcal{C}_{20}$	42	58	51	49.72	4.8460	0.2641	79.32

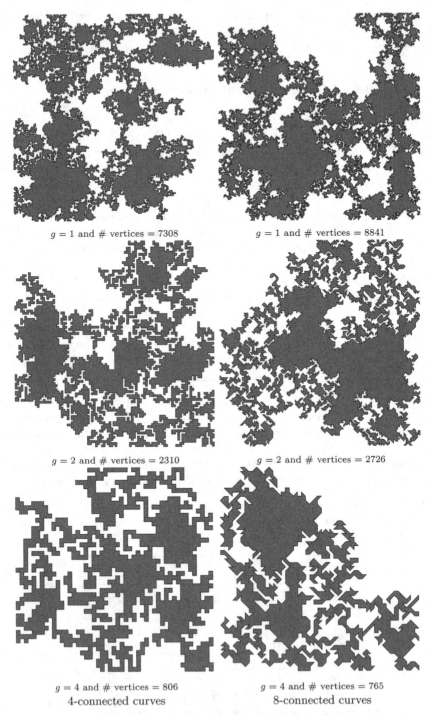

$g = 1$ and # vertices = 7308

$g = 1$ and # vertices = 8841

$g = 2$ and # vertices = 2310

$g = 2$ and # vertices = 2726

$g = 4$ and # vertices = 806
4-connected curves

$g = 4$ and # vertices = 765
8-connected curves

Fig. 14. Instances of 4-connected and 8-connected digital random curves on a canvas of size 200×200

6   Conclusions

The work presents a fast linear-time algorithm for generation of random simple orthogonal digital curves based on some combinatorial technique which avoids backtracking. It is also presented that with modified constraints the same algorithm can generate 8-connected curves. The experimental results establish the randomness of the curves generated. The algorithm can be further tuned to generate random paths inside a digital object which can be useful for practical applications.

References

1. Auer, T., Held, M.: Heuristics for the generation of random polygons. In: Proc. Canadian Conference on Computational Geometry, pp. 38–44 (1996)
2. Bhowmick, P., Bhattacharya, B.B.: Fast polygonal approximation of digital curves using relaxed straightness properties. IEEE Trans. PAMI 29(9), 1590–1602 (2007)
3. Bhowmick, P., Pal, O., Klette, R.: Linear-time algorithm for the generation of random digital curves. In: Proceedings of the 2010 Fourth Pacific-Rim Symposium on Image and Video Technology, pp. 168–173 (2010)
4. Dailey, D., Whitfield, D.: Constructing random polygons. In: Proceedings of the 9th ACM SIGITE Conf., Information Tech. Education, pp. 119–124 (2008)
5. Epstein, P.: Generating geometric objects at random. Master's thesis, CS Dept., Carleton University, Canada (1992)
6. Klette, R., Rosenfeld, A.: Digital Geometry: Geometric Methods for Digital Picture Analysis. Morgan Kaufmann, San Francisco (2004)
7. Richard, I.: Levin. Statistics for Management. Prentice Hall India (1986)
8. Mason, R.D.: Essentials of Statistics. Prentice-Hall (1986)
9. Rennesson, I., Luc, R., Degli, J.: Segmentation of discrete curves into fuzzy segments. Electronic Notes in Discrete Mathematics 12, 372–383 (2003)
10. Rourke, J., Virmani, M.: Generating random polygons, TR:011, CS Dept. Smith College, Northampton (1991)
11. Tomas, A.P., Bajuelos, A.L.: Generating random orthogonal polygons. In: Conejo, R., Urretavizcaya, M., Pérez-de-la-Cruz, J.-L. (eds.) CAEPIA/TTIA 2003. LNCS (LNAI), vol. 3040, pp. 364–373. Springer, Heidelberg (2004)
12. Zhu, C., Sundaram, G., Snoeyink, J., Mitchell, J.S.B.: Generating random polygons with given vertices. CGTA 6, 277–290 (1996)

Author Index